Informatik-Fachberichte 275

Herausgeber: W. Brauer
im Auftrag der Gesellschaft für Informatik (GI)

D. P. F. Möller O. Richter (Hrsg.)

Analyse dynamischer Systeme in Medizin, Biologie und Ökologie

4. Ebernburgerer Gespräch
Bad Münster, 5.-7. April 1990

Proceedings

Springer-Verlag

Berlin Heidelberg New York London Paris
Tokyo Hong Kong Barcelona Budapest

Herausgeber

Dietmar P. F. Möller
Drägerwerk AG, Geschäftsgebiet Anästhesie
Moislinger Allee 53-55, W-2400 Lübeck

Otto Richter
Institut für Geographie und Geoökologie
Technische Universität Braunschweig
Langer Kamp 19c, W-3300 Braunschweig

CR Subject Classification (1991): B.1.2, B.1.3, B.2.1, B.3.3, B.5.2, C.1.2, D.3.2, F.1.2, F.2.1, G.1.6, G.1.7, G.1.8, G.1.9, H.1.1, K.3.2

ISBN-13: 978-3-540-54669-6   e-ISBN: 978-3-642-77020-3
DOI: 10.1007/978-3-642-77020-3

Vorwort

Der vorliegende Tagungsband enthält die drei eingeladenen Hauptvorträge und die 30 ausgewählten Beiträge, die während der "Erwin Riesch-Arbeitstagung Analyse dynamischer Systeme in Medizin, Biologie und Ökologie" vom 5. April bis 7. April 1990 auf der Ebernburg im Rahmen des 4. Ebernburger Gespräches gehalten wurden.

Die Hauptvorträge hielten der Träger der Max-Planck-Medaille Professor Dr. H. Haken über Synergetik, einer umfassenden Methode zur Beschreibung nichtlinearer vermaschter dynamischer Systeme (Beitrag lag zum Druck nicht vor), Professor Dr. R. Kaehr über die philosophischen Ansätze zur Beschreibung selbstorganisierender dynamischer Prozesse und Dr. J.P. Schloeder über die Identifikation unbekannter Parameter nichtlinearer Differentialgleichungssysteme, wie sie zur Beschreibung dynamischer Prozesse verwendet werden (Beitrag lag zum Druck nicht vor).

Die Beiträge des vorliegenden Tagungsbandes sind in thematisch geschlossenen Abschnitten zusammengefaßt, um dem Leser eine bessere Übersicht zu ermöglichen: Medizin, Biologie, Ökologie, Umweltqualität und Mathematik.

Die Arbeitstagung wurde vom Arbeitskreis Simulation in Medizin, Biologie und Ökologie (AK 4.5.2.1) des Fachausschuß 4.5 Simulation (ASIM) in der Gesellschaft für Informatik (GI) organisiert, in Zusammenarbeit mit der Arbeitsgruppe Mathematische Modelle in Biologie und Medizin, der Gesellschaft für Medizinische Dokumentation und Statistik (GMDS) und der Deutschen Region der Biometrischen Gesellschaft.

Die Beiträge der Arbeitstagung wurden von einem internationalen Programmkomitee sorgfältig begutachtet und ausgewählt.
Dem Programmkomitee gehörten an: H.G. Bock, Univ. Augsburg; R.P. van Wijk van Brievingh, TH-Delft; B.A. Gottwald, Univ. Freiburg; S.S. Hacisalihzade, Univ. of Berkeley; L. Mathäus, ZKI Ost-Berlin; D.P.F. Möller, Univ. Mainz; O. Richter, TU-Braunschweig; R. Rudolph, ZKI Ost-Berlin; H.G. Schuster, Univ. Kiel; H.P. Schwefel, Univ. Dortmund; H.E. Wichmann, Univ. Wuppertal; A. Sydow, ZKI Ost-Berlin.

Mit der Bezeichnung "Erwin Riesch-Arbeitstagung" soll wieder die großzügige Unterstützung der Tagung durch die Erwin Riesch-Stiftung gewürdigt werden.

Bad Münster am Stein-Ebernburg mit der auf einem wuchtigen Porphyrfelsen gelegenen Ebernburg ist eine hübsche kleine Kurstadt mit einer romantischen Felsenlandschaft, ausgedehnten Wäldern und der anmutigen Flußlandschaft an der Nahe. Die Naheregion ist insbesondere wegen des wesentlich wärmeren Klimas und der Böden für erstklassige Weinbergslagen und damit für die Naheweine bekannt.

Die Ebernburg, sie wird 1212 zum ersten Mal unter dem Namen
Heberenburg erwähnt, wurde 1448 vom Geschlecht der Sickinger
erworben, die später durch Ulrich von Hulten in die Gedanken-
welt des Humanismus und der Reformation eingeführt wurden.
So wurde die Ebernburg als Herberge der Gerechtigkeit zum
Wahrzeichen humanistischer und religiöser Erneuerung und dazu
eine der wehrhaften Burgen Deutschlands ganz in der Nähe der
Mitte und des Kräftezentrums des alten Heiligen Römischen
Reiches Deutscher Nation. Jedoch wurde die Ebernburg 1523 von
den Heeren des Erzbischof von Trier, dem Kurfürsten von der
Pfalz und dem Landgrafen von Hessen belagert und zerstört.
Sie wurde nach 1542 wieder von den Sickingern aufgebaut.
1698 verlangten die Franzosen vom Deutschen Kaiser die
Schleifung der Burg, d.h. ihre völlige Zerstörung.

1914 wurde die Ebernburgstiftung gegründet mit dem Ziel, die
Burg zu erhalten und auszugestalten. Nachdem die Gebäude in den
Ruinen der Burg im Jahre 1945 durch Artilleriebeschuß der
Alliierten erneut stark gelitten hatten, begann man von 1954
bis 1971 und von 1974 bis 1981 die Ebernburg in ihrer historischen Form des 16. Jahrhunderts wieder herzustellen und die
freigeistige Tradition fortzuführen.

Die besondere politische Entwicklung in der ehemaligen DDR hat
es ermöglicht, daß 14 Wissenschaftler aus der ehemaligen DDR an
der Tagung auf der Ebernburg teilnehmen konnten und so den
Anfang für eine neue gemeinsame freigeistige wissenschaftliche
Tradition eröffnet wurde. Die Veranstalter danken in diesem
Zusammenhang der VW-Stiftung, die in großzügiger Weise die
finanzielle Unterstützung der Wissenschaftler aus der ehemaligen DDR und der eingeladenen Hauptreferenten übernommen hat.

Herrn Stadtbürgermeister W. Schaust sowie Herrn Kurdirektor
R. Bolfing und Herrn K. Gattung von den Kurbetrieben in Bad
Münster am Stein-Ebernburg danken die Veranstalter für die
Mitwirkung bei der Gestaltung des gesellschaftlich attraktiv
ausgewählten Rahmenprogrammes während der Tagung und die
Unterstützung für einen reibungslosen und, wie wir hoffen, alle
Tagungsteilnehmer zufriedenstellenden Ablauf. So wurde zur
Eröffnung der Tagung am Donnerstagabend nach der Ebernburger
Wildtafel im Gewölbekeller der Ebernburg eine Weinprobe mit
12 Naheweinen durchgeführt.

Den gesellschaftlichen Höhepunkt der Tagung bildete das
festliche Gelage an der Tafelrunde von König Artus im Ritter-
saal der Altenbaumburg. Die Gruppe Chamelot untermalte das
festliche Gelage mit Musik, Gauklertum, Minnesang, Balladen und
gar manchen historischen Begebenheiten von König Artus Tafelrunde. Herzlicher Dank gebührt diesen edlen Helden.

Dem Ehepaar Rauschenplat und seinen Mitarbeitern von der
Evangelischen Familienfeiern- und Bildungsstätte Ebernburg
danken wir für den unermüdlichen Einsatz bei der Beherbergung
und Bewirtung der Tagungsteilnehmer auf der Ebernburg.

Dem Ehepaar Prietz und seinen Mitarbeitern von der Altenbaumburg danken wir für die gelungene Ausrichtung der festlichen Tafelrunde.

Ebenfalls danken wir der Boehringer Ingelheim für die Unterstützung der Tagung mit Schreibmaterial und Namensschildern.

Schließlich gilt unser Dank allen Vortragenden, Sitzungsleitern und Diskutanden und dem Springer Verlag, der sich bereit erklärt hat, den Tagungsband im Rahmen der Reihe "Informatik Fachberichte" zu veröffentlichen.

Abschließend möchten die Herausgeber ihren Familien danken für deren Geduld und Nachsicht, die sie uns während der Vorbereitung der Arbeitstagung entgegengebracht haben.

Mainz/Lübeck, Ostern 1991   Dietmar P.F. Möller

Braunschweig, Ostern 1991   Otto Richter

Inhaltsverzeichnis Seite

Seite

Biologie

Ökologie

H A U P T V O R T R Ä G E

SYNERGETIK
(H. Haken, Univ. Stuttgart)

(der Beitrag lag bei Drucklegung
des Bandes nicht vor)

PROBLEMS OF AUTONOMY AND DISCONTEXTURALITY
IN THE THEORY OF LIVING SYSTEMS
(E. von Goldammer and R. Kaehr,
Med. Univ. Lübeck)

PARAMETERIDENTIFIKATION IN NICHTLINEAREN
DIFFERENTIALGLEICHUNGEN
(J.P. Schloeder, Uni-. Augsburg)

(der Beitrag lag bei Drucklegung
des Bandes nicht vor)

Problems of Autonomy and Discontexturality in the Theory of Living Systems

E. VON GOLDAMMER and R. KAEHR
Medizinische Universität Lübeck
Universität Witten/Herdecke

ABSTRACT: In the theory of living systems any description of self-organizing processes is confronted by a very central problem concerning the role of the system's boundary, i.e., there is the necessity of a simultaneous formal representation of the inside and the outside of a system.

On the other hand, in a theory of self-organization restricted to changes of states within a system, which may be defined by some physical state variables, the question of the boundary has been eliminated and the distinction between a system and its environment (its inside and outside) generally is interpreted as an information process between both, the system and the environment.

In the theory of autopoietic systems (TAS), on the other hand, it is the autonomy of a system which plays a fundamental role and therefore the TAS represents a theory of self-organization in relation to a system and its environment and not primarily a theory of self-organization of states within a system. This, however, results in the logical problem of circularity as an immediate consequence of the postulated closure of any living system.

As a result of the closure principle, the distinction between a system and its environment (the boundary of a system) interpreted as an information transfer in the theory of self-organization cannot be established any longer as a primarily relevant process in the theory of autopoietic systems.

For an adequate description of closed systems it is the discontexturality between autonomous and non-autonomous systems which takes the place of the 'system-environment-relation'. On the basis of the theory of poly-contextural logic discontexturality between a system and its environment results as an explication and conceptual precision of the 'structural coupling concept' as introduced in the theory of autopoietic systems.

— HISTORICAL NOTES

The scientific concept of both 'cybernetics' and 'general systems theory' was founded in the early forties within the biological sciences, and the declared aim of cybernetics was the modelling, simulation, and technical reproduction of living processes [Wiener, 1943]:

> "...a uniform behavioristic analysis is applicable to both the mechanistic and living organisms, regardless of the complexity of their behavior."   (1)

The correlation between biology and technique was established by a common scientific approach [Wiener, 1943]:

"Given any object, relatively abstracted from its surroundings for
study, the behavioristic approach consits in the examination of the
output of the object and of the relation of this outputs to the input.
By output is meant any change produced in the surroundings by the
object. By input, conversely, is meant any event external to the object
that modifies this object in any manner... By behavior is meant any
change of an entity with respect to its surroundings."

(2)

The attempt to develop a scientific description of living systems in the sense of
a holistic (non-reductionistic) 'theory of the living' (not based on the terminology of physical sciences) has resulted in a fundamental change of the scientific paradigm concerning the role of an observer during the process of observation [von Foerster, 1980]:

"A living organism is an independent autonomous organizationally
closed unity;"
and
an observing organism is part, partner, and participant in its world of
observations."

(3)

It is the inclusion of the observer into the description of living systems as claimed
by modern cybernetics which causes several basic scientific and logical problems.

— AUTONOMY and CLOSURE

According to the postulate (3) given above, all living systems are autonnomous,
i.e., self-regulating organisms. If the prefix "self-" is substituted by the corresponding noun, the meaning of "autonomy" becomes synonymous with "regulation of regulation". In the terminology of cybernetics this means:

"A living system regulates its own regulation."

(4)

This statement stipulates operational closure in the sense that systems have to
be described with no inputs and no outputs in order to emphasize their autonomous constituents [Varela, 1979]:

"Closure Thesis:
Every autonomous system is organizationally closed."

(5)

This point of view is alien to the Wienerian idea of feedback *simpliciter*.

— COGNITION and AUTOPOIESIS

Parallel to the discovery of the special role of an observer in the 'theory of living systems' great importance is attributed to the relation between a system
and its environment which has to be seen under the aspect of the cognitive
abilities as primordial attribute of 'The Living' [Maturana, 1980]:

> "Living systems are cognitive systems, and living as a process is a
> process of cognition. This statement is valid for all organisms, with (6)
> and without nervous systems."

A milestone on the way to a theory of living systems is given by the concept
of 'autopoiesis' introduced by Maturana and Varela [Maturana, 1980; 1985; Varela, 1979]:

> "An autopoietic machine is a machine organized (defined as unity) as a
> network of processes of production (transformation and destruction)
> of components that produces the components which:
> i) through their interactions and transformations continuously regene-
>   rate and realize the network of processes (relations) that produced
>   them; (7)
> ii) constitute it (the machine) as a concrete unity in the space which
>   they (the components) exist by specifying the topology domain of its
>   realization as such a network.
> ... autopoietic machines are autonomous ...
> ... autopoietic machines have individuality ...
> ... autopoietic machines do not have inputs or outputs...
> ... autopoisesis is necessary and sufficient to
>   characterize the organization of living...
> ... a physical system, if autopoietic, is living."

Again it is the conception of 'autonomy' and 'closure' which is of importance in
characterizing autopoietic machines, i.e., living systems. It should be emphasi-
zed that in the theory of autopoietic systems the concept of 'information' has
been excluded for the description of living systems, because information does
not exist independent of an organization that generates a cognitive domain,
from which an observer-community can describe certain elements as informati-
onal and symbolic. In other words, information does not exist *sui generis*.

— MATHEMATICAL CONSEQUENCES

The basic epistemological point in the 'Theory of Autopoietic Systems' results
from the insight that 'closure' and 'autonomy' of living systems are incompati-
ble with any representation of a system from its system-environment relationship. The system's boundaries defined by an observer of a system (and its environment) always differ from the boundaries generated by an autonomous system itself in relation to all other systems. It is this different description of a system, i)
from a point outside the system (from the view of an observer) and ii) from the
inside of its autonomy, which is of fundamental importance for any theoretical
description of the living.

The transition from so-called 'first order cybernetics' (the cybernetics of observed systems) to 'second order cybernetics' (the cybernetics of observing systems) is demonstrated by comparison of the postulates (1)&(2) representing the
classical (first order) situation with the postulates (3)-(7) reflecting the position
of 'second order conceptions'.

In order to demonstrate the fundamental difference of both positions a topoplogical analysis is given in figure 1.

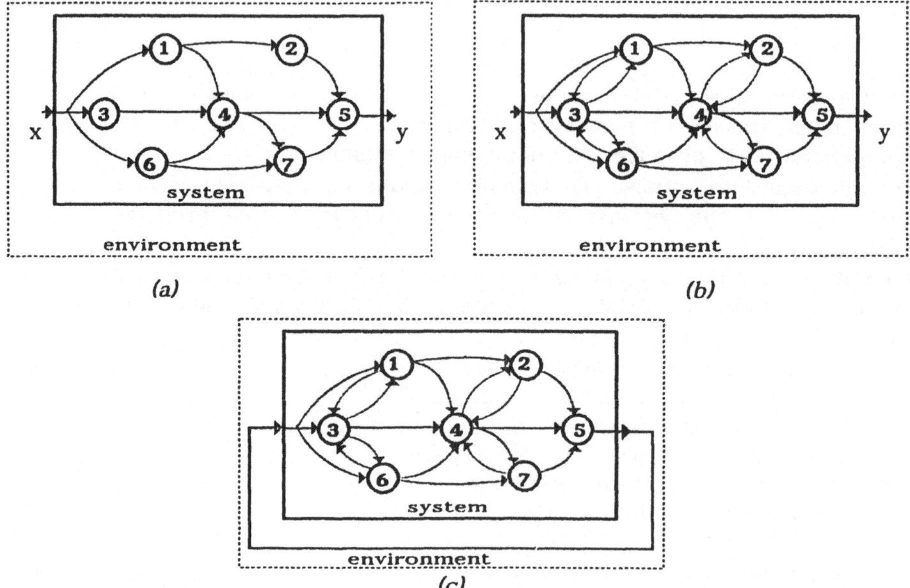

Figure 1: A system as a set of elements and relations in interaction with its environment.
  a) a classical input-output system;
  b) a classical input-output system with closed loops and recurrent connections;
  c) a autonomous closed system with no inputs and no outputs.

$$x : \text{input variable,} \quad y: \text{output variable,} \quad n_i: \text{state of element n}$$

The mathematical description of the system in fig.1a is given as:

$$\frac{d}{dt} n_i = f_i \left[x; u_1, u_2, \ldots u_n \right]$$

$$y = g \left[x; u_1, u_2, \ldots u_N \right] \qquad \text{with} \quad i = 1, 2, \ldots n-1, n, n+1, \ldots, N$$

$$[\ 8a \ ]$$

If there are closed loops caused, for example, by mutual interactions such as indicated in fig.1b, the mathematical corresponding description becomes,

$$\frac{d}{dt} n_i = f_i \left[x; u_1, u_2, \ldots u_N \right]$$

$$y = g \left[x; u_1, u_2, \ldots u_N \right]$$

$$[\ 8b \ ]$$

The difference between eq.(8a) and (8b) is given by the indices. While eqs.(8a) can be solved under certain conditions, eqs.(8b) cannot be reduced any further which means that the system in fig.1b has to be described by a different model.

For a closed system defined in the sense of the closure thesis with no inputs and no outputs as it is shown in fig.1c the corresponding differential equation becomes:

$$\frac{d}{dt} u_i = f_i (u_1, u_2, \ldots u_N) \qquad\qquad (\,8c\,)$$

Because of its recursive form eq.(8c) cannot be solved unless an input-output function is introduced which, however, is in contradiction to the definition of the closure condition for an autonomous system. In other words, on the basis of the 'closure thesis' a mathematical description of an autonomous system cannot be given if the closure of the system is to be maintained within the theoretical description.

— COGNITION AND VOLITION

In the previous postulates, the process of cognition was attributed to be an essential feature of all living systems. Combined with the idea of computation, cognition appears as self-reference which means that an autonomous (cognitive) system must be able to produce an image of the system (itself) and its environment (inside the system).

As a result of the closure thesis, however, a paradoxical situation emerges concerning the relationship between an operationally closed system and its environment:

> "The more closed an autonomous system appears, the more open is its
> relation to the outside world." $\qquad\qquad (\,9\,)$

In other words, the concept of cognition alone turns out to be inadequate for any consistent representation of the living able to generate a cognitive domain where information is imposed on the environment and not picked up from it, as it is demanded categorically by 'second order cybernetics' [Varela, 1979 ; p.238].

That means, no self-reference is possible unless a system acquires a certain degree of freedom. But any system is only free insofar as it is capable of interpreting its environment and choosing between different interpretations for regulation of its own behavior [Günther, 1968; p.44]. Therefore, decision making processes (volition) also have to be considered for an adequate description of systems with the capability of self- generation of choices and the ability *to act* in a decisional manner upon self-generated alternatives.

— SELF-ORGANIZATION

In the following a distinction between two completely different processes of self-organization will be demanded:

> a) self-organization of data (elements, components, objects, processes) inside a system from the view of an observer of the system, and
>
> b) self-organization of the system (itself) in relation to its environment from the view of the autonomous system itself.

In other words, for the description of self-organization 'first order conceptions' again have to be distinguished from 'second order conceptions'.

a) Self-Organization of 1st Order

In this category a system is defined by an observer, i.e., a clear distinction between the system and its environment exists from the view of the observer of the system and its environment. However, this boundary between the system and the environment will not be reconstructed by the (autonomous) system itself. The boundary only exists for the external observer and both domains of distinction are well defined: what belongs to the system, belongs to the system, what belongs to the environment, belongs to the environment. Both tautologies are dualistic, i.e., what does not belong to the system, belongs to the environment, and what does not belong to the environment, belongs to the system.

In other words, the unambiguity of the difference between the system and the environment as defined by an observer does not affect the laws of the classical (mono-contextural) logic. On the contrary, the uniquness of the difference confirms the validity of the logical identity principle. The changes of the system described by the (external) observer are changes within the system represented by a set of parameters chosen by the observer for the definition and adequate description of the system under consideration.

All non-linear theories of dynamical systems and processes such as the 'theory of dissipative structures', 'synergetics', or the 'theory of determined chaos' belong to the concept of '1st order self-organization'.

b) Self-Organization of 2nd Order

Self-organization in the sense of autonomy is the self- realization of an autonomous system in its environment by at least two simultaneously interacting processes:

> i) a *volitive process* structuring the environment by determination of relevances and a corresponding context (e.g., a cognitive domain), and $\qquad$ (10a)
>
> ii) a classification and abstraction of the data within the context chosen in i) by *cognitive process* producing a representational structure of content $\qquad$ (10b) and meaning (e.g., an interpretation of the data).

Both processes are complementary to each other, i.e., neither of the two can be considered or described separately. The situation may be visualized by the following scheme [Kaehr, 1989]:

$$\left\{ \begin{array}{ll} \text{DISTINCTION 1} : (\text{SYSTEM } \underline{O} \mid \text{ENVIRONMENT } O) & (11a) \\ \text{DISTINCTION 2} : (\text{SYSTEM } \underline{O} \mid (\text{SYSTEM } \underline{O} \mid \text{ENVIRONMENT } O))) & (11b) \end{array} \right.$$

The braces in relation (11) symbolize the complementarity of the two simultaneous processes in the sense of a parallelism which cannot be linearized without describing a completely different process. Thus the operator (program) of the volitive process (11a) becomes the operand (data structure) of the cognitive system and what has been the operator of the cognitive process (11b) may change during the process into the operand of the volitive system. The logical criterion for an adequate formal description of such closely interwoven processes is the existence of a logical system which allows several simultaneous successions of deductive steps in different logical domains mediated to each other.

Only if the representation of the system is restricted to some aspects of itself as in the project of 'computational reflection' [Maes, 1988],

> "...a reflective system is a system which incorporates structures representing aspects of itself..." (12)

no logical problem will appear. However, since the autonomy of a living system is not related only to some parts of the system but to its wholeness, an unambiguous self-explication of living systems is not possible in the linguistic framework of the classical sciences.

– TOWARDS MODELLING

It is this irreducible difference, the dis-contexturality, between an autonomous system and its environment which has to be realized in modelling a system possessing its self- determined boundaries. In other words, this fundamental structure necessarily has to be repeated in the tools (the methods of notation) for any (formal) description of these processes, i.e., the system of notation must reflect the dis-contextural structure of the process if modelling is to avoid assimilation of the difference between the system/environment which is constitutive for the self-organization of an autonomous system.

A theoretical framework which offers the complexity necessary for an adequate and unambigous modelling is given by the 'Theory of Poly-Contexturality' introduced by Günther [Günther, 1980] and Kaehr [Kaehr, 1981]. This theory represents a formal and operative system of mathematical logic. Fig.2 summarizes not only the main arguments of the foregoing discussion but also gives a qualitative impression of the idea behind a poly- contextural logical system. For details refer to the literature [Kaehr, 1981; 1988; 1989].

Fig.2a, which corresponds to fig.1c, illustrates the circularity *(circulus vitiosus)* arising from any representation of self-reference, i.e., for cognitive processes if described on the basis of a mono-contextural logical system. It is well known that the name or the image (operator) of an object belongs to another logical type than the corresponding object (operator), i.e., Russell's theory of logical types (cf. fig.2d) suffices for any classification. For an operative modelling of processes such as the one given by relation (11), however, transitions between the different logical types or domains in fig.2d are necessary as indicated in fig.2b.

Fig.2b depicts two logical domains (contextures) $L_{1,2}$ in such a way that circularity (compared to fig.2a) is distributed among the two different logical domains provided the meaning of the terms are retained during the transition from one domain to another. On the other hand, the relationship between the operators and operands distributed on different logical domains escapes any circularity (or ambiguity) if the individual process is discriminated during transitions between different contextures as indicated by the indices in fig.2b. Fig.2c shows the composition of fig. 2b and fig.2e gives a graphical

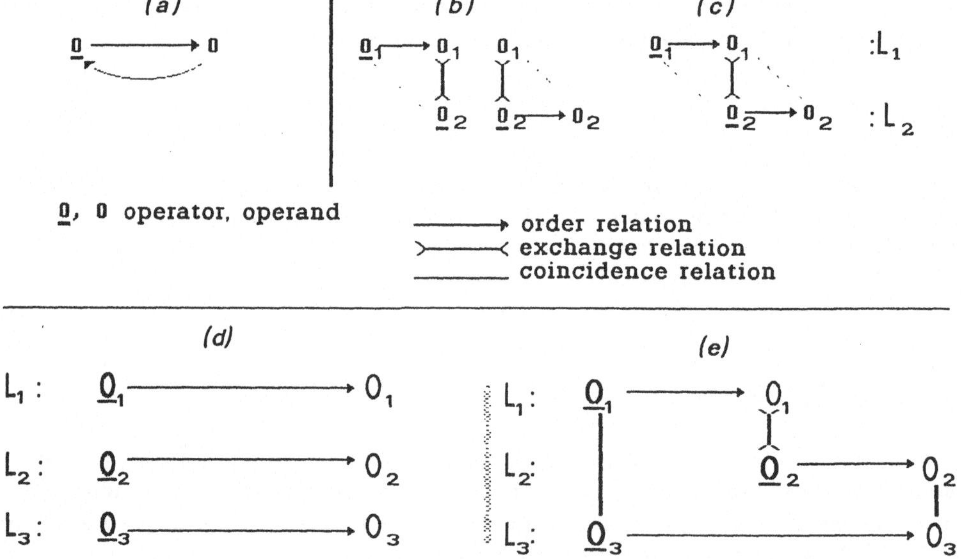

Figure 2: Graphical representation of mono- and poly-contexturality.

       a) circularity caused by self-reference in a mono-contextural system;
       b) circularity distributed on two logical (mediated) domains;
       c) composition of b)
       d) three logical domains $L_{1,2,3}$ which are isolated (not mediated), the indices may be attributed to three different types in Russell's theory of logical types;
       e) three logical contextures (a three-contextural system) $L_{1,2,3}$ as smallest (irreducible) unit in of a poly-contextural system.

representation of the smallest irreducible unit (three contextures) in the theory of poly-contexturality. Such an interchange, i.e., the distribution and mediation of domains is designed as 'heterarchy'. Heterarchically organized structures or processes belong to the category of autonomous and not to the class of input/output-systems. In the terminology of poly- contexturality, heterarchy is constituted inter-contextural, whereas intra-contexturally all descriptions (of systems or processes) are hierarchically structured. Intra-contexturally, i.e., within the logic of the contexture, the transitivity law holds rigorously, as do all classic logical rules.

The essential point of 'poly-contexturality' results from the mediation by order and exchange relations between different (at least three) contextures which is achieved by new (non- classical) logical operators such as the 'transjunction'. This allows the modelling of bifurcation from one logical domain into at least two parallel, simultaneously existing contextures. Thus a parallelism is constituted by a distributed circularity of operator and operand, which is no longer reducible to linearity as would be demanded for an adequate (formal) description of volitive and cognitive processes according to relation (11). For more details concerning the tranjunctional or the multi- negational operations it is referred to the literature [Kaehr, 1981; 1989].

— CONCLUSION

The 'Theory of Autopoietic Systems' represents the scientific attempt of a purely semantic, i.e, non-formal theory of living systems with the declared intention to develop a biological conception – this is its merit. What cannot be achieved on this basis is a sysmbiosis of computer and bio-logical sciences – the declared aim of cybernetics.

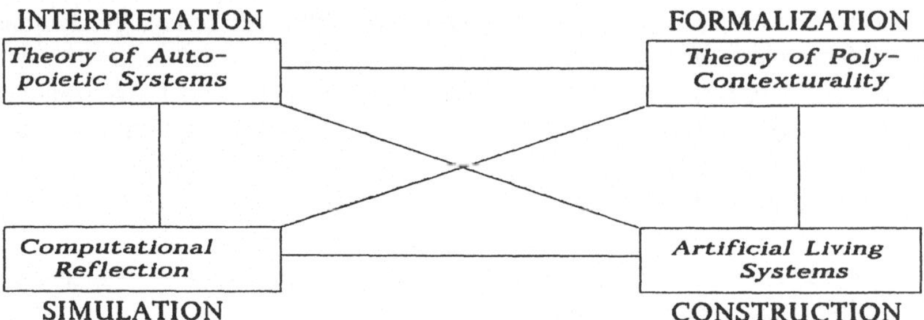

Figure 3: A scientific allocation of the different approaches towards a theory of living systems. Cybernetical research, which is founded by its interdisciplinary and transdisciplinary methodology, has to be allocated between the fields represented by the four corners. This is symbolized by the interconnections in the diagram.

Although the principial logical difficulties arising from the description and modelling of cognitive [self-referential] processes have been recognized, no efforts have been undertaken to overcome these limitations in the sense of an extension of the logical axiomatic basis. Any technical reconstruction of cognitive processes, however, necessarily requires their unambiguous representation.

The 'Theory of Poly-Contexturality' offers a semiotic framework with the degree of complexity as demanded for a non- reductionistic representation of living processes.

The project of 'Computational Reflection' for the simulation of reflective processes only describes partial autonomy since for simplicity's sake all senso-motoric aspects of the living have been excluded.

The situation is presented in fig.3, where the four corners represent 'formalization', 'interpretation' or 'modelling', 'simulation', and 'construction' resp. It is the region of the interconnections between these different scientific activities which is of special interest for modern cybernetical research.

ACKNOWLEDGEMENT
The study was supported ba a grant from the Stiftung Volkswagenwerk.

— REFERENCES

FOERSTER, VON H. (1980): 'Kybernetik einer Erkenntnistheorie', in: Sicht und Einsicht,
       Vieweg Verlag, Braunschweig.
GÜNTHER, G.(1968): 'Many-Valued Designations and a Hierarchy of First Order Ontologies'
       Akten des XIV.Interantionalen Kongresses für Philosophie, Wien.
GÜNTHER, G.(1980): Beiträge zur Grundlegung einer operationsfähigen Dialektik, Vol.I–III,
       Felix Meiner-Verlag, Hamburg.
KAEHR, R. (1981) in: Idee und Grundriss einer nicht–Aristotelischen Logik (G.Günther, ed.),
       Felix Meiner-Verlag, Hamburg.
KAEHR, R. & VON GOLDAMMER, E. (1988): 'Again Computers and the Brain', in: Journal
       of Molecular Electronics, Vol.4, p.31–37.
KAEHR, R. & VON GOLDAMMER, E. (1989): 'Poly-contextural modelling of heterarchies
       in brain functions', in: Models of Brain Function (R.M.J.Cotterill, ed.) Cambridge
       University Press, Cambridge, p. 463–497.
MAES, P. (1988): 'Computational Reflection', in: The Knowledge Engineering Review,
       Vol. 3, p. 1–19.
MATURANA, H.R. & VARELA, J.F. (1980): 'Autopoiesis: The Organization of the Living',
       in: Autopoiesis and Cognition, Boston Studies in Philosophy of Science (R.S.Cohen,
       M.W.Wartofsky, eds.) Vol. 42, D.Reidel Publ., Dodrecht.
MATURANA, H.R. (1985): Erkennen: Die Organisation und Verkörperung von Wirklichkeit,
       Vieweg Verlag, Braunschweig.
VARELA, J.F.(1979): 'Principles of Biological Autonomy', in: General Systems Research
       (G.Klir, ed.) Vol. 2, North Holland Publ., New York.
WIENER, N. & ROSENBLUTH, A. (1943) in: Philosophy of Science, Vol. 10, p.18.

MEDIZIN

USE OF TRANSPUTERS IN BIOMEDICAL SIMULATION

F.J. Pasveer
Institute of Technology, HR&O
Dpt. of technical computer science
Rotterdam. The Netherlands

Summary. The mathematics of the electrical analogon of catheter-manometer systems is completely known by partial differential equations. The equations are solved by the method of characteristics, leading to a set of coupled normal differential equations with wave solutions in forward and backward direction. The equations are solved with the help of parallel operating transputers, in order to save computation time. This article presents a first attempt to parallelize the differential equations, with data exchange through channels, the synchronous communication path for coupled concurrently operating processes. The program has been written in OCCAM, a known language for parallel processing with transputers.

1. Introduction.
In biomedical simulations often the degree of mathematical complexity plays an important role in model setup, and in restrictions due to insufficient computation power.
We tried to implicate a transputer set in the simulation of a catheter-manometer-system, a biomedical problem with interesting aspects for simulation and for measurements with the model [4].
The transputer set is attached to a pc AT. The Host is programmed to control the simulation and to process measurement data from the simulation. The actual catheter-manometer-system simulation is implemented in the transputer set, attached to the Host. So, in fact two entirely different items have been coupled, just as if they were physical entities.

2. Transputer.
Recently, the transputer has become much more important in various fields of science, especially where many computations have to be done. Picture processing, multi-dimensional Fourier transforms and advanced digital signal processing are known examples for usage of transputers. Basically the transputer is a single chip computer, designed for purposes of parallel processing. The architectonic layout implies a processor (32 bits) connected to an on-chip-RAM (2 K byte), to 4 link interfaces, to an extended memory interface (4G byte addressing space) as well as to various other on-chip functions. These units are integrated on a single wafer and are connected through a 32 bits wide internal bus. Apart from the 4 link interfaces, we can think of a normal single chip microprocessor, when dealing with a transputer. However, the link interfaces guarantee a synchronous communication with other transputers attached to the one in discussion.

3. OCCAM.
The transputer is relative easily programmed in the language OCCAM. This language yields special features for parallellism in processes. Arbitrary whether these processes run in the same transputer or in different transputers. Synchronous communication is guaranteed through the mechanism of high speed channels. Details about programming in Occam are found in [1] and in [2].
Some elements of this programming language are further explained below to give the reader a better insight in the use of this interesting method of parallel programming.

Basically OCCAM knows three primitive processes:
```
    1. assignment process:        jim := one + two
    2. input process:             chann ? memvar
    3. output process:                    chann ! memvar
```
The assignment process in OCCAM looks like a statement in other programming languages. The input and output processes need a channel, (called chann above). In case of an input process, the contents of the channel is transported to a memory location, called memvar (memory variable) above. Similarly, the output process transports the contents of the location in memory to the channel with the name chann. Input and output primitive processes are recognized by the query (?) and the exclamation point (!) respectively.

Declarations of all variable and channel names must explicitly be done in OCCAM, as will be shown in examples below.

A second feature of OCCAM is the possibility for sequential and parallel execution of primitive processes. We must explicitly force OCCAM to perform primitive processes in series. Note that this is considered to be "normal" in familiar languages. Similarly, we must explicitely tell OCCAM to perform primitive processes concurrently. Within a list of sequential processes data exchange occurs through variables in memory, as is recognized in our familiar languages. But, data exchange between parallel operating processes can only be done through channels. This is obvious if you think in terms of parallel processes operating in different transputers. The different transputers can only communicate through their links!

Consider the following program segment:
```
    CHAN OF REAL32 realchann:
    CHAN OF INT integchann:
    INT j,k :
    REAL32 fred,puck,nextvar:
    SEQ
       fred := fred + puck
       nextvar := nextvar * puck
       realchann ! nextvar
       j := ( k + 32 ) - 12
       integchann ! j
```
In this program segment we observe the declaration of two channels. One channel for real data transfer, 32 bits wide. The addition of 32 in the specification for real munbers must be given in OCCAM. The second channel is for integer data transfer, 32 bits wide if not specified. All variables have been declared too. Each of these declarations end with a colon (:) indicating that they belong to the process immediately following, i.e. the SEQ process. Indentation is obligatory in OCCAM. It operates as a sort of delimiter for so called constructs. The keyword SEQ introduces a construct, yielding the sequence of primitive assignment and output processes to be executed sequentially.

Since only dyadic operations are allowed, the primitive process calculating j in the example, is forced to that operation by the brackets in the appropriate line of source text.

The program segment above needs a simultaneously operating process, attached to it, in order to be able to communicate with. Simply stated: each exclamation point (!) in a process must be accompanied by a query (?) in some other parallel operating process and vice versa.

The program segment below shows two sequential processes operating concurrently. The parallellism is governed by the construct PAR, and both sequential processes by SEQ.
```
    CHAN OF REAL32 chann:
    PAR
```

```
      REAL32 varabove:
      SEQ
        SQRTP(3.1415926(REAL32) *  2.0(REAL32),varabove)
        varabove := 1.0(REAL32) / varabove
        chann ! varabove
      REAL32 varbelow:
      SEQ
        varbelow ? chann
        varbelow :=  etcetera
```
These two processes operate concurrently. The first sequential
process computes the reciprokal of the square root of two times π.
The answer is used in the second SEQ process. Data exchange takes
place through the channel chann, declared for transfer of 32 bits
wide real numbers above and in line with the keyword PAR since this
channel belongs to the PAR construct.

4. Catheter-manometer-system.

In the catheter-manometer-system, pressure waves are transmitted from
heart side to the distal end, where a manometer is for observation of
the blood's pressure behaviour inside the heart. The analogon used
for the catheter-manometer is the electrical transmission line,
terminated by a capacitor. The voltage along the line is calculated
by superposition of forward and backward travelling wave fronts. The
basic partial differential equations for the transmission line are:

$$C \frac{\delta v}{\delta t} + \frac{\delta i}{\delta x} = -G.v \qquad \text{and} \qquad L \frac{\delta i}{\delta t} + \frac{\delta v}{\delta x} = -R.i$$

R, L, C and G are the resistance, inductance, capacitance resp:
conductance per unit of length. The variables v and i are the voltage
and current along the line, i.e. v(x,t) resp: i(x,t). By introducing
two auxilliary variables fie and psi, the voltage v(x,t) and i(x,t)
is found by:

$$v = \frac{fie + psi}{2} \qquad \text{and} \qquad i = \frac{fie - psi}{2} . SQRT(C/L)$$

In the equations above fie(j) and psi(j) form the forward, resp.
backward travelling wave fronts. Their normal differential equations
with respect to time become:

$$\frac{d\ fie}{dt} = d_1 * fie + d_2 * psi \qquad \text{and} \qquad \frac{d\ psi}{dt} = d_1 * psi + d_2 * fie$$

in which the constants d1 and d2 are determined by the parameters R,
L and C of the transmission line.
The conductance G can be neglected. Note here that fie and psi are
not written in their Greek characters. Introducing difference
equations by discretization in time and space, the accumulation
algorithms for the j-th compartment become:

$$fie_n(j) = d_1 * fie_o(j-1) + d_2 * psi_o(j-1) \quad \text{and}$$
$$psi_n(j) = d_1 * psi_o(j+1) + d_2 * fie_o(j+1)$$

Details about these equations can be found in [3] and [4].
Observe the forward travelling effect in fie(j). Previous values are
taken at the (j-1) th discretisation to compute the recent value.
Similarly for psi(j) in opposite direction. Figure 1 explains this
aspect once more, by two layers at two successive time steps. At
layer t_n in Figure 1 all fie and psi data are known. These data form
the distribution of the wave behaviour along the catheter at specific
instant of time t_n. At layer t_{n+1} the fie and psi distribution has
still to be estimated according to the difference equations. This
means that the contribution to fie must be derived from the values at
time t_n, of the left layer in Figure 1, indicated by the arrow

pointing to the right. Similarly for the estimation of the new value of psi from the next compartment in sequence.

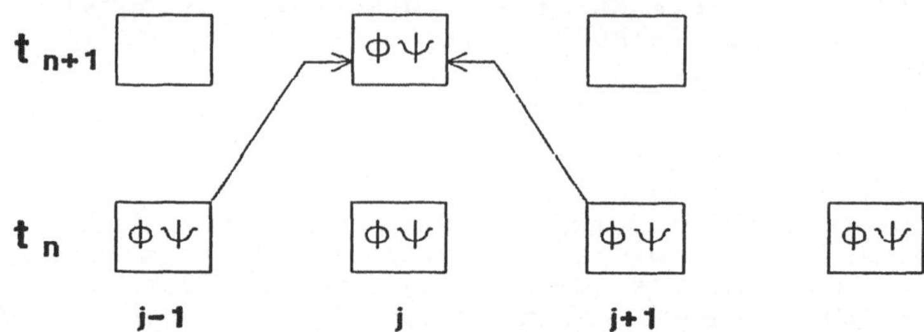

Figure 1. fie, psi forward and backward going wave fronts.

5. Implementation on a transputer.

Apart from the first and the last compartment in the discretization of the transmission line all numerical actions are entirely the same. They can be brought in separate parallel working sequential processes. The data transport between the separate sequential processes takes place through channels. Once the data transport has been done in forward and backward direction the individual computations per compartment can start simultaneously. In the first compartment the "heart excitation" is simulated by a step function. The reflection at the heart side is only a change in phase, so the reflection coefficient is minus 1.

At the manometer side a capacitor is present, introducing a separate difference equation to be solved. See ref: [3] for details. In order to be able to observe wave behaviour along the line each voltage in all compartments in each time step is fed to a separate process, called the measurement process. This final process is fed by the other processes via a channel too.

The parallel setup is shown in Figure 2.

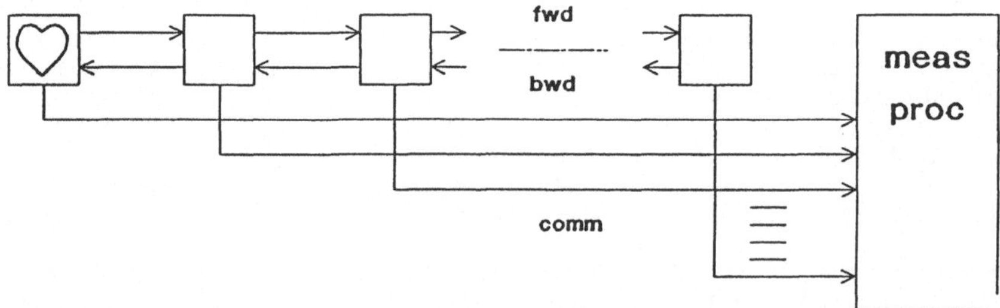

Figure 2. setup parallel process together with measurement process.

The source text of the parallel setup in figure 2 in an OCCAM program is partially shown below. Three separate procedures have been made. Two procedures for the excitation and manometer segment. A third procedure has been made for all intermediate parts. However, this procedure "parallelizes" internally!

Important lines of source text are given below to show the program flow.

```
PROCEDURE catheter(CHAN OF ANY keyboard,screen)
-- addition of auxilliary libraries
VAL cathlength IS 10:
PROTOCOL datastream IS REAL32;REAL32:
[cathlength-1]CHAN OF datastrean fwd:
```

```
[cathlength-1]CHAN OF datastream bwd:
[cathlength]CHAN OF REAL32 comm:
[cathlength]REAL32 fie,psi:
PROC leftsegment()
  REAL32 bfie,bpsi,hufie,hupsi:
  SEQ
    fwd[0] ! fie[0];psi[0]
    bwd[0] ? bfie;bpsi
   -- solve difference equations and reflection
    comm[0] ! fie[0] + psi[0]
:
PROC intersegment()
  PAR m = 1 FOR (cathlength-2)
    REAL32 ffie,fpsi,bfie,bpsi,hufie,hupsi:
    SEQ
       PAR
         SEQ
            fdw[m-1] ? ffie;fpsi
            bdw[m-1] ! fie[m];psi[m]
         SEQ
            fdw[m]    ! fie[m];psi[m]
            bdw[m]    ? bfie;bpsi
       SEQ
         hufie := (d1*ffie)+(d2*fpsi)   -- compute new fie
         hupsi := (d1*bpsi)+(d2*bfie)   -- and new psi value
         fie[m] := hufie
         psi[m] := hupsi
         comm[m] ! fie[m] + psi[m]
:
PROC rightsegment()
  REAL32 ffie,fpsi,hufie,hupsi,huv:
  SEQ
    fwd[cathlength-2] ? ffie;fpsi
    bwd[cathlength-2] ! fie[cathlebgth-1];pri[cathlength-1]
    -- solve difference equations and reflection
    -- due to the manometer capacity
    fie[cathlength-1] := hufie
    psi[cathlebgth-1] := hufie
    comm[cathlength-1] ! fie[cathlength-1]+psi[cathlength-1]
:
PROC measurement()
  [cathlength]REAL32 vout:
  SEQ
    -- read all comm[] values and store in an
    -- intermediate array
:
```

Up to now the main procedures have been mentioned.
The variable names ffie, fpsi stand for forward fie and psi values,
to be read from the next higher neighbouring process. Similarly for
bfie and bpsi in the opposite direction. The variables hufie and
hupsi are intermediate names for new values of fie and psi, which
have to be written in the arrays fie and psi which are globally
declared here! Observe that the procedures begin with the keyword
PROC and end with the colon(:).
The program source text continues with computation of all constant
values like d1 and d2. Finally, the parallel loop in the OCCAM
program looks like:

```
    iks := 0
    WHILE iks < timelength    -- timelength, previously declared
```

```
    PAR
      leftsegment( )
      intersegment( )
      rightsegment( )
      measurements( )
    SEQ
      post processing of data
```

6. Preliminar result with one transputer.

Restricting this catheter-manometer simulation to 10 compartments, it is easy to run the entire model in a single transputer, on the transputer evaluation board attached to the Host.
Of the first 20 time steps the numerical results are shown in the table below. The model had been excited by a step function with a height of 10.

```
10.0    0.0    0.0    0.0    0.0    0.0    0.0    0.0    0.0    0.0
10.0   10.0    0.0    0.0    0.0    0.0    0.0    0.0    0.0    0.0
10.0   10.0    9.9    0.0    0.0    0.0    0.0    0.0    0.0    0.0
10.0   10.0    9.9    9.9    0.0    0.0    0.0    0.0    0.0    0.0
10.0   10.0   10.0    9.9    9.8    0.0    0.0    0.0    0.0    0.0
10.0   10.0    9.9    9.9    9.8    9.8    0.0    0.0    0.0    0.0
10.0   10.0   10.0    9.9    9.9    9.8    9.8    0.0    0.0    0.0
10.0   10.0    9.9    9.9    9.8    9.8    9.8    9.7    0.0    0.0
10.0   10.0   10.0    9.9    9.9    9.8    9.8    9.7    9.7    0.0
10.0   10.0    9.9    9.9    9.9    9.9    9.8    9.8    9.7    2.6
10.0   10.0   10.0    9.9    9.9    9.8    9.8    9.7    2.7    4.9
10.0   10.0    9.9    9.9    9.9    9.9    9.8    2.8    4.9    6.8
10.0   10.0   10.0    9.9    9.9    9.8    2.8    5.0    6.8    8.5
10.0   10.0    9.9    9.9    9.9    2.9    5.0    6.9    8.5    9.9
```

Discussion.
This presentation is only a first experiment in modelling of a catheter-manometer-system by a transputer. The aim is to implement completely the catheter model in a multi-transputer setup.
At this stage only one transputer has been used. With more coupled transputers available, it can be imagined that heart beat data will be fed in a separate transputer, excitating the rest of the model. In principle the same program code can be used. Some more text is necessary to place parts of code into appropriate transputers. Then they really communicate through their channels. In a single transputer the same concept of channel communication is used between the (quasi) parallel operating processes in order to be compatible with multi transputer programs. And in fact this is a powerful feature of OCCAM. One can prepare and test multi transputer programs on a single transputer!
The computational work has been taken away from the Host, leaving the Host free for other work, for example compensation of manometer data. This last aspect is still a topic of great interest, but could not yet be handled adequately due to lack of sufficient computation power, even in an AT. It is the author's opinion that with the pc set of T800 transputers (with a floating point processor on board) computation times can be reduced considerably.
Some aspects for further research.
A point of further research is to examine the possibility of real time simulation with a set of T800 transputers attached to a AT Host computer.
If real time simulation would be possible one can try to perform catheter-manometer-compensation by such a transputerised catheter.
This compensation will be initialized within a couple of months,

together with some students at the HR&O in Rotterdam.
Inclusion of an air bubble in the simulation does not introduce more
complexity, since it is only another process in the parallel system.
This supplement of the catheter-manometer-system is point of interest
too.
A further point of interest is the Wormersly correction. A separate
process parallel to the catheter manometer simulating processes may
be programmed to find running spectra of the excitation function.
From the running spectrum varying values for the catheter parameters
R, L and C can be derived and can be transported to the parallel
operating processes of the catheter itself. This laborious task is
beyond the possibilities of a normal AT and must be parallelized
adequately.

Acknowledgement.
The transputer set, applied for this research work has been put at my
disposal at the HR&O (Institute of Technology in Rotterdam) by Mr.
A. Kleinendorst of TME at Heeswijk Dinther near s'Hertogenbosch in
the Netherlands.
The Host computer has been put at my disposal at the HR&O by Mr. L.
Kuipers of Kuipers Electronics (telemonitoring systems and industrial
microcomputers) in Zwijndrecht near Rotterdam.
This preliminary work has encouraged me to involve students at the
HR&O who I have found to be very interested in parallel processing
with transputers. It provides them with a new tool to become familiar
with in a broad sense of applications, both in hardware and in
software.
Furthermore this initial transputer implemented catheter-manometer-
system is a following basis for further cooperation with the
Laboratory for Biomedical Techniques of Mr. R van Wijk van Brievingh
at the Delft University of Technology.

References.

[1]   Dick Pountain and David May.
       A tutorial introduction to Occam programming.
       Inmos. BSP professional books.
[2]   OCCAM 2. Reference manual.
       Inmos. Limited.
[3]   S.W. Brok. and E.E.E. Frietman.
       Parallel implementation of a catheter manometer system
       on the DPP84.
       Proceedings European Simulation Multiconference. June 1988.
       Nice, France. pp 376 - 379.
[4]   R.P. van Wijk van Brievingh. and F.J. Pasveer.
       Simulation of Catheter-Manometer Dynamic Response.
       Advances in System Analysis. Volume 5: System Analysis
       of Biomedical Processes. Edited by Dietmar P. F. Möller.

Simulation of the effects of potassium-channel inhibiting drugs on the duration of cardiac action potentials

Hafner, D., Berger, F., Borchard, U., Stöcker, K.
Institute of Pharmacology
University of Düsseldorf
Moorenstr.5, 4000 Düsseldorf, FRG

Summary: The mathematical model of DIFRANCESCO and NOBLE (6) is a complex summary of experimental knowledge about electrophysiological processes in cardiac cell membranes. A simulation series was performed describing the role of theoretical model support in a concrete pharmacological problem giving rise to possible interpretation of our own experimental results compared to contradictory data from the literature. Model simulation suggest that the effects of potassium channel inhibiting substances on cardiac action potentials may be dependent on the duration of the action potential itself or the underlying individual constellation of ionic current components.

Zusammenfassung: Das mathematische Modell von DIFRANCESCO und NOBLE (6) stellt eine komplexe Zusammenfassung experimenteller Erkenntnisse von den elektrophysiologischen Vorgängen an Herzzellmembranen dar. Zu einer konkreten pharmakologischen Fragestellung wird eine Simulationsstudie beschrieben, die zur möglichen Erklärung eigener und scheinbar widersprüchlicher Resultate aus der Literatur geführt hat. Dabei ergab sich, daß die Wirkung von Substanzen, welche die Kaliumionenkanäle blockieren, auf das kardiale Aktionspotential verschiedenartig ausfällt in Abhängigkeit von der Dauer der Aktionspotentiale bzw. der aktuellen Konstellation verschiedener Stromkomponenten eines Zellsystems.

Introduction :

HODGKIN and HUXLEY (9) established the first mathematical formulation of the electrical activities of the nerve cell 40 years ago. Their theoretical approach gave a fundamental insight into the interaction of different ionic currents involved in the depolarizing and repolarizing processes and their dual dependence upon time and membrane potential.

A number of further models were developed as experimental results grew and only in the 1970´s the first mathematical formulations were applied to the more complex situation in muscle cells of different cardiac tissues such as: Ventricular muscle (1), Purkinje fibre (11) and sinus node (4). These models were still specialized to the corresponding cardiac tissue and had only few current components in common.

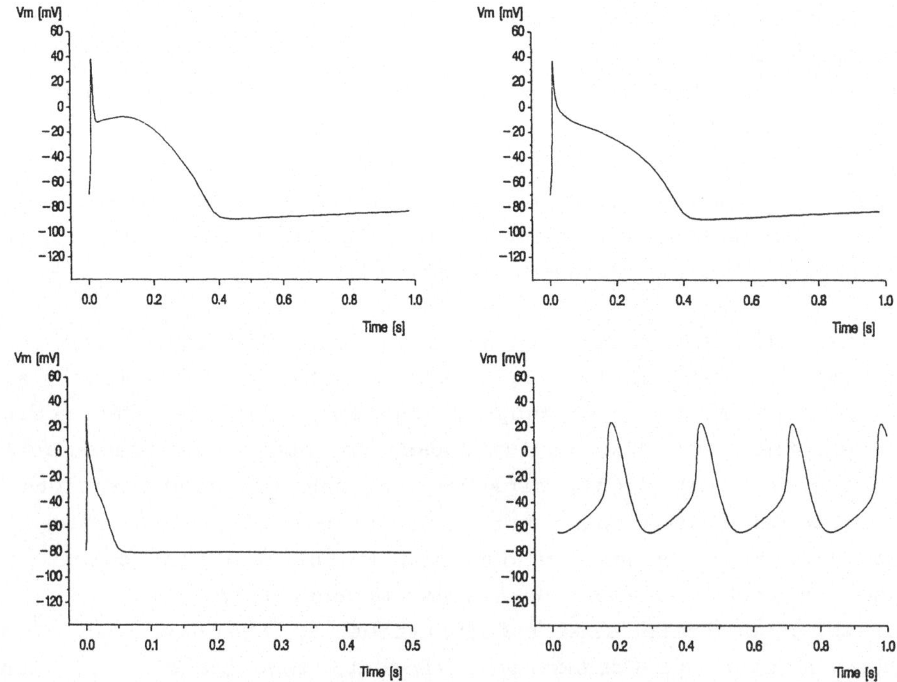

Fig. 1: Action potential time courses simulated with OXSOFT HEART
using 4 different parameter sets : PURKINJE with "notch" (**top left**),
and without "notch" (**top right**), rat ventricular cell (**bottom left**)
and sinus-node (**bottom right**).

Accumulated experimental results led to the assumption, that there may
be a number of specific and ion-selective ion channels and transport-
systems which can be identified in many different cardiac structures.
The fact of not finding a special current in a certain tissue, e.g.
the fast sodium inward current in the AV-node, was interpreted in
terms of silent or inactive ion channels.

Recently DIFRANCESCO and NOBLE (6) developed a more general mathematical model (DN-model) which may be applied to different cardiac
tissues of different biological species by only initializing the
underlying equations with different parameter sets. It has been the
aim of the authors and many other working groups to summarize the actual experimental knowledge on the basis of the DN-model and to dis-
tribute further updates to the interested electrophysiologists.

Some major new properties of the model are briefly summarized: It is
available as a source code PASCAL-program which can be implemented on
many computers ranging from PC to large main frame machines. It is
rather easy to change the model equations or to add new ones. The
current edition (OXSOFT HEART 2.0) contains parameter sets for 5
different cardiac preparations as obtained from Purkinje fibre, ven-

tricle, SA-node and atrium. Fig.1 shows simulated action potential time courses resulting from some of these parameter sets. Two different configurations of Purkinje fibre action potentials (AP) (with and without "notch") are available, exhibiting both diastolic pacemaker potentials, whereas the extremely short rat ventricular AP shows stable diastolic resting potential. A sinus node AP demonstrates elevated resting potential, rather slow depolarisation velocity but high spontaneous activity of nearly 4 Hz.

While former model equations only dealt with passive ion fluxes being modulated by Hodgkin-Huxley-type channel gates the DN-model also comprises electrogenic ion-transport systems such as the sodium-calcium exchanger and the sodium-potassium pump. As these active processes are not only highly dependent on the ion concentrations in intra- and extracellular space but also strongly influence them, the model has taken into account the dynamics of intracellular calcium and sodium and extracellular cleft potassium concentration.

In order to demonstrate some possibilities of the new model Fig.2 shows the effects of changing different conductances of ionic currents. These simulations were performed using the parameter set for the cardiac Purkinje fibre because the following simulation study deals with experimental data from this tissue.

Method :

The mathematical model of DIFRANCESCO and NOBLE (6) (OXSOFT HEART, Lic.-No. 2.024, Version 2.0) programmed in PASCAL, was implemented on a large main frame computer (SIEMENS 7580-P). Computation intervals always comprised at least a 400 s real time period, which sufficed to attain steady-state data. CPU-time was about 800 - 1000 s. Output data were transmitted to a local PC for graphical representations using SAS/GRAPH procedures.

The experimental data were obtained from isolated sheep cardiac Purkinje fibres and standard microelectrode intracellular recordings of membrane potential or two microelectrode voltage clamp experiments. For details of the experimental setup see (2).

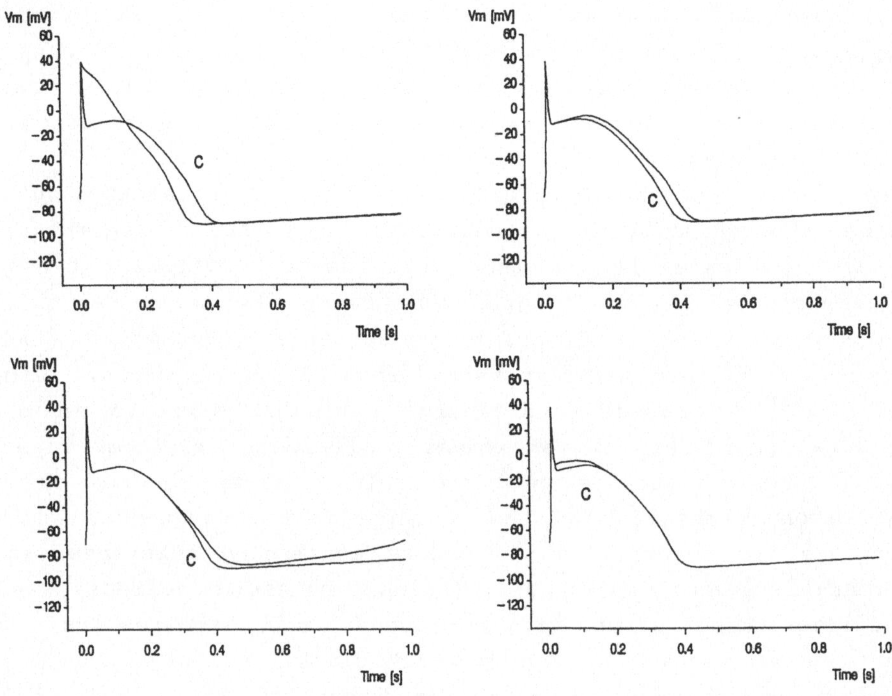

Fig. 2: Simulation of the effects of single current conductance changes on the contour of the action potential (PURKINJE with "notch"). **Top left:** Reduction of i_{to}-conductance (to 1 %). **Top right:** Reduction of I_K-conductance (to 80 %). **Bottom left:** Reduction of I_{K1}-conductance (to 60 %): **Bottom right:** Increase in I_{si}-conductance (to 200 %). C = control.

Results :

Our working group has been studying electrophysiological effects and mechanisms of antiarrhythmic drugs in cardiac tissues. Classical antiarrhythmic therapy was concentrated on drugs which interact with the cardiac sodium channel (quinidine, lidocaine) in order to reduce depolarisation- and conduction-velocity in atrial and ventricular cardiac fibres. Some effects of these drugs had also been subject to simulation studies (8). Another therapeutic principle was based on the ability of certain drugs (verapamil, diltiazem) to reduce calcium inward current thus slowing spontaneous sinus rhythm or conduction velocity in the atrio-ventricular node. As antiarrhythmic therapy still suffers from many shortcomings as e.g. unwanted side-effects or pro-arrhythmic potency of the drugs used and many unexplicabel cases of non-responders, interest is growing now for a new sort of pharmacological agents which inhibit cardiac potassium channels. Blockade of potassium outward current should lead to a prolongation of action potential and increase of refractory period thus reducing the probability of premature ectopic activity.

The study of electrophysiological mechanisms of potassium channel inhibiting substances and the discussion of their therapeutical potency is complicated by the fact, that several different types of potassium channels are existing in the heart. We performed voltage clamp investigations of effects of 4 different potassium channel blockers and found that each substance produced a different pattern of interaction with at least 4 different potassium currents in sheep cardiac Purkinje fibres. It is extremely difficult to either forecast or explain the consequences of these voltage clamp results on free running action potentials and the therapeutical consequences as long as there are even contradictory results concerning substances which interact highly selective with only a single potassium current.

From the literature (10) and from our own voltage clamp results (3) 4-aminopyridine was known as an agent which blocks in a highly selective way i_{to}-current in the concentration range of 10 - 500 µmol/l. Although these findings are generally accepted, contradictory results can be found in the literature and in our own experimental studies concerning the effects of 4-aminopyridine (4-AP) on action potential duration (APD) which was found either to be prolonged or reduced under the influence of the drug. It seemed worth while studying this situation in more detail as the question of selective current blockade upon AP-configuration is an important question in the context of antiarrhythmic treatment especially as several other potassium channel blocking substances being studied until now (e.g.(2)) show inhibition of the i_{to}-current on top of effects on other potassium channels.

In contrast to other authors (5, 10) we found that increasing concentrations of 4-AP (0 - 500 µmol/l) led to successive AP-prolongation at all frequencies of electrical stimulation tested (0.05, 0.25, 1 Hz; Fig. 3). At this point of our investigation we started simulations using the OXSOFT HEART model and looked for effects on APD when i_{to}-current was stepwise reduced. Fig.4 (right) shows the time courses of Purkinje fibre APs (stimulation frequency: 1 Hz) where i_{to}-conductance (GTO =100 µS) was reduced to 50, 25 and 1% of its control value. It can clearly be seen that APD (measured at the level of -70 mV) decreases monotonically thus showing the opposite effect to our experimental findings (Fig. 3). Similar results were obtained under

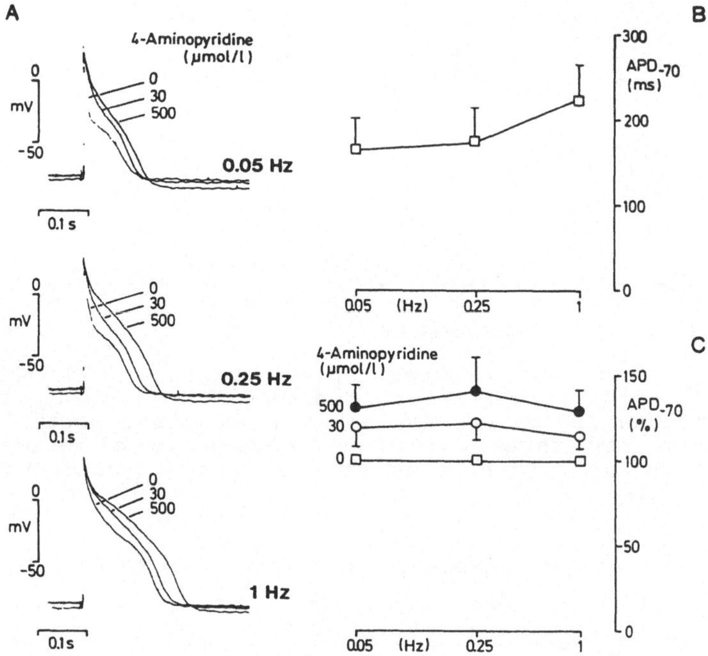

Fig. 3: Experimental results: Effects of 4-Aminopyridine (0-500 µmol/l) on duration of sheep cardiac Purkinje fibre action potentials under different frequencies of electrical stimulation (0.05 - 1 Hz). **A:** Original action potential recordings. **B:** APD in the absence of 4-AP as dependent on stimulation frequency. **C:** Frequency-dependent effects of 4-AP on APD. (n = 5).

different frequencies of electrical stimulation. Under control conditions APD shows a biphasic contour as dependent on frequency of stimulation with a maximal APD at 1 Hz , which is typical for cardiac Purkinje fibre (7). Compared to the experimental studies, we used slightly different stimulation frequencies in the model simulation, because it was intended to study frequency dependence of the 4-AP effect on both sides of the maximum. Reducing i_{to}-conductance from 100 µS down to 1 µS at 0.25, 1 and 3 Hz always led to subsequent AP-shorting (Fig.4,left). APD reduction under 3 Hz of stimulation is due to frequency-dependent effects on time-dependent membrane currents. For details of the experimental results see (3).

In a further step i_{to}-conductance was increased beyond its normal control value (GTO = 100 µS) and Fig. 5 shows the effects: Starting from GTO = 100 µS (control), GTO = 300 µS led to AP-prolongation whereas GTO = 600 µS and 1000 µS caused successive shortening of the AP even below the control value. Thus the model revealed a biphasic

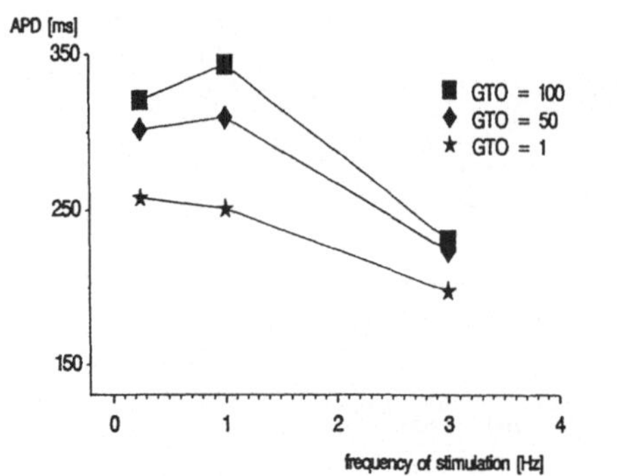

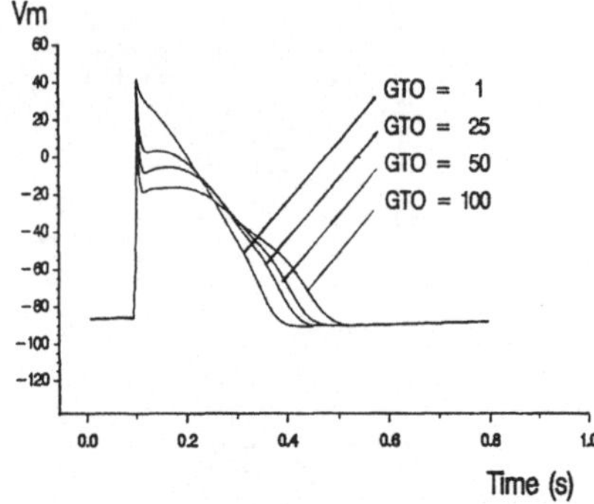

Fig. 4: Effects of reduced i_{to}-conductance GTO (μS) on APD of simulated action potentials as dependent on frequency of stimulation. **Left:** Absolute APD values evaluated at membrane potential $V_m = -70$ mV are plotted for three different frequencies of stimulation. **Right:** Action potential time courses at 1 Hz for i_{to}-conductance GTO = 100, 50, 25, 1 μS.

behaviour of APD as dependent on conductance GTO in the range of 1 - 1000 μS. It could be seen from this result that there was a possibility to reproduce AP-prolongation upon successive i_{to}-blockade: Starting with a conductance value of GTO = 1000 μS further decrease of GTO down to 600 and 300 μS was accompanied by an increase in APD. Only at lower GTO-values (< 300 μS) APD reversed its direction of change i.e. AP-shortening was observed.

In order to attain closer reproduction of our experimental data which showed APD reduction under 4-AP concentrations up to 500 μmol/l, we tried to find additional modifications of the DN-model parameter set which might shift the peak of biphasic APD-GTO-relation in the model down to lower values of GTO. This would widen the range of i_{to}-inhibition resulting in AP-prolongation as observed in the experiments.

Some aspects were considered: Duration of AP is highly influenced by repolarizing potassium currents (I_K and I_{K1}). The delayed rectifier I_K, as a time-dependent current, is activated during the AP-plateau phase. Its degree of activation increases both with more depolarized plateau levels or longer plateau durations of the action potential. Plateau level of the AP is strongly determined by the extent of i_{to}-current: With increasing i_{to}-amplitudes the triggered AP is repolarized instantaneously to lower plateau levels thus modulating the subsequent activation of I_K-current. These considerations suggested to vary some parameters important for activation and deactivation of the delayed potassium outward current I_K and to study APD changes unter i_{to}-blockade: Different methods to either enhance or diminish I_k influence

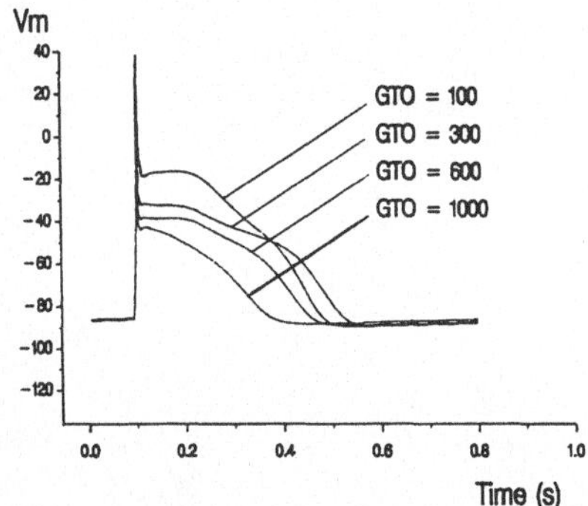

Fig. 5: When the range of i_{to}-conductance was enlarged (1 - 1000 µS) APD showed a biphasic dependence: Increase of GTO = 100 µS to 300 µS led to AP-prolongation, further increase to GTO = 600 µS and 1000 µS to subsequent AP-shortening (frequency of stimulation 1 Hz).

during the AP-plateau were considered: Increasing the activation time constant of the voltage-dependent activation gate (x) or shifting the voltage-dependent steady-state activation curve (x_∞) to less negative values of the membrane potential leads to a weakening of I_K influence and vice versa.

Membrane potential of I_K half-activation is reported to be -25 mV: This value was shifted by 10 mV to -15 mV or -35 mV (SHIFTX = +10 mV, SHIFTX = -10 mV). Potential-dependent maximal time constant of I_K is about 600 ms. This value was either doubled or halved (SPEEDX = 0.5, SPEEDX = 2.0).

Frequency-dependent AP-simulations were now performed under the extended i_{to}-conductance range (GTO = 1 - 1000 µS) and all 4 above mentioned isolated I_K parameter modifications: Summary of results for a stimulation frequency of 1 Hz are shown in Fig. 6: Reduction of I_K by using SHIFTX = +10 mV or SPEEDX = 0.5 did not shift the peak of APD-GTO relation into the desired direction. Under some GTO values action potentials did no longer return to the resting potential between two pulses of stimulation. On the whole, very long APs resulted from the reduction of I_K and most values of APD > 400 ms were very much higher than those observed in our experimental data.

Enhancement of I_K-current produced good simulations of our experimental results. Increase of I_K-current by two different approaches (SPEEDX = 2.0, SHIFTX = -10 mV) led to nearly identical results (GTO $\geq$ 300 µS). The shift of potential dependence of activation, for example, produced continuous APD reduction, as a result of decreasing I_{to}-conductances, down to GTO = 100 µS (10% of the starting value GTO = 1000 µS). Fig. 6 (right) demonstrates the corresponding time courses of simulated APs (SHIFTX = -10 mV) for GTO values of 1000 to 1 µS, where AP prolongation down to GTO = 100 µS can be observed.

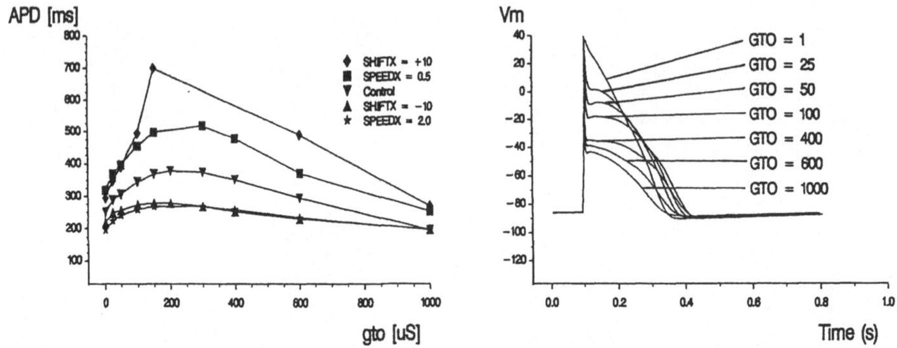

Fig.6: Summary of results from simulation data. **Left:** APD as dependent on i_{to}-conductance GTO (1 - 1000 µS): Influence of repolarizing I_K was either enhanced (SPEEDX = 2 or SHIFTX= -10) or reduced (SPEEDX = 0.5, SHIFTX = 10). For details see text. **Right:** Action potential time courses simulated for decreasing i_{to}-conductance values GTO = 1000, 600, 400, 100, 50, 25, 1 µS and enhanced influence of potassium outward current I_K (SHIFTX = -10). APs are prolonged down to GTO = 100 µS, only further decrease of GTO results in subsequent shortening of the AP.

Discussion :

Contradictory results obtained by different authors who studied the effects of 4-AP on action potential duration in cardiac Purkinje fibres were examined in experimental and simulation work. The experimental data have been published elsewhere (3). Shortening of APD was reported when 4-AP blocked transient outward current i_{to} (5, 10) while our own data (3) document prolongation of AP-duration. The former results were in accordance with initial model simulation of OXSOFT HEART using standard PURKINJE parameter sets and different frequencies of electrical stimulation (0.25 - 3 Hz). Discrepancies of this kind are often discussed in the literature using only global arguments concerning the applied experimental conditions or the different animal species used. It was the aim of our simulation study not to give an exact quantitative reproduction of the experimental findings but to see if in the context of the mathematical model there may be special ionic current constellations which may qualitatively explain AP prolongation on i_{to}-blockade. In summary, we deduced from the model that only two minor changes were sufficient to reproduce the phenomenom under consideration. Furthermore, model simulations gave also hints as to understand apparently contradictory results from similar experiments: AP-prolongation over a wide range of i_{to}-blockade could only be produced by the model if influence of repolarizing I_K-

current was enhanced. The consequence of elevated I_K in general leads to APD shortening, so that our simulated effects took place in a lower range of APD values (Fig.6, left). Weakening of I_K, in contrast, produces longer APs and shifts the peak of APD-GTO relation to higher GTO values, thus increasing the probability to observe AP-shortening on i_{to}-blockade in longer APs. This, actually, is reality in the data of KENYON and GIBBONS (10) who reported shortening of AP on i_{to}-blockade: Their shortest APs (under control conditions) are longer than our longest ones.

There are still some limitations interpreting the results of our simulations: Comparing simulated and real biological APs slight deviations in the initial repolarisation phase can be observed which is dominated by the transient outward current i_{to}. It seems that real i_{to}-current in our preparations seems to inactivate somewhat slower than the computed current. First voltage clamp based analysis suggest that the time constant of inactivation has to be reduced to about 40% of the model value. There will be, however, no principally different qualitative behaviour of the model compared to the results shown here when a reduced time constant for i_{to}-inactivation is used.

Another aspect which has to be incorporated into further model simulation studies is concerned with the nature of i_{to} itself: In sheep Purkinje fibres i_{to}-current consists of several components of which only the dominant one is represented in the model. There exists a smaller i_{to}-component, called i_{bo} in the literature, which is calcium-activated and insensitive to 4-AP, i.e. which cannot be blocked even at very high 4-AP concentrations (1 mmol/l). Therefore, until we lack precise data for incorporating this second i_{to}-component into the model, we may be consolated by the simulated results, where AP-prolongation is reproduced by reduction of GTO values in the range of 1000 μS to 100 μS. The i_{to}-current remaining when GTO is reduced to 10% of its starting value (GTO = 1000 μS) may represent the 4-AP insensitive component of i_{to}.

Thus, further model studies have to follow more closely voltage clamp analysis of single ionic current components in order to adjust the most important model parameters to the experimental reality. With these informations a more quantitative model synthesis of AP-experiments will be possible.

Besides these shortcomings of the model equations, the simulation studies have taught us, that pharmacological interventions (e.g. application of 4-AP) are not always due to exert the same unidirectional effects (either AP-prolongation or AP-shortening) under similar experimental conditions. It may rather be observed that the same interventions lead to apparently contradictory results, which are produced by different complex constellations of ionic currents in the individual preparations.

As to the therapeutic relevance of the effects described above it may be beneficial to the heart if a drug tends to shorten relatively long APs and to prolong the shorter ones, thus enhancing homogenity of refractory periods in the cardiac tissues and reducing the probability for ectopic activities to find unphysiological pathways.

Literature:

1. Beeler, G.W., Reuter, H. Reconstruction of the action potential of ventricular myocardial fibres. J.Physiol., Lond., 268, 177-210 (1977).
2. Berger, F., Borchard, U., Hafner, D. Effects of (+)- and (±)-sotalol on repolarizing outward currents and pacemaker current in sheep cardiac Purkinje fibres. Naunyn-Schmiedeberg's Arch.Pharmacol. 340, 696-704 (1989)
3. Berger, F., Borchard, U., Hafner, D., Kammer, T., Weis, T. Transient outward current blockade and correlated effects on action potential duration in sheep cardiac Purkinje fibres. Submitted for publication: Naunyn-Schmiedeberg's Arch.Pharmacol. (1990)
4. Bristow, D.G., Clark, J.W. A mathematical model of primary pacemaking cell in SA node of the heart. Am.J.Physiol., 243, H207-H218 (1982).
5. Coraboeuf, E. , Carmeliet, E. Existence of two transient outward currents in sheep cardiac Purkinje fibres. Pflügers Archiv 392, 352-359 (1982).
6. DiFrancesco, D., Noble, D. A model of cardiac electrical activity incorporating ionic pumps and concentration changes. Phil.Trans.R. Soc.Lond.B 307,353-398 (1985).
7. Fedida, D., Boyett, M.R. Mechanism underlying the shortening of the action potential at high and low stimulation rates in sheep Purkinje fibres. Proc.R.Soc.Lond.B 225, 481-502 (1985).
8. Hafner, D., Berger, F., Borchard, U. Simulation in electrophysiological pharmacology: Specific interactions of antiarrhythmic agents with ion channels of the cardiac cell membrane. In: Advances in System Analysis, D.P.F.Möller (Ed.), Vieweg-Verlag Braunschweig/ Wiesbaden (1987).
9. Hodgkin, A.L., Huxley, A.F. A quantitative description of membrane current and its application to conduction and excitation in nerve. J.Physiol.,Lond.,117, 500-544 (1952).
10. Kenyon, J., Gibbons, W.R. 4-Aminopyridine and the early outward current of sheep cardiac Purkinje fibres. J.Gen.Physiol. 73, 139-157 (1979).
11. McAllister, R.E., Noble, D., Tsien, R.W. Reconstruction of the electrical activity of cardiac Purkinje fibres. J.Physiol.,Lond., 251, 1-59 (1975).

Modern Control Theory as a Tool to Describe the Biomathematical Model of Granulocytopoiesis

E.P. Hofer*, B. Tibken* and T.M. Fliedner**
*Department of Measurement, Control and Microtechnology
**Institute of Clinical Physiology, Occupational and Social Medicine
University of Ulm, D-7900 Ulm, P.O. Box 4066

Introduction

Pertubations of the human granulocytopoiesis for which many different reasons can be named have been described with the aid of a biomathematical model by Fliedner & Steinbach [1]. Their model consists of a set of coupled nonlinear differential equations describing the time evolution of the cell content of 7 different cell compartments. The most fundamental of these compartments is the stem cell pool because in case of a pertubation the number of remaining stem cells influences the whole process of recovery. According to Fliedner & Steinbach only a small amount of surviving stem cells is sufficient for complete restauration. Thus, it is a very important issue to estimate the stem cell pool size in order to predict the future behaviour of the granulocytopoiesis. Work towards this direction has already been done by Fliedner, Steinbach and Szepesi [2].

The main issue of this report is to further analyse the biomathematical model and to show up the links to modern control theory. It turns out that the model of Fliedner & Steinbach can be decomposed into bilinear subsystems which are coupled both directly and via dynamic nonlinear regulators. Bilinear systems are a special class of nonlinear systems and play a major role in modern control theory because every nonlinear input-output system can be approximated arbitrarily well by a bilinear system. In addition for bilinear systems many structural properties are known.

Fliedner & Steinbach Model of Granulocytopoiesis

Fliedner and Steinbach have published a biomathematical model of granulocytopoiesis [1] which is based on the scheme in Figure 1 where the basic compartments considered by Fliedner & Steinbach are shown. In this model a stem cell pool (S) continuously delivers stem cells into the cell cycle which consists of two compartments for the progenitor cells (CBM/CBL), one compartment for the precursor cells (P), reserve cells

(R), maturing cells (M) and the granulocytes in the functioncompartment (F). The stem cells differentiate during this cell cycle into granulocytes. Two hormon compartments (Reg.I) and (Reg.II) are introduced to control the whole process. More details of the model can be found in [1],[2]. In the next section we will give the differential equations governing the dynamical behaviour of the model. These equations will be interpreted in terms of modern control theory. A block diagram shows the structure of the feedback control scheme.

State Space Model and Control Scheme of Granulocytopoiesis

The derivation of the state space model needs the introduction of the so called state variables x_i, input variables u_j and output variables y_k. Every cell content of the Fliedner & Steinbach model obeys a differential equation of first order. This leads to the following definition of the state variables

$$x_1 = S \ , \quad x_2 = CBM_1 \ , \quad \ldots \ , \quad x_{11} = CBM_{10} \ ,$$
$$x_{12} = CBL_1 \ , \quad \ldots \ , \quad x_{21} = CBL_{10} \ , \quad x_{22} = P_1 \ , \quad \ldots \ , \quad x_{31} = P_{10} \ , \quad (1)$$
$$x_{32} = M \ , \quad x_{33} = R \ , \quad x_{34} = F \ , \quad x_{35} = Reg.I \ , \quad x_{36} = Reg.II \ .$$

Auxiliary input and output variables are defined by

$$u_1 = \gamma_1 \exp(-\nu_1 x_1) + \gamma_2 \exp(-\nu_2 [x_2 + \ldots + x_{11} + x_{22} + \ldots + x_{33}]) + \gamma_3 \ ,$$
$$u_2 = \gamma_4 - \gamma_5 \exp(-\nu_3 x_{35}) \ , \quad u_3 = 2(1 - \rho) u_1 x_1 \ , \quad u_4 = f(x_2 + \ldots + x_{11}) \ ,$$
$$u_5 = \lambda_c x_{11} \ , \quad u_6 = \gamma_6 - \gamma_7 \exp(-\nu_4 x_{35}) \ , \quad u_7 = \lambda_p x_{31} \ ,$$
$$u_8 = \gamma_8 - \gamma_9 \exp(-\nu_5 x_{36}) \ , \quad u_9 = u_8 x_{33} \ ,$$
$$u_{10} = \gamma_{10} \exp(-\nu_6 [g_1 x_1 + g_2 (x_2 + \ldots + x_{11}) + g_3 (x_{22} + \ldots + x_{34})]) \ ,$$
$$u_{11} = \gamma_{11} \exp(-\nu_7 x_{34}) \ ,$$
$$y_1 = x_1 \ , \quad y_2 = \lambda_c x_{11} \ , \quad y_3 = x_2 + \ldots + x_{11} \ ,$$
$$y_4 = \lambda_p x_{31} \ , \quad y_5 = x_{22} + \ldots + x_{31} \ ,$$
$$y_6 = x_{32} + x_{33} \ , \quad y_7 = x_{33} \ , \quad y_8 = x_{34} \ ,$$
$$y_9 = x_{35} \ , \quad y_{10} = x_{36} \ .$$

$$(2)$$

Using the previous definitions the state space model for granulocytopoiesis is given by

$$\dot{x}_1 = (2\rho - 1)u_1 x_1 \ ,$$

$$\dot{x}_2 = u_3 - \lambda_c x_2 + u_2 x_2 - u_4 x_2 + \phi x_{12} \ ,$$

$$\vdots$$

$$\dot{x}_i = \lambda_c x_{i-1} - \lambda_c x_i + u_2 x_i - u_4 x_i + \phi x_{i+10} \ , \quad i = 3,\ldots,10 \ ,$$

$$\vdots$$

$$\dot{x}_{11} = \lambda_c x_{10} - \lambda_c x_{11} + u_2 x_{11} - u_4 x_{11} + \phi x_{21} \ ,$$

$$\dot{x}_{12} = u_4 x_2 - \phi x_{12} \ ,$$

$$\vdots$$

$$\dot{x}_i = u_4 x_{i-10} - \phi x_i \ , \quad i = 13,\ldots,20 \ ,$$

$$\vdots \tag{3}$$

$$\dot{x}_{21} = u_4 x_{11} - \phi x_{21} \ ,$$

$$\dot{x}_{22} = u_5 + u_6 x_{22} - \lambda_p x_{22} \ ,$$

$$\vdots$$

$$\dot{x}_i = \lambda_p x_{i-1} + u_6 x_i - \lambda_p x_i \ , \quad i = 23,\ldots,30 \ ,$$

$$\vdots$$

$$\dot{x}_{31} = \lambda_p x_{30} + u_6 x_{31} - \lambda_p x_{31} \ ,$$

$$\dot{x}_{32} = u_7 - \lambda_M x_{32} \ , \quad \dot{x}_{33} = \lambda_M x_{32} - u_8 x_{33} \ ,$$

$$\dot{x}_{34} = u_9 - \lambda_F x_{34} \ ,$$

$$\dot{x}_{35} = u_{10} - \lambda_{Reg.I} x_{35} \ , \quad \dot{x}_{36} = u_{11} - \lambda_{Reg.II} x_{36} \ .$$

For the parameters in this model the following data are used [3]

$$\gamma_1 = 3.0E-2 \ , \quad \gamma_2 = 9.0E-3 \ , \quad \gamma_3 = 3.0E-2 \ , \quad \gamma_4 = 1.0E-1 \ ,$$

$$\gamma_5 = 7.85E-2 \ , \quad \gamma_6 = 1.0E-1 \ , \quad \gamma_7 = 7.5796E-2 \ , \quad \gamma_8 = 8.0E-2 \ ,$$

$$\gamma_9 = 7.0E-2 \ , \quad \gamma_{10} = 2.0 \ , \quad \gamma_{11} = 1.993 \ ,$$

$$\nu_1 = 2.2507286E-9 \ , \quad \nu_2 = 6.138163E-12 \ , \quad \nu_3 = 2.49123047E-3 \ ,$$

$$\nu_4 = 1.3422069E-5 \ , \quad \nu_5 = 6.31789013E-3 \ , \quad \nu_6 = 3.66235E-4 \ , \tag{4}$$

$$\nu_7 = 3.70554E-12 \ ,$$

$$\phi = 1.1 \ , \quad \lambda_c = 8.0E-2 \ , \quad \lambda_p = 1.0E-1 \ ,$$

$$\lambda_F = 8.0E-2 \ , \quad \lambda_M = 8.0E-2 \ .$$

A block diagram associated with equations (1), (2), (3) is shown in **Figure 2**. The blocks labeled as in the **Fliedner & Steinbach** model represent dynamical systems which are interconnected by the inputs and outputs. In this scheme the basic control loops containing (**Reg.I**) and (**Reg.II**) are easily identified. An important special point of the equations (3) is that they are linear with respect to the x_i and the u_j, but not jointly linear. In the literature systems of this structure are called bilinear [4],[5]. Bilinear Systems have been investigated in detail during the recent years [6],[7],[8]. The next section introduces this class of systems from a control viewpoint and gives a theorem which is useful in further investigating the granulocytopoiesis model.

Bilinear Systems

General bilinear systems are given by the state space representation

$$\dot{\underline{x}} = \underline{A}\,\underline{x} + \underline{B}\,\underline{u} + \sum_{i=1}^{m} u_i\,\underline{N}_i\,\underline{x} \ , \quad \underline{x}(0) = \underline{x}^0 \ ,$$

$$\underline{y} = \underline{C}\,\underline{x} \ ,$$

(5)

where $\underline{x} \in R^n$ represents the state vector, $\underline{u} \in R^m$ the input vector and $\underline{y} \in R^p$ the output vector. The rectangular matrices $\underline{A}, \underline{B}, \underline{N}_i, \underline{C}$ are of appropriate dimensions. Bilinear systems have been used to model a number of technical and nontechnical systems. As an overview the article of Mohler and Kolodziej [9] is recommended. These systems are especially important because every nonlinear system can be approximated to any desired accuracy by a bilinear system of appropriate dimension [10].

In the previous paragraphs the Fliedner & Steinbach model has been divided into interconnected bilinear systems as shown by the block diagram. The bilinear structure of the mathematical model (3) is the basis for a theorem to prove the nonnegativity of the solution to this system. This guarantees the nonnegativity of the cell content in all compartments of the model under the assumption that the initial cell contents are nonnegative and the input functions $u_i(t)$ are nonnegative for all $t \geq 0$. Although this is evident from a biomedical viewpoint it is a nontrivial result from the mathematical standpoint.

Theorem: *If all nondiagonal elements of the matrices $\underline{A}, \underline{N}_i$ and all elements of the matrix $\underline{B}$ are nonnegative the solution to (5) corresponding to $x_i(0) \geq 0$, $i = 1,\ldots,n$, $u_i(t) \geq 0$, $i = 1,\ldots,m$, $\forall t$ obeys $x_i(t) \geq 0$, $\forall t \geq 0$, $i = 1,\ldots,n$.*

Using results from [11] for timevarying linear systems the proof of the forgoing theorem is rather easy.

Simulation Results

In practice not all of the initial conditions for model (3) are known and only a few of the state variables are measurable directly. This complicates the prediction of the future behaviour of the model. Therfor by using (3) it is an important task to estimate the unknown initial states on the basis of given measurements. The most important initial state is the stem cell pool size given by $x_1(0)$. It has major influence on the future behaviour of the system. In order to estimate this a performance index given by the sums of the quadratic deviations between simulated model data and real data is minimized with respect to the initial conditions for stem cells, CBM/CBL cells, P cells and the two hormon compartments Reg.I and Reg.II. The results are shown in Figure 3. The first 15 measurements have been used to estimate the initial conditions. The estimation algorithm uses the Nelder-Mead method. As seen from the Figure 3 the future behaviour is predicted extremely well. This is surprising in view of not knowing the tolerances of the measured data.

Conclusions

In this paper the biomathematical model of granulocytopoiesis has been identified as a system consisting of bilinear subsystems. Thus bilinear system theory is applicable and the nonnegativity of the solution to the biomathematical model can be proved as a first application.

The next step is the prediction of the future behaviour of the granulocytopoiesis on the basis of real measurements taken from a patient with disturbed granulocytopoiesis. Note that the data from the first 15 measurements taken in approximately 17 days provide enough information to simulate the future behaviour up to the 70th day. This is an important result in view of further treatment of real patients.

References

[1] Fliedner, T.M. and Steinbach, K.-H.: *Simulationsmodelle von Perturbationen des granulozytären Zellerneuerungssystems*; in Doerr, W. and Schipperges, H. (Ed.); *Modelle der Pathologischen Physiologie*; Springer; Berlin; 1987.

[2] Fliedner, T.M., Steinbach, K.-H. and Szepesi, T.: *Hematological Indicators in the Determination of Clinical Management Strategies in Radiation Accidents*; International Conference on Biological Effects of Large Dose Ionizing and Non-Ionizing Radiation; Hangzhou; Society of Radiation Medicine and Protection; Chinese Medical Association; Bejing; 1988.

[3] Steinbach, K.-H.: private communication.

[4] Mohler, R.R.: *Bilinear control processes*; Academic Press; New York, London; 1973.

[5] Isidori, A.: *Nonlinear Control Systems: An Introduction*; Springer; Berlin, Heidelberg; 1985.

[6] Hofer, E.P. and Tibken, B.: *An Iterative Method for the Finite-Time Bilinear-Quadratic Control Problem*; Journal of Optimization Theory and Applications; (1988)57; pp. 411-427.

[7] Tibken, B. and Hofer, E.P.: *Systematic Observer Design for Bilinear Systems*; Proceedings of the International Symposium on Circuits and Systems; Portland; 1989.

[8] Tibken, B. and Hofer, E.P.: *A novel computer approach to optimal feedback control of bilinear systems*; Proceedings of the IFAC Symposium on Nonlinear Control Systems Design; Capri; 1989.

[9] Mohler, R.R. and Kolodziej, W.J.: *An overview of bilinear System Theory and applications*; IEEE Transactions on Systems, Man and Cybernetics; SMC-10(1980)10; pp. 683-688.

[10] Krener, A.J.: *Linearization and Bilinearization of Control Systems*; Proceedings of the 1974 Allerton Conference on Circuit and Systems Theory; Urbana Ill.; 1974.

[11] Bellman, R. and Bekenbach, F.: *Inequalities*; Springer; New York; 1977.

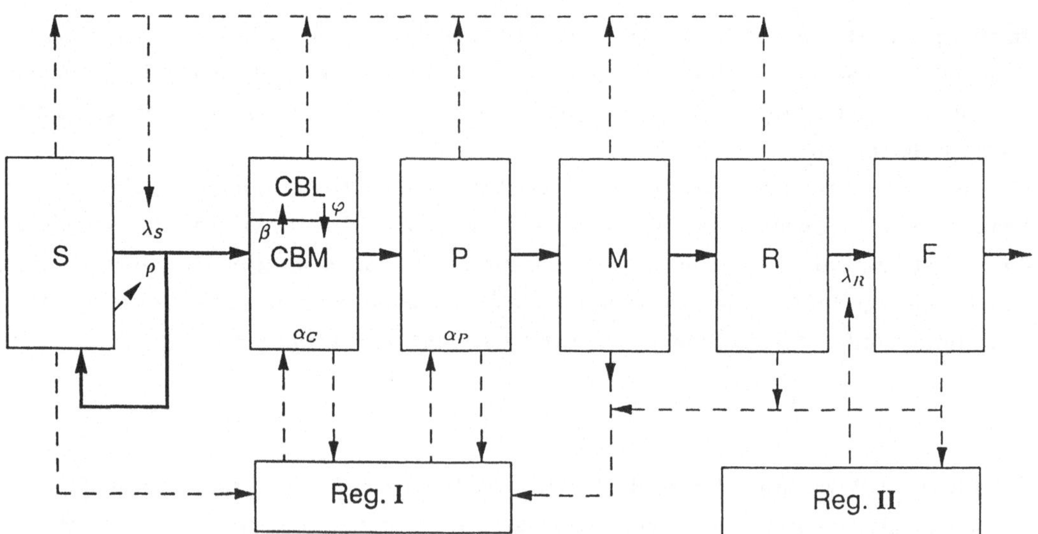

Figure 1. Fliedner & Steinbach Model of Granulocytopoiesis

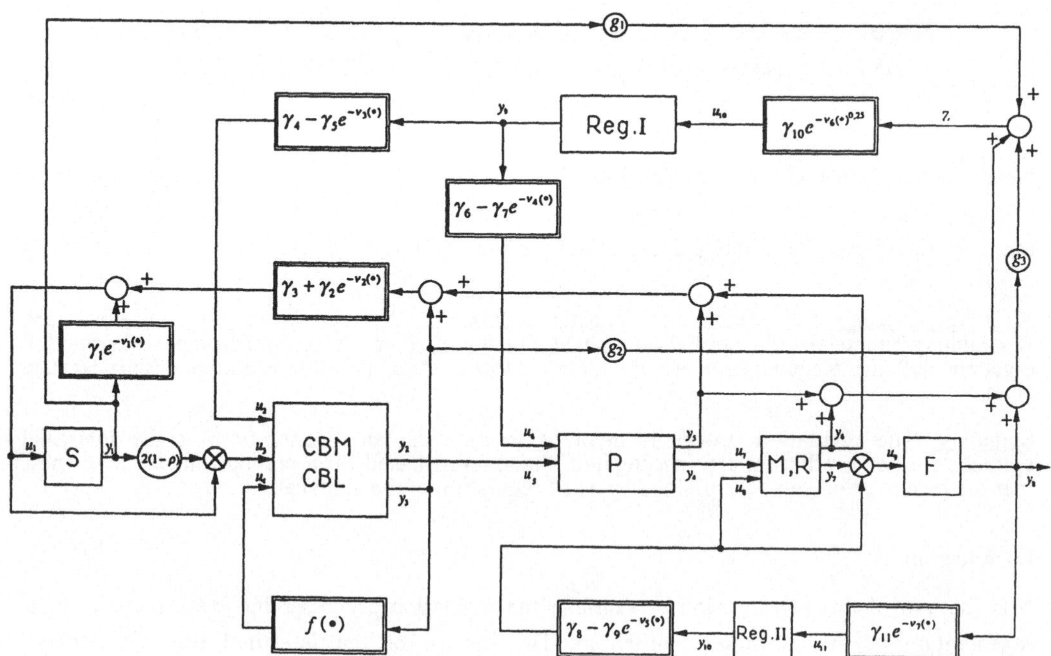

Figure 2. Control Scheme of Granulocytopoiesis

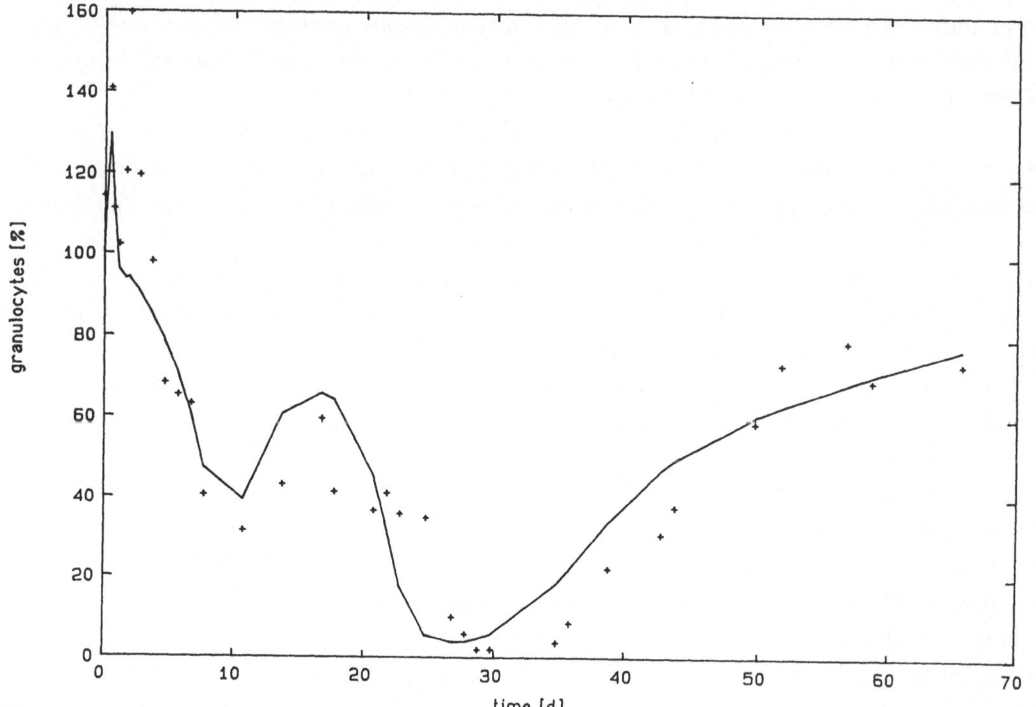

Figure 3. Simulation Results (-) and Patient Data (+)

Strategies for Identification of Regulation Processes in the Intestinal Epithelium after Perturbation

Ursula Paulus, Jürgen Glatzer, Markus Loeffler

Medizinische Universitätsklinik I, Joseph Stelzmannstr. 9, D-5000 Köln 41

Zusammenfassung: Es werden die Regulationsprozesse der zellulären Regeneration nach Bestrahlung im intestinalen Epithel untersucht. Die Simulationen mit einem Kompartmentmodell ergeben, daß die Regeneration allein durch Autoregulation der Stammzellen erklärt werden kann.

Summary: The regulation processes involved in the cellular regeneration of the intestinal epithelium after irradiation are investigated. Simulations based on a compartment model show that the regeneration can be explained by a self regulation of the stem cells.

Introduction

It is the aim of this investigation to examine the regulation processes involved in the cellular regeneration of the intestinal epithelium. The surface of the intestinal wall is a folded structure which exhibits fingerlike protrusions into the lumen of the gut called villi and baglike structures embedded in the wall called crypts. These structures are covered by a one layer thick lining of cells. There is a directed motion of cells from the crypts where all the cell production takes place to the villi where they are finally shed into the lumen of the gut. These villus cells are responsible for the digestion and resorption of the food.

In the following our interest will focus on the crypts. Besides a few non-proliferative Paneth cells at the bottom of the crypt the majority of all other cells are morphologically not distinguishable. Biologists have undertaken numerous efforts to investigate functional heterogeneity among these cells. They are nowadays classified into three functional groups. The upper third of the crypt contains about fifty cells, which are considered as mature (M). Beneath these there are one hundred and fifty cells which undergo division. These proliferating cells can be labelled by radioactive nucleotides which are incorporated in the DNA during DNA synthesis. The proliferating cells can be further subdivided. Cells in the middle region of the crypt differentiate after each cell division and generate the mature cells. These cells are not able to maintain the epithelium over long time periods. Therefore they are called transient cells (T). They number over one hundred. In contrast only a few cells in the lower crypt region are capable of selfmaintenance. They continuously regenerate the epithelium in steady state and are therefore called actual stem cells (A). They can not be morphologically identified at present but their number is estimated to be less than 16. A closely related term is "clonogenic cells". These cells are able to regenerate a crypt entirely after damage. Their number can be estimated from crypt survival studies after damage. Under the assumption that a single clonogenic cell can repopulate an entire crypt the number of clonogenic cells has been estimated to lie between 30 and 50. In another set of experiments aiming at a localized cell kill at the bottom of the crypt lower estimates for the number of the clonogenic cells of 4 to 16 have been obtained (Potten 1990).

In a previous model of the steady state situation taking single cell behaviour into account we have characterized the organisation of the crypt and the cellular developments including the spatial arrangements (Loeffler et al. 1986/1988). From such model considerations the existence of 4 to 16 actual stem cells in steady state was concluded. These obvious discrepancies point to the fact that the stem cells in steady state may not be the same as those during perturbation. There may be a certain flexibility within the operational status which results in different estimates after different perturbations.

Working Hypothesis about Organisation of the Crypt

With our steady state model it was possible to simulate the spatial arrangements of proliferating cells and their kinetic development. The basic model assumptions were as follows: There is a fixed number of symmetric cell divisions of transient cells (4-5). The stem cells divide strictly asymmetrically (exact steady state). Consequently there is a strict distinction of stem cells and transient cells. The succession of cell divisions and the maturation and differentiation program determine the "stem tree". The spatial arrangement of cells is organized by selection processes on the basis of local next neighbour interactions suggesting an entirely local regulation process in the system.

Experimental Data after Irradiation with 8 Gy

In a comprehensive series of experiments Potten and collaborators (Potten 1990) have examined the recovery of the intestinal crypt after 8 Gy. After such an exposure there is a mitotic inhibition which results in no cell divisions taking place for at least 8 hours in the middle of the crypt while at the bottom of the crypt no cell division can be observed for 20 hours. One can identify only a very small number of about 5 dead cells which are all located at the bottom of the crypt. Although the number of dead cells is small the total number of cells in the crypt is reduced up to 50% by day 2 after irradiation which is followed by an overshoot to 150% on day 4. On day 7 the steady state value is reached again.

The time behaviour of the labelled cells (**Figure 1**) which are the proliferating cells is similar to the pattern of the total cell numbers. After an initial reduction more than twice the normal number of labelled cells are found on day 3. The cell cycle times in the middle of the crypt are reduced from 12 to 9 hours during this phase. The cell production (**Figure 2**) in the crypt can be calculated from the migration velocities of labelled cells at the upper edge of the crypt and the change of the crypt cellularity. The cell production comes to an entire stop on day 1 but overshoots beyond normal levels on day 3 and 4. The drastic reduction of the cell production shows that the number of proliferating cells on day 1 after irradiation has decreased to less than 10%.

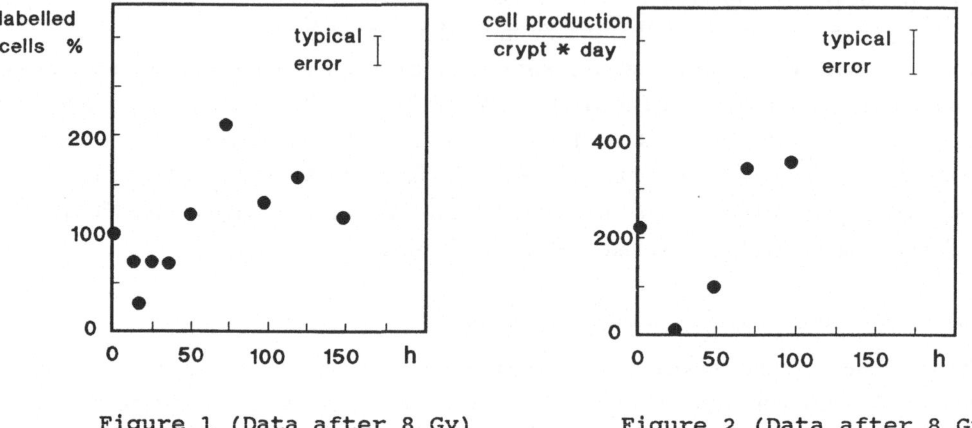

Figure 1 (Data after 8 Gy)   Figure 2 (Data after 8 Gy)

On day 1 and 2 after irradiation the number of clonogenic cells falls dramatically from about 40 to 3. However, on day 4 normal values are reached again. During this time the cell cycle time in the lower crypt region is reduced from 24-40 hours to about 8 hours i.e. at least three fold.

The leading questions of this examination are :

Can the steady state working hypothesis of a fixed number of transient cell generations be maintained? Is it possible to describe the crypt population with a fixed stem tree also after irradiation?

Can the working hypothesis of a local regulation also be applied to post irradiation recovery? Can one explain the regeneration phenomena without the assumption of a long range feed back mechanism from the villus to the crypt? Such models have been postulated by various authors (Kicherer 1983, Britton et al. 1982, Jakovlev 1988) but lack experimental proof.

Model

To answer these questions we suggest a compartment model. Cells of equal generation are summarized in compartments, which have an internal age structure. Transient cells are characterized by an obligatory transition from one to the next compartment. In the steady state only stem cells can selfmaintain. In each compartment the cell cycle phases (G1, S, G2, M) are considered. A variance of the cell cycle is attributed to a variance in the G1 phase.

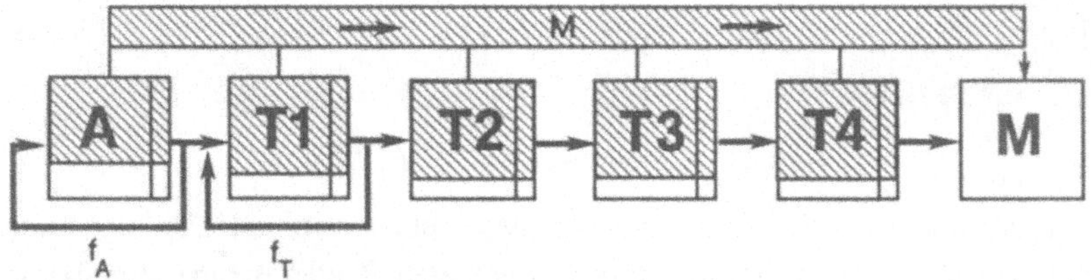

Figure 3 (Model scheme)

The following assumptions about the damage process are made **(Figure 3)**:

a) 90% of all cells leave the cell cycle after damage. This is necessary to understand the breakdown of the cell production. The cells are considered to survive as non-proliferative, mature cells and are therefore counted as M cells (see shades in Figure 3). The number of stem cells is reduced from 16 (normal value) to values between 2-12 (various scenarios).

b) The cell cycle is blocked in the G2 phase for 8-10 hours for T-cells and for 20-24 hours for stem cells (see vertical lines inside the compartments in Figure 3).

The following assumptions about the regulation process governing regeneration are made:

a) The cell cycle time of stem cells is assumed to be a monotonous function of the stem cell number (A):

$$T_c(A) = \begin{cases} A^n & \text{for} \quad A^n \geq \dfrac{1}{m} \\[2mm] \dfrac{1}{m} & \text{for} \quad A^n < \dfrac{1}{m} \end{cases}$$

The minimum value is 1/m i.e. the shortest cell cycle time measured for reduced numbers of stem cells. In this description the number of stem cells and the cycle times are normalized to the steady state value. Realistic values for m range between 2 and 5.

b) The selfmaintainance rate f_A is also assumed to be a function of the number of stem cells. f_A is defined as the fraction of stem cells that is added to the stem cell pool after division. $f_A = 1$ implies that both daughter cells become stem cells and that no differentiation to T1 cells takes place; $f_A = 0$ implies asymmetric divisions and therefore steady state; $f_A = -1$ implies that all cells differentiate to the next compartment. For biological reasons the maximum values of f_A are 1 and -1. This justifies the following approach:

$$f_A(A) = \begin{cases} f_A^{min} & \text{for} \quad s_A*(1-A) \leq f_A^{min} \\[2mm] s_A*(1-A) & \text{else} \\[2mm] f_A^{max} & \text{for} \quad s_A*(1-A) \geq f_A^{max} \end{cases}$$

In addition to a selfmaintenance of A-cells a similar selfmaintenance rate is assumed for the transient T1 cells which is given in the following formula:

$$f_T(A) = \begin{cases} -1 & \text{for} \quad s_T*(1-A)-1 \leq -1 \\ s_T*(1-A)-1 & \text{else} \\ f_T^{max} & \text{for} \quad s_T*(1-A)-1 \geq f_T^{max} \end{cases}$$

This description implies that the strict distinction between stem cells (asymmetric division) and transient cells (symmetric division) in the steady state is weakened. Under strong stimulation T1 cells can take up properties ($f_T^{max}=0$) which only steady state A-cells had before (**Figure 4**).

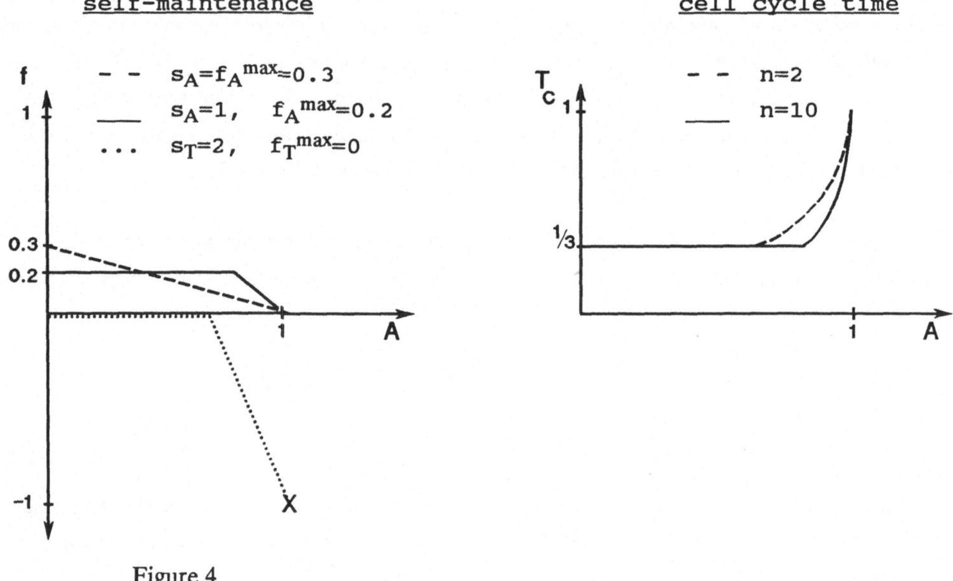

Figure 4

Choice of Parameters

In this model approach the parameters n, m, s, f^{max} have to be determined. From experimental data we can estimate m to lie between 3 and 5. For the choice of n and s a qualitative model argument helps to narrow down the options. For this purpose we simplify the model to a two compartment model describing only stem cells and the differentiated cells:

$$\frac{dA}{dt} = \frac{f_A(A)}{T_c(A)} * A$$

$$\frac{dT}{dt} = K* \left(\frac{1-f_A(A)}{T_c(A)} * A - T \right)$$

In this dimensionless system of ordinary differential equations K is a scaling factor while f_A and T_c are the functions defined above. These equations can be solved analytically. One can determine fixpoints and their local and global stability properties and it is possible to calculate phase space trajectories. Comparing these trajectories with experimental data (eliminating the actual time scale) one obtains estimates for n ($n \geq 2$) and s ($s \leq 0.3$). With $n = 2$ and $s = 0.3$ one can return to the more complex model and perform simulations. Although the recovery pattern for the proliferative cells is qualitatively correct the time scale exhibits a recovery which is far too slow. This is not surprising because the shape of the phase space trajectory did not imply quantitative statements about time development. A possibility to correct the time scale in the complex model is to modify the functions T_c and f_A by bending the functions towards a more rectangular shape (**Figure 4**).

Results

Various model scenarios have been calculated with these functions. They varied in the number of proliferating stem cells surviving the perturbation (initial value) and the degree to which the T1 population exhibits selfmaintenance (f_T^{max}). **Figures 5** and **6** show an example in which the regulation of T1 selfmaintenance was introduced ($f_T^{max} = 0$). These model calculations exhibit a time behaviour of the labelled cells and the cell production which is comparable to the data (see Figure 1 and 2).

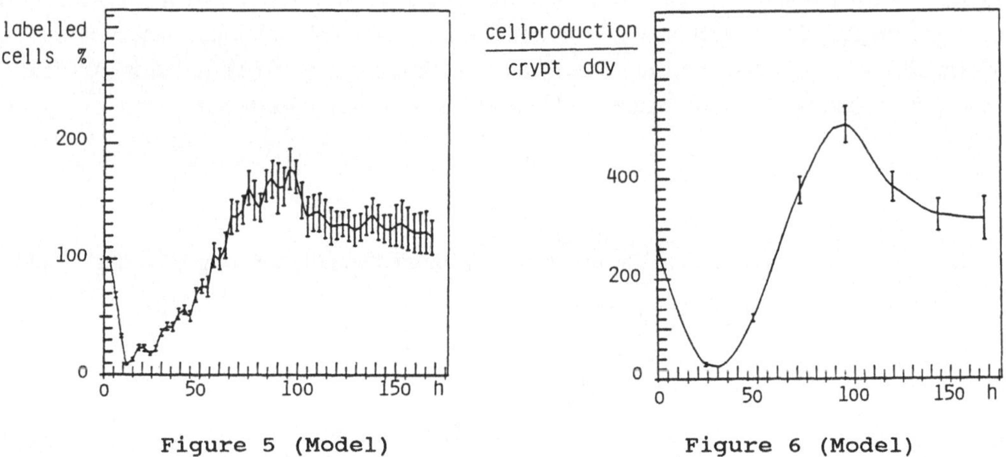

Figure 5 (Model)   Figure 6 (Model)

By varying the model parameter m and the T1 regulation (f_T^{max}) one finds that the simulations are consistent with the data provided that the A-cell number has at least the minimum value as shown in the table:

minimal number of A-cells after damage
(steady state value: 16)

T1 regulation

	no ($f_T^{max}=-1$)	yes ($f_T^{max}=0$)
m=3	11	3
m=5	6	2

This table indicates how one can select appropriate models once a precise measurement of the stem cell damage becomes available. If the damage can be shown to be strong leaving less than 10% (2/16) of the stem cells reproductively alive one can simulate the data by using $m=5$ with T1 regulation. If the damage to the stem cells is small leaving more than 70% (11/16) reproductively alive then models with $m=3$ and no T1 regulation are appropriate. On the basis of the present data one cannot exclude any of these scenarios. It will be necessary to narrow down the number of reproductively active stem cells and in particular the value for m by further experiments.

Conclusions

1. The hypothesis of an internally determined maturation and differentiation program (stem tree) can be maintained for the description of the crypt after damage.
2. The strictly asymmetric stem cell division of the steady state model has to be abandoned and can be replaced by a functional dependence of selfmaintenance on the stem cell numbers.
3. A model based on local regulation mechanisms (self regulation of stem cells) is sufficient to explain the regeneration. A long range feed back is not a necessary assumption.

References

1. Britton, N.F. et al: J. Theor. Biol. (1982) 98, 531-541
2. Jakovlev, A.Y.; Zorin, A.V.: Lecture Notes in Biomathematics, Springer Verlag (1988) Vol. 74
3. Kicherer, G.: Dissertation Stuttgart (1983)
4. Loeffler, M. et al: Cell Tissue Kinet. (1986) 19, 627-645
5. Loeffler, M. et al: Cell Tissue Kinet. (1988) 21, 247-258
6. Potten, C.S.: Int. J. Rad. Biol. (in press)

A Stochastic Branching Model of the Steady State Growth of Intestinal Crypts

Markus Loeffler and Bernd Großmann

Medizinische Klinik I, Dept.Biometrie, LFI-5, Josef-Stelzmann-Str.9, D-5000 Koeln 41, FRG
Institut für Theoretische Physik, Philosophenweg 19, D-6900 Heidelberg, FRG

Zusammenfassung: Ein Modell des Gleichgewichtswachstums intestinaler Krypten wird vorgestellt. Es basiert auf einer Überlagerung zweier stochastischer Verzweigungsprozesse, welche die Teilung der epithelialen Stammzellen und die Kryptteilung beschreiben. Parametersätze können identifiziert werden, welche bekannte Daten über Größenverteilungen, Extinktions- und Teilungsraten von Krypten erklären.

Abstract: A Model of the steady state growth of intestinal crypts is presented. It is based on a superposition of two stochastic branching processes which describe division of the epithelial stem cells and crypts. Sets of parameters can be identified which explain measured data on crypt size distribution, extinction and crypt division rates.

INTRODUCTION

The intestinal epithelium is one of the most rapidly regenerating tissues in mammals. Cell production takes place in the intestinal crypts which are imbedded in the wall of the gut. Only a small minority of the 250 proliferating cells in a crypt is able to maintain the epithelium over a long period of time. However, attempts to identify these stem cells on grounds of histology or genetics were unsuccessful. Little is known about the growth properties of the stem cells. Their number may range between 1 and 60 (Potten, 1990).

Crypts are dynamic structures. Their life starts by a longitudinal fission of existing crypts and ends by fission or extinction. This process is diplayed in **Figure** 1. On the left a section of a normal crypt is shown. Crypt doubling occurs by a longitudinal fission of a crypt starting at the bottom (the presumable stem cell location) and continuing upwards (middle). Crypts disappear by shrinking in length and circumference until they are integrated into the surface lining (right). It is assumed that extinction occurs only if a crypt has lost all selfmaintaining stem cells. It is believed that a crypt can stay alive as long as it has at least one active stem cell in it (Potten 1990). In addition crypt size distributions have recently been measured (Totafurno et al., 1987).

Scheme of Birth and Death of Crypts

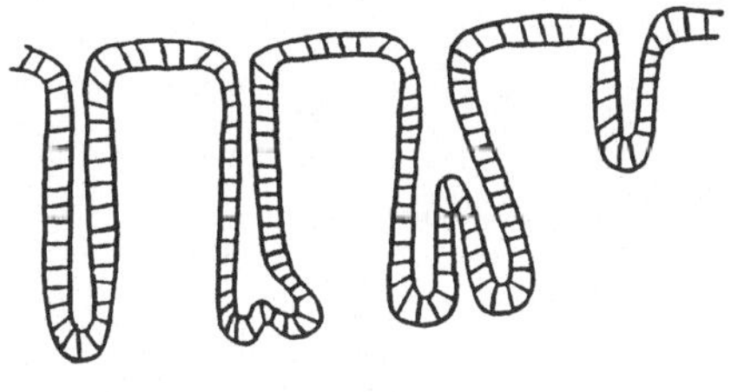

Normal crypt   Crypts in fission   Dying crypt

Figure 1

The basic data on the life cycle of crypts and their size distribution can be summarized as follows (from Totafurno et al., 1987; Potten, 1990):
1) The number of crypts increases continuously with age with a doubling time in the order of a life time of a mouse; 2) The crypt fission rate can be estimated to be between 0.0013 and 0.006 per day; 3) The crypt extinction rate can be estimated to be considerably lower than 0.001 per day; 4) Small crypts are very rare and the mode of the crypt size distribution is about 1/2 of the size of the biggest crypts and the coefficient of variation is about 0.25.

Here we demonstrate a model that can quantitatively explain the life cycle of crypts on the basis of growth characteristics of individual stem cells.
This implies that one can relate the dynamics of the macroscopic crypt structure to the microscopic cellular process of stem cell selfrenewal and differentiation.

MODEL ASSUMPTIONS AND FORMALISM
The following model assumptions are introduced (see also Vogel et al., 1969):
Assumption 1 (Cell division):
If the number S of stem cells in a crypt is below a critical value S_f stem cells grow according to a supracritical time independent Markov process of the Galton Watson type with probabilities p,r, and q of producing 2, 1, and 0 new stem cells (p+q+r=1).

Assumption 2 (Crypt division):
If the number S of stem cells in a crypt exceeds S_f two new crypts are formed and the stem cells are distributed according to binomial statistics with probability v (B(v,S)). With respect to the old crypt this is equivalent with the view that either 0 or 1 stem cells remain.(For practical reasons we will assume v = 1/2 later on.)

These two assumptions enable us to write down the transition probability $P_{ij}(v)$ for a crypt of size S_n = i in generation n to one of size S_{n+1} = j in generation n+1. It is given by

$$[1a] \quad P_{ij}(v) = P_v(S_{n+1} = j \mid S_n = i) = \begin{cases} P^*_{ij}(v) & \text{für } i > 0,\ j \geq 0 \\ \delta_{ij} & \text{für } i = 0,\ j \geq 0 \end{cases}$$

where $P^*_{ij}(v)$ stands for the i-fold convolution of $\{p_{ik}(v)\}$ at point j and

$$[1b] \quad p_{ik}(v) = \begin{cases} q & \text{for } i < S_f & \text{and } K = 0 \\ r & \text{for } i < S_f & \text{and } K = 1 \\ p & \text{for } i < S_f & \text{and } K = 2 \\ 1-v & \text{for } i \geq S_f & \text{and } K = 0 \\ v & \text{for } i \geq S_f & \text{and } K = 1 \\ 0 & \text{else} \end{cases}$$

In order to obtain experimentally meaningful results one has to consider an additional assumption which leads to conditional probabilities:

Assumption 3 (Extinction of subunits):

Crypts with zero stem cells are considered extinct (absorbing state). They are eliminated from consideration.

Furthermore the experimental observation of the evolution of individual crypts over time is not possible. In experiments one rather observes expectation values averaged over certain time periods and populations of crypts. This justifies the following assumption:

Assumption 4 (Effective Crypt Approximation):

The evolution of a population of crypts (with all the possible combinatorial calculus) can be approximated by consideration of an effective probability distribution for one average crypt and a separate counting of the total number of crypts.

This assumption has the advantage that the dynamics of the number of crypts and the cell distribution per (effective) crypt can be separated. It requires a particular renormalisation procedure of the probability distributions. Let $P_{eff}(S_n \mid S_n > 0)$ denote an appropriately defined effective distribution of stem cells in step n. Then one has to assume fission of those crypts with $S_n \geq S_f$ cells. The fraction of crypts which is added by this doubling process is given by

$$[2] \quad g = \sum_{i \geq S_f} P_{eff}(S_n = i) \Big/ \Big\{ \sum_{i < S_f} P_{eff}(S_n = i) + 2 \cdot \sum_{i \geq S_f} P_{eff}(S_n = i) \Big\}$$

g can be used as a renormalisation factor. After generation of the transition from n to n+1 via equations [1] one can return to a properly normalized effective probability distribution

$$[3] \quad P_{eff}(S_{n+1} = i) = P(S_{n+1} = i \mid S_f \geq 0) \cdot (1-g)$$
$$+ P(S_{n+1} = i \mid S_n \geq S_f) \cdot g$$

With this effective probability distribution one can derive all other quantities of interest. The fission probability (rate) is defined by

$$[4] \quad \phi := \sum_{i \geq S_f} P_{eff}\, (S_n = i \mid S_n \neq 0)$$

and the extinction probability (rate) is

$$[5] \quad \Omega := P_{eff}\, (S_{n+1} = 0 \mid S_n \neq 0)$$

From here the iteration continues with equations [2] and [3] substituting P_{eff} by the conditional probability $P_{eff}\, (S_{n+1} \mid S_{n+1} > 0)$

Scheme of the model and numerical approximation

A. Model

1. Cellular branching process

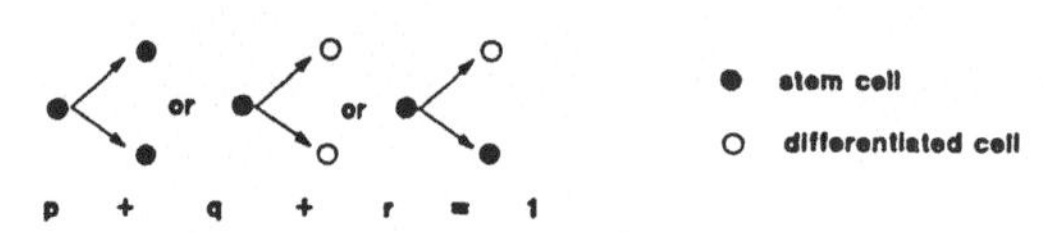

2. Branching process of subunits

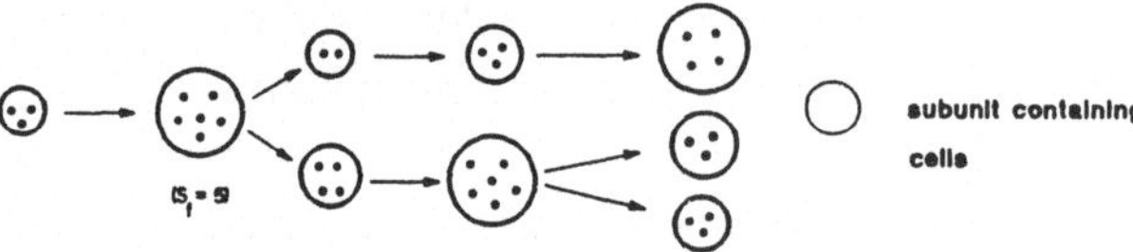

B. Approximation (effective probability distribution)

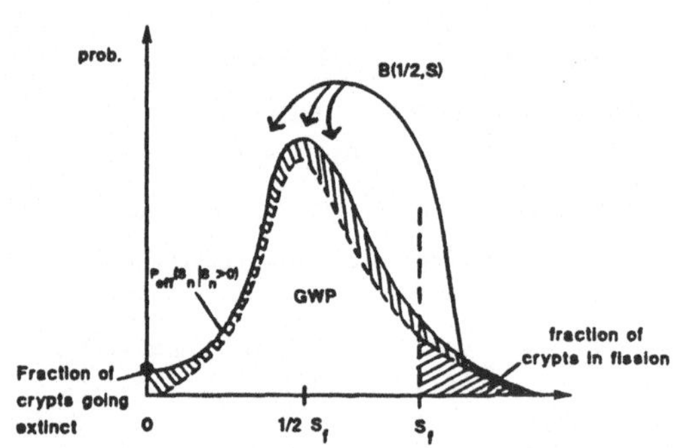

<u>Figure 2</u>

Figure 2 gives a schematic summary of the model concept and the approximation strategy. In order to compare the probability distribution of stem cells with macroscopic observables a fifth assumption holds.

Assumption 5 (Proportionality): The size of a crypt is proportional to the number of stem cells in it.

NUMERICAL PROCEDURE AND CHOICE OF PARAMETERS

There are only three free parameters in the model. One is the fission threshold S_f. Based on data it is assumed to range between 4 and 64. For numerical purposes values of 4,8,16,32, and 64 are used. The second parameter is the doubling time T of the number of crypts. It relates to the microscopic parameters p,q, and r via the formula $(2p+r)^T = 2$ and $p+q+r=1$. For numerical purposes values of 12, 25, 50, 100, 200, and 400 are used. The probability for asymmetric stem cell division r is chosen as third parameter. It will be varied between 0 and the maximum value r_{max} which is given by $r_{max} = 2 - 2^{1/T}$. Typically r=0, 0.6, 0.9, 0.97, 0.99, r_{max} are used.

MODEL RESULTS

STATIONARY PROBABILITY DISTRIBUTION

Numerical calculations of the effective probability distribution (P_{eff}) were undertaken.

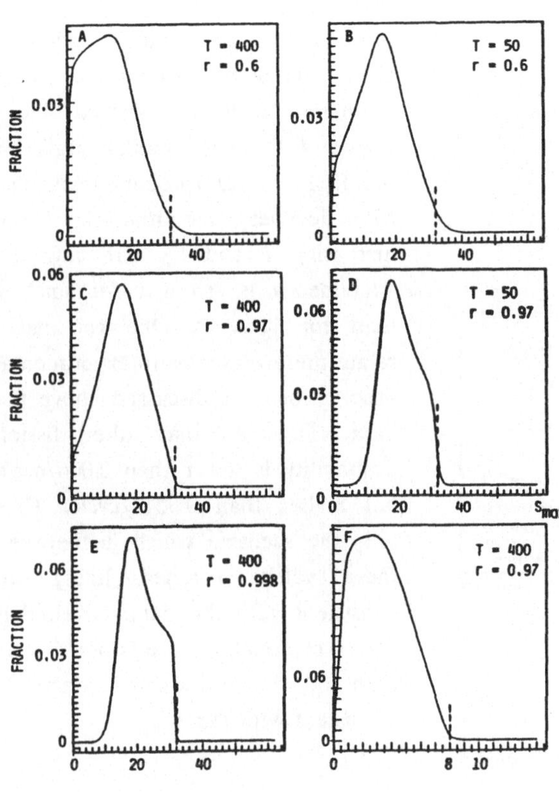

Figure 3 gives examples. The distributions in the left three panels (A,C,E) are obtained for the same threshold parameter S_f and doubling time T. It is apparent that variation of the asymmetric division probability r drastically influences the probability distributions. For low r (0.6) the distributions are broad and crypts with very few and many stem cells are likely. The coefficient of variation exceeds 0.5. For high r (>0.9) the distribution is shifted to the right and the mode gets close to 1/2 S_f. Coefficients of variation can be as small as 0.25. Comparison of the effective probability distribution of stem cells with the measured size distribution of crypts suggests that the broad distributions associated with small r, high T and low S_f are qualitatively not consistent with the data. In particular distributions with a flat left hand side cannot exist for $S_f < 8$.

<u>Figure 3</u>

EXTINCTION AND FISSION PROBABILITIES

A quantitative insight can be gained by considering the extinction probabilities (EP) and fission probabilities (FP) over the three dimensional parameter space. Extensive numerical investigations were undertaken to explore the behaviour over a wide range of reasonable parameters. Two–dimensional projections of the results are displayed in **Figure 4** (fixed T) with variation of the other two parameters. **Figure 4 A** shows the extinction probabilities. Each point in the diagram represents one calculation for a fixed combination of parameters. The lines interpolate between model scenaria of equal values of r. It becomes apparent that for given T and r the EP declines monotonously with increasing S_f. This dependency is, however, weak if r is small (<0.6) but strong for large r. One can relate these diagrams

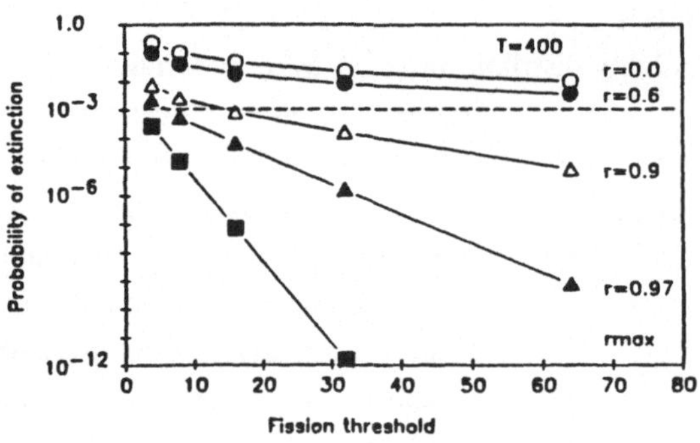

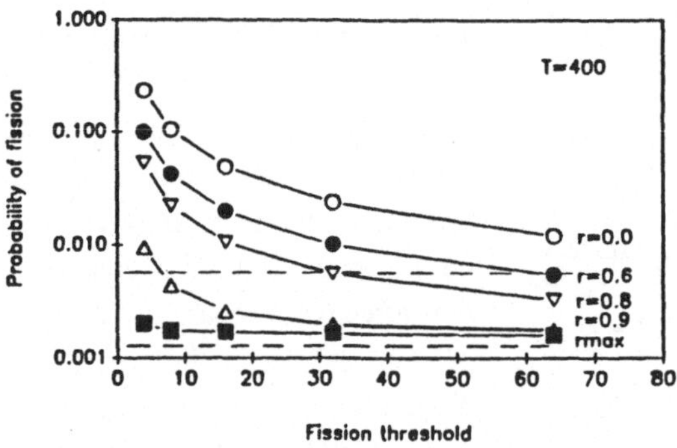

<u>Figure 4</u>

directly to experimental observations. As discussed above, the data indicate that the extinction rate is realistically lower than 0.001/cell cycle. Apparently only models with sufficiently large r and S_f can give EPs consistent with the data (scenaria below the dashed lines). **Figure 4 B** shows similar plots for the fission rates (probabilities). The FP declines monotonously with increasing r and S_f, but now the dependency is stronger for small r than for large r. One can again relate these diagrams to experimental observations. As discussed above the data indicate that the fission probability is lower than 0.006/cycle but higher than 0.0013/cycle. Thus only the scenaria which lie between these values (dashed lines) are consistent with the data. Given that the crypt doubling time is not shorter than 100 stem cell cycles models consistent with the data on can only be obtained if $r \geq 0.8$ and $S_f \geq 8$.

DISCUSSION

In this paper we have discussed properties of a new type of stochastic branching processes. The basic process is a microscopic cellular renewal process of stem cells which is basically described by a conventional Galton Watson process. The new component is a threshold dependent segregation of the cells into subunits.

Biologically the investigation was undertaken to understand the birth and death process of macroscopic intestinal crypts and relate it back to the microscopic birth and death process of a few epithelial stem cells which are the constituents of the crypt. Consideration of the segregation process allowed us to introduce a new (weaker) concept of stability. We found that even in the case of supracritical growth of the entire cell number the probability distributions (per effective crypt) become stationary. Furthermore, this concept allows for a meaningful definition of a fission rate and provides estimates for the extinction process.

The particular system of intestinal crypts can quantitatively be explained by the model proposed. It was designed to explain four sets of data: the size distribution of crypts, the low extinction rate, the apparent fission rate and the growing number of crypts. We conclude that these phenomena can quantitatively be explained by a stochastic branching process of epithelial stem cells if the following conditions hold: (1) The probability for asymmetric stem cell division r is at least 0.8; (2) The number of stem cells grows with a doubling time of 100 or more cell divisions; (3) The threshold value S_f is at least 8 or larger. The high value of r is biologically remarkable. For short time scales one can neglect any symmetric cell divisions. This is interesting because it gives a justification for a previous model of our group in which the short term steady state of single crypts was descibed on the basis of strictly asymmetric divisions of stem cells (Loeffler et al 1986, Potten and Loeffler 1987).

The biological basis for the asymmetric division is unclear. Several explanations are conceivable. Stem cells might be defined by their attachment to some supportive neighbour cell, or they might be determined by some internal marker attached to the mother chromosomes. Discrimination of these possibilities is presently not possible. This makes clear that future experiments should centre around understanding these cells and their control in more detail.

Acknowledgements: We like to thank C.S.Potten (Manchester), Ursula Paulus, Heiner Stroetmann and Dirk Hasenclever (Koeln) for helpful discussion.

References

Loeffler,M.; et al.: Cell Tissue Kinet. 19 (1986) 627-645

Potten,C.S.: Int.J.Radiat.Biol. (1990) (in press)

Potten,C.S.; Loeffler,M.: J.theor.Biol. 127 (1987) 381-391

Totafurno,J.; et al.: Biophys.J.52 (1987) 279-294

Vogel,H.; et al.: J.Theoret.Biol. 22 (1969) 249-270

Modell zur Simulierung der pulmonalen Eliminierung von Benzol bei der Maus nach intraperitonealer Bolusgabe

H. Kindler, L.Weber, T.M. Fliedner

FAW Ulm, Postfach 2060, D-7900 Ulm, 0731/501480, KINDLER@DULFAW1A; Institut für Arbeits- und Sozialmedizin, Albert Einstein Allee 11, D-7900 Ulm, 0731/1763445

0. Kurzfassung

Intraperitoneale Injektion von lipophilen Lösemitteln gelöst in Öl ist ein gutes Mittel zur Verabreichung ohne Inhalationsapparaturen. Durch den dabei erwirkten langsamen Übertritt der Lösemittel ins Blut kann trotz starker kapazitativer Begrenzung der metabolischen Abbauleistung (z. B. der Leber) ein großer Lösemittelanteil in Metaboliten umgewandelt werden, die quantitativ im Urin erfaßt werden können. Um den exhalierten Teil des Lösemittels, der nur für 50 Minuten gemessen werden konnte, grob abschätzen zu können, wurde ein Modell zur Simulierung der Lösemittelabatmung entwickelt. Das Versuchstier war die Maus und das Lösemittel Benzol. Dieses Modell erfordert neben der Berücksichtigung der Diffusion ins Blut (basierend auf Partitionskoeffizienten, Fiserova-Bergerova 86) auch die strukturelle Modellierung des Kreislaufsystems und seiner Konvektionsströme.

1. Einleitung

Neben dem inhalativen Angebot von Lösemitteln an Probanden bzw. Versuchstiere, was großvolumige technische Geräte zur Exposition und Überwachung voraussetzt, bietet die intraperitoneale Applikation von lipophilen Lösemitteln in öligen Medien eine einfache, schnell dosier- und wiederholbare Möglichkeit, eine Exposition zu simulieren und die Verteilungsprozesse sowie Stoffwechselprodukte dosisabhängig zu bestimmen. Die Injektion von flüssigem Benzol, gelöst in Maiskeimöl, in das Peritoneum ist eine geläufige Applikationsart zum Studium von Lösemittelbelastungen im tierischen Organismus. Mit dieser Applikationsart lassen sich verschiedene Parameter des biologischen Monitorings abgreifen. Einerseits können die Metabolite im Blut, Urin und Kot bestimmt werden, andererseits kann die Ausscheidung des unveränderten Lösemittels in der Atemluft bestimmt werden nach Resorption. Am Ende einer Exposition mit einem Lösemittel ist die Eliminierung durch pulmonale Exhalation ein anteiliger Prozeß. Am Beispiel der intraperitonealen Gabe von Benzol in Öl soll unter Zugrundelegung   der gemessenen Benzolkonzentrationen in der abgeatmeten Luft ein Modell zur Simulierung der Eliminationsverläufe erstellt werden. Die Versuche, die die Exhalationskurven liefern, wurden zur quantitativen Bestimmung von Benzol-Metaboliten im Urin durchgeführt. Die Idee der Modellierung der Benzolexhalation als Differentialgleichungsmodell kam erst nach der Durchführung der Versuche, was wie alle retrospektiven Studien nicht optimal sein kann. Durch die ungewöhnlich niedrigen Fallzahlen für Mäuseversuche ist keine Basis für eine richtig abgesicherte quantitatve Aussage möglich. Die Ergebnisse bei der Modellierung sind eher qualitativer Natur, zeigen aber doch, daß die Berücksichtigung der anatomischen Struktur und der physiologischen Funktion für die Erklärung und Vorraussage des Systemverhaltens unabdingbar sind. Die durch die exaktere Nachbildung größere Anzahl von unbekannten Parametern läßt sich durch Werte aus der Literatur, eigene Messungen und vernünftige Analogieschlüsse reduzieren. Durch die Struktur des Modelles bedingt war die Zahl der wirklich sensitiven Parameter sehr gering.

2. Material und Methoden

2.1. Applikation des Lösemittels

Am Beispiel von in Öl gelöstem Benzol wird die pulmonale Eliminierungskinetik von Benzol untersucht. Die einzelne BDF-1 Maus (Gewicht: 22 - 28 g) erhält 110, 220 oder 440 mg Benzol gelöst in 10 μl Maiskeimöl (220 mg Benzol/kg auch in 5 μl Öl) je Kilogramm Körpergewicht. Die ölige Lösung wird mit einer Insulinspritze in einen der beiden unteren Quadranten des Bauchraums gespritzt. Die ölige Lösung verteilt sich bei 39 Grad Körpertemperatur über den gesamten Bauchraum. Über das Pfortadersystem wird das resorbierte Benzol ins venöse Stromgebiet der Leber und von dort in das Herz-Lungen-Kreislaufsystem verteilt. Durch Diffusions- und Konvektionsprozesse gelangt das Benzol in die Ausatemluft.

2.2. Aufbau der Meßvorrichtung

Die mit Benzol in Öl belastete Maus exhaliert in eine Meßkammer mit 160 ml Volumen (Abb. 1). Die Meßkammer besteht aus transparentem Polystyrol, der Seitendeckel rechts und der Mittelkonus sind rot eingefärbt. Die bei Eintritt in die Meßkammer benzolfreie Raumluft (25 Grad Celsius, 50 % rel. Feuchte) wird durch zwei Pumpen mit einer Luftwechselrate von 160 ml/ Minute durch die Exhalationskammer geführt. Im konischen Mittelteil der

Meßkammer liegt der Einsaugort für die Pumpe, die einen Teilstrom (65 ml/Min) der Meßkammerluft an einem Einlaßeil für einen Gaschromatographen vorbeiführt. Alle 3 Minuten werden über eine Probenschleife von 1 ml Inhalt dem Gaschromatographen Meßkammerluftproben zur Analyse zugeführt. Die Dauer eines gaschromatographischen Analysenlaufs der Luftproben dauert 2,7 Minuten. Eine zweite Pumpe saugt einen Teilstrom (95 ml/min) der Meßkammerlauft in eine Aktivkohlesammelphase. Hier können Sammelfraktionen über 15 Minuten gewonnen werden. Die Pumpvolumina der beiden Präzisionspumpen werden nach jedem Meßlauf über 60 Minuten mit einem Seifenblasenmesser kalibriert.

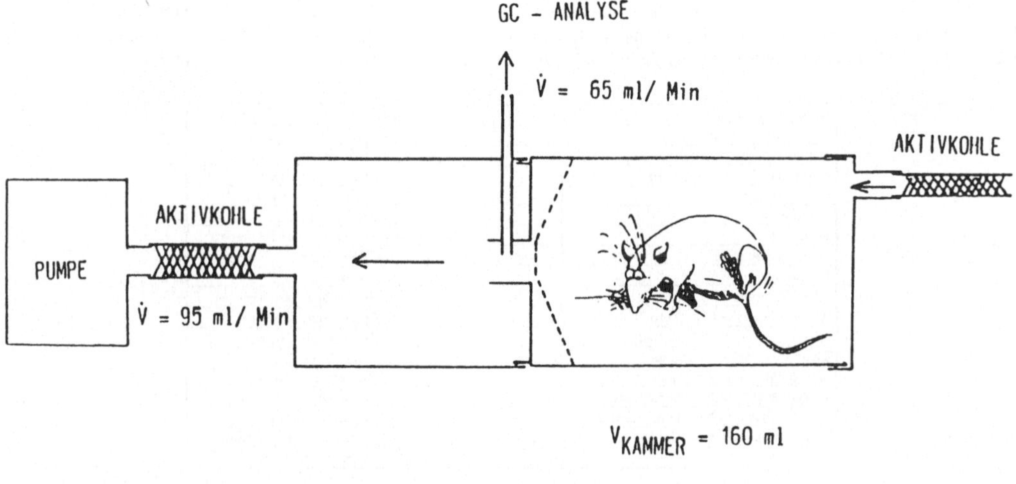

ABB.1 SCHNITTBILD DER EXHALATIONSKAMMER

2.3. Verhalten der Maus in der Meßkammer

Die Maus wird mit dem Kopf voraus von rechts in die Meßkammer eingeführt. Die Dimensionen der Meßkammer erlauben der Maus eine geringe horizontale Bewegungsmöglichkeit, sowie eine Drehung um die vertikale Körperachse. Die Maus ist nach der intraperitonealen Applikation für etwa 2 Minuten erregt, gewöhnt sich dann aber schnell an die neue Umgebung in der Meßkammer und die Geräuschkulisse durch die Meßgeräte. Die Maus erkundet während der einstündigen Meßdauer ihren Aufenthaltsort durch Belecken, Benagen und Beriechen und befindet sich dabei in körperlicher Bewegung in der Meßkammer. Mit einer gewissen Zufälligkeit werden auch gaschromatographische Proben gezogen, bei denen sich der Kopf der Maus in unmittelbarer Nähe der Einsaugstelle befindet. Ein Mindestabstand wird aber durch den mit Radiallamellen aufgefüllten Mittelkonus der Meßkammer immer eingehalten.

3. Modellbeschreibung der Benzolexhalation mit Blut als zentralem Kompartiment

Die Exhalationsverläufe (Abb. 5, 6, 7 und 8, dort die verbundenen Punktreihen) weisen die Form von γ -Funktionen auf, wie sie bei der Hintereinanderreihung von Kompartimenten entstehen. Es lag nahe, beim ersten Modellbauen eine solche Struktur zu verwenden.

Die Struktur des Modells ist in Abb. 2 zu sehen. Das Öl im Bauchraum, das Blut und die Luft in Käfig/Lunge bilden eine Kompartimentreihe, an die über das Blut das Fettgewebe und das Gewebe mit niedrigerer Benzolaffinität parallel angekoppelt sind. Das Blut wird als ein Zentralkompartiment mit homogener Durchmischung angesehen.

Die zu bestimmenden Parameter sind die in Abb. 2 eingetragenen Flüsse. Die Partitionskoeffizienten zwischen den einzelnen Kompartimenten sind bekannt (Fiserova-Bergerova 86). Die Modellierung eines Diffusionsprozesse zwischen zwei Kompartimenten erfolgt als ein Paar von Flüssen, deren Vorzeichen unterschiedlich sind, deren Stärke untereinander vom Partitionskoeffizienten abhängt, so daß das Verhältnis der Kompartimentkonzentrationen des Benzols im "steady-state" eines isolierten Kompartimentpaares dem Partitionskoeffizienten gleich sind. Die konstanten Konzentrationen im "steady-state" werden durch ein Fließgleichgewicht erreicht. Nur für die Dynamik des Übergangs in den "steady-state" muß die Flußstärke der Flußpaare noch geschätzt werden.

Durch Grobabschätzung mit Hilfe des 1. Fickschen Diffusionsgesetzes ($Q/t = D*F*\Delta C/I$) über die Kapillarflächen (F) und die je nach Gewebe verschiedenen mittleren Diffusionsstrecken (I) kann zwischen den vier Flußkonstantenpaaren (o/b2, f/b1, t/b3, a1/b4) der relative Unterschied festgelegt werden. Die Flußkonstante (D) wird für alle Körperewebe als gleich angenommen. Die Flußkonstante (D) für die Luft geht gegen Null im Vergleich zu den anderen Flußkonstanten. Die Kapillarfläche für die Lunge entstammt der Literatur (Kaplan 83), die anderen Kapillarflächen wurden durch Analogie zum Menschen geschätzt (Dedrick 80, Silbernagel 83). Die Grobabschätzung der mittleren Diffusionsstrecken erfolgte grob über ein Kugelmodell (einer Blutkugel mit Kapillardurchmesser wird eine zu versorgende Gewebekugel zugeordnet), so daß für $I_{o/b2}$ und $I_{a1/b4}$ untere Schranken angenommen werden.

Danach erfolgt das Schätzen der absoluten Flüsse über einen Parameter, von dem alle Flüsse abhängig sind.

Die Diffusionsstrecken mußten beim Feinabschätzen später noch untereinander und absolut modifiziert werden (vergleiche Tab. 1 und Tab. 2, Veränderung der D/I beim Übergang vom ersten Modell zum genauen Modell).

Bei der Parameteranpassung sollte die Summe der Quadrate der Abweichungen des ppm-Wertes bei 50 min, des ppm-Wertes des Maximums der Kurve und der Zeitabweichung des Maximums der Kurve minimiert werden.

TAB. 1

Parametername	Wert	Quelle
Part.-Koeff. Blut/Fett	66,4	Fiserova-Bergerova 86
Part.-Koeff. Blut/Nichtfettgewebe	2,5	Fiserova-Bergerova 86
Part.-Koeff. Blut/Ol in Periton.	66,4	Fiserova-Bergerova 86
Part.-Koeff Blut/Lunge	0,15625	Fiserova-Bergerova 86
Fettmasse/g	3	Präparation
Ölmasse/g	0,198 oder 0,099	bekannt je nach Ölmenge
Gewebemasse/g	20	Präparation
Blutmasse/g	2	Präparation
D/I(Öl,Blut) pro D/I(Luft,Blut)	0,273	geschätzt, siehe Text
D/I(Nichtfettgewebe,Blut) pro D/I(Luft,Blut	0,282	geschätzt, siehe Text
D/I(Fett,Blut) pro D/I(Luft,Blut)	0,122	geschätzt, siehe Text
Kapillarfläche(Blut,Luft)/qmm	256	Kaplan 83
Kapillarfläche(Blut,Fett)/qmm	19,2	geschätzt, siehe Text
Kapillarfläche(Blut,Peritoneum)qmm	96	geschätzt, siehe Text
Kapillarfläche(Blut,Nichtfettgewebe)/qmm	268,8	geschätzt, siehe Text
Parameter zur Anpassung von D/I	kein Wert für erstes Modell	Optimierung hier sinnlos
a2	0,0166666	errechnet

Bei geringen Flußkonstanten ist das Ergebnis eine γ-Funktion (nur auf ca. 20 ppm ansteigend, flache untere Kurve in Abb. 3), die bei weiterem Erhöhen der Flußraten in heftiges Schwingen übergeht. Dieses Modell ist von seiner Strukter her also nicht geeignet, die beobachteten Benzolexhalationskurven zu reproduzieren.

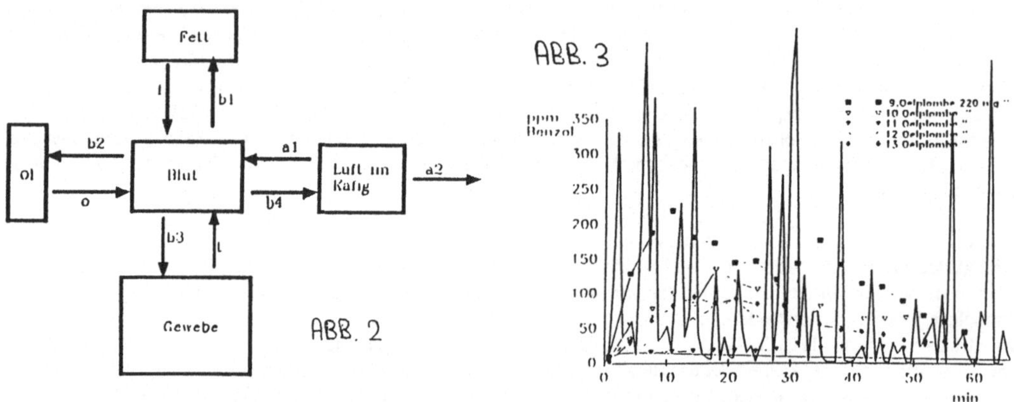

4. Berücksichtigung des Mischungsprozesses im Blut durch ein nichtlineares Modell mit dem Blut als Zentralkompartiment

Die Injektion in das Peritoneum führt dazu, daß das Blut, das im Bauchraum Benzol aufnehmen kann, durch die Leber zuerst in die Lunge und danach erst in die anderen Gewebe transportiert wird. Ein Ansatz, dies zu modellieren, ist, die Diffusionsströme zwischen Blut/Fett und Blut/Nichtfettgewebe erst langsam nach Benzolverabreichung zu steigern. Dies führt zu einem nichtlinearen Modell.

Mit diesem Modell können γ-Funktionen mit einem Maximum bei höchstens 60 ppm erzeugt werden. Bei weiterer Erhöhung der Flußraten zeigt auch dieses Modell Oszillationen.

5. Modell der pulmonalen Benzolexhalation mit Berücksichtigung der Diffusions- und Konvektionsprozesse

Das genaue Modell (Abb. 4) berücksichtigt die Konvektionsströme des Kreislaufs. Die Kompartimente des arteriellen und venösen Blutes verbinden die Kapillarkompartimente und haben eine Pufferfunktion. Zwischen den Kapillarkompartimenten und den Geweben oder der Lungenluft erfolgt der Lösemittelaustausch über Diffusionsvorgänge.

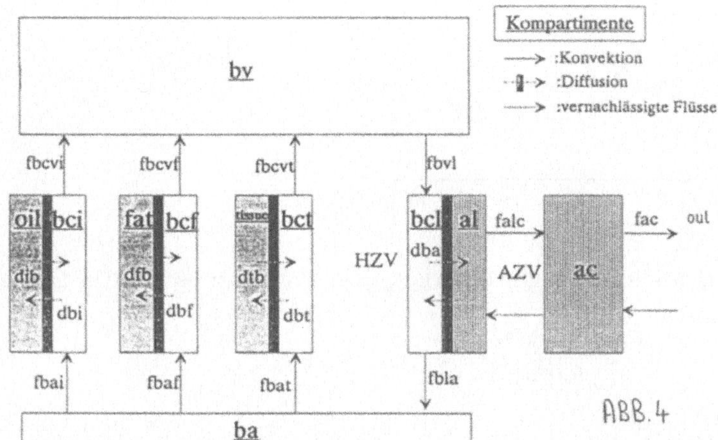

Die Gewebsmassen der Kompartimente, die Partitionskoeffizienten, die Grobschätzungen der Diffusionsflußverhältnisse, die Kapillarflächen und die Perfusionsraten der verschiedenen Kompartimente wurden aus dem ersten Modell übernommen (Tab. 1). Die weiteren Parameter wurden selbst gemessen oder der Literatur entnommen (Tab. 2), die Verteilung des Blutes auf die verschiedenen Kompartimente wurde vom Menschen übertragen (Silbernagel 83).

Die Verlaufskurve für 220 mg Benzol in 10 μl Öl pro kg Maus ist in Abb.5, die für 220 mg Benzol in 5μ Öl pro kg Maus ist in Abb.6 zu sehen. Bei genauer Betrachtung können leichte Oszillationen, die den Grundverläufen aufgelagert sind, beobachtet werden. Die Diffusionflüsse wurden mit der Zielfunktion wie beim ersten Modell neu fein abgeschätzt.

TAB. 2

Parametername	Wert	Quelle
HZV in g/s	0,155	Kaplan 83
AZV in g/s	0,00051	Kaplan 83
Verteilung des Blutes auf Kompartiments		siehe Text
D/l(Öl,Blut) pro D/l(Luft,Blut)	0,105	feiner angepaßt
D/l(Nichtfettgewebe,Blut) pro D/l(Luft,Blut	0,217	feiner angepaßt
D/l(Fett,Blut) pro D/l(Luft,Blut)	0,0938	feiner angepaßt
Parameter zur Anpassung von D/l	0,00000037	geschätzt, Zielfunktion siehe Text
fac	0,0166666	errechnet

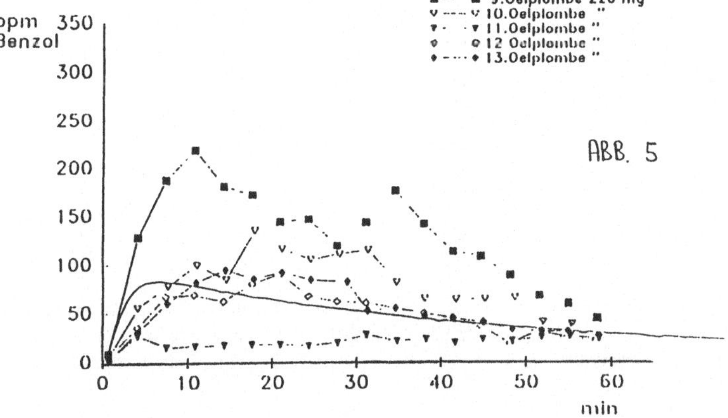

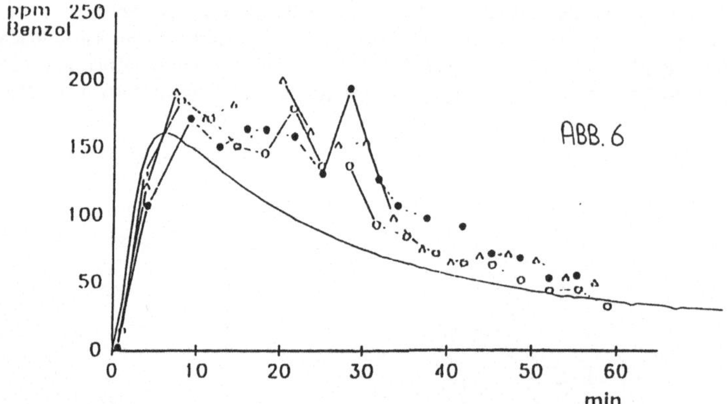

Die Verlaufskurve der Benzolexhalation bei 110 mg in 10 μl Öl pro kg Maus ist in Abb.7 und die bei 440 mg in 10 μl Öl pro kg Maus in Abb.8 zu sehen.

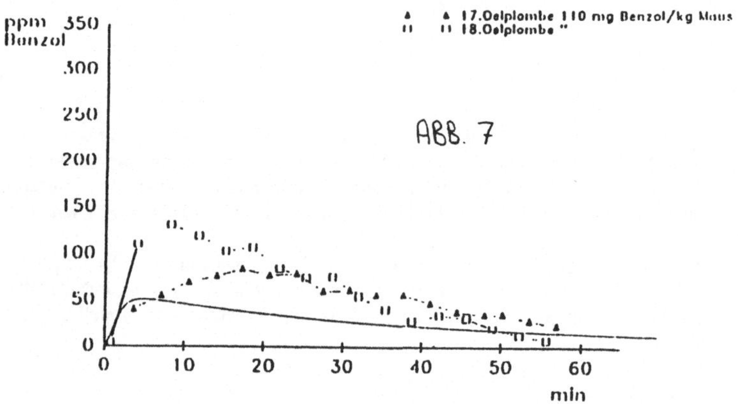

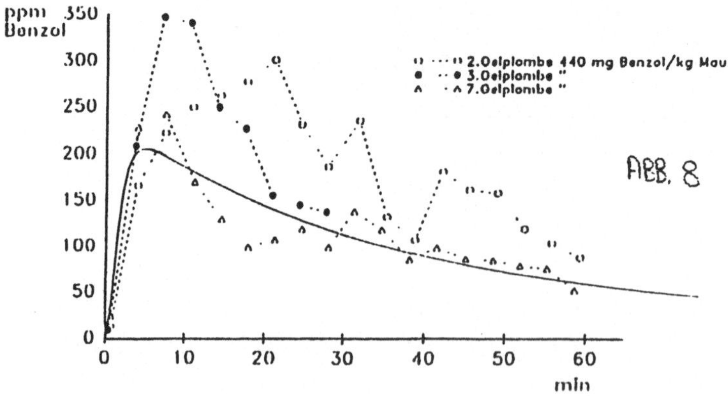

In Abb. 9 ist der Verlauf der Verteilung des Benzols während 50 Minuten (bei einer Exhalationskurve wie in Abb. 5) auf die verschiedenen Kompartimente zu sehen. Die Menge des in die Außenluft ausgeschiedenen Benzols stimmt in etwa mit der mit Hilfe von Aktivkohle aufgefangenen Benzolmenge aus der Abluft überein (Abb. 1, auf der linken Seite der Aktivkohle-"Integrator").

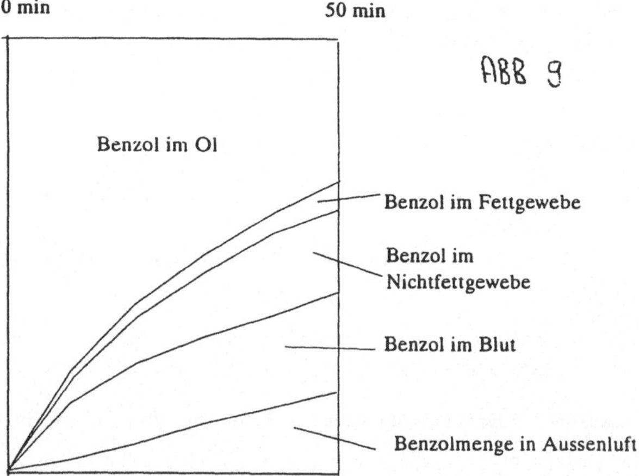

Bei der Sensitivitätsanalyse stellte sich heraus, daß bezüglich Herz- und Atemzeitvolumen das Modell eine geringe Sensitivität besitzt. Mit Änderungen der Diffusionsflüsse und der Perfusionsraten im Kapillarbereich des Körpers ist nur eine schwache Adaption des Systemverhaltens möglich. Durch Veränderung des Parameters der Diffusionsflüsse zwischen Kapillaren und Lungenluft kann das Systemverhalten am meisten beeinflußt werden.

6. Simulationsmethoden
Simuliert wurde auf einer VAX vo DEC mit CSMP. Als Integrationsverfahren wurde "Runge-Kutta-Prediktor-Korrektor" verwandt.

7. Konklusion
Das genaue Modell ist gut geeignet, um die Benzolexhalation bei intraperitonealer Bolusgabe qualitativ zu beschreiben und zu erklären. Eine gute Übereinstimmung der gemessenen/gerechneten Kurven erhält man bei Lösemittelapplikation in weniger Öl als Trägermedium. Bei größeren Ölmengen treten deutliche individuelle Schwankungen in den Exhalationskinetiken auf.
Bei der Applikation von Lösungen in den unteren Bauchraum mit Hilfe von kleinen Injektionsnadeln (z.B. Insulinspritzen) weichen die Darmschlingen der Nadelspitze im Normalfall reflexartig aus. Unter ungünstigen Bedingungen kann die Deponierung der Ölmenge aber auch ausnahmsweise in den Darm bzw. in eine Fettstruktur im Bauchraum erfolgen. Selten sollten Injektionen in Oberbauchorgane bei der gewählten Spritzregion vorkommen. Besteht ein Verdacht dahin, kann dies nach Versuchsende geprüft werden.
Durch die schlechte Qualität der Daten der Originale - retrospektive Studie - ist keine hochwertige quantitative Schätzung der gewebsspezifischen Diffusionskonstanten und mittleren Diffusionswege im Kapillarbereich möglich. Neben der geringen Fallzahl und der niedrigen Abtastfrequenz sind weitere Nachteile bei der intraperitonealen Bolusgabe bei der Maus die geringe Größe des Versuchstiers, die den Zugang zu anderen Meßgrößen als der Exhalationsrate verwehrt, und die unbekannten Diffusionsverhältnisse im Peritoneum.
Bei genauer Kenntniss der Diffusionskonstanten und der mittleren Diffusionswege bei Säugern, die wahrscheinlich klassenspezifisch sind (z. B. Kapillardurchmesser, Schmid-Schönbein 80) könnte das genaue Modell, das struktur- und funktionsanalog für alle Säuger ist, nach Beschaffung der weiteren individuellen anatomischen, physiologischen Parameter (Fettmasse, Körpergewicht, Herzzeitvolumen, Atemzeitvolumen, Kapillarflächen, quantitative Strömungsverhältnisse, Partitionskoeffizienten für verschiedene Lösemittel etc.) zur Vorhersage von Lösemittellabatmung und -verteilung benutzt werden.
Eine Versuchsanordnung zur Schätzung der Diffusionskonstanten und mittleren Diffusionwege könnte wie folgt aussehen: intravenöse Injektion (Vorschlag von H. E. Wichmann, was dazu führt, daß eine definierte Lösemittelmenge auf einen Schlag ins venösen Kompartiment eingebracht wird und die unbekannten Diffusionsverhältnisse im Peritoneum nicht mehr beachtet werden müssen) in größere Säuger (bei größeren Säugern ist neben der Exhalationsrate auch ein Zugriff auf das venöse und arterielle Blutkompartiment möglich). Diese Versuchanordnung hat den Vorteil, daß zur Schätzung von drei Parametern (D/I für die Diffusionsströme zwischen Lungenblut/Luft, Körperblut/Fett und Körperblut/Nichtfettgewebe) drei meßbare Größen zur Verfügung stehen. Durch Erhöhen des apparativen Aufwands kann die Meßfrequenz erhöht werden (z. B. Parallelschalten von Gaschromatographen). Durch Versuche an verschiedenen Säugerarten müßte sichergestellt werden, daß die Diffusionsparameter klassen- und nicht

nur artspezifisch sind (Vorschlag von G. Wünscher, solche Modelle in größerem Rahmen zu vergleichen). Nach Kenntnis dieser Parameter könnte das Diffusionverhalten im Peritoneum genauer untersucht werden.

8. Literaturverzeichnis

[1] Crank, J.: The Mathematics of Diffusion, Clarendon Press, Oxford, 1975

[2] Dedrick, R. L., Bischoff, K. B.: Species Similarities in Pharmacokinetics, Fed. Proc. 39, 1980, 54-59

[3] Fiserova-Bergerova, V., Diaz, L. M.: Determination and Prediction of Tissuegas Partition Coefficients, in: Int. Arch. Occup. Environ. Health 58, 1986, 75 - 87

[4] Fiserova-Bergerova, V., Tichy, M., DiCarlo, F. J.: Effects of Biosolubility on Pulmonary Uptake and Disposition of Gases and Vapors of Lipophilic Chemicals, in: Drug. Metab. Rev. 15, 1984 , 1033 - 1070

[5] Frumin, M.J., Salanitre, E., Rackow, H.: Excretion of Nitrous Oxide in Anesthetized Man, in: J. Appl. Physiol. 16, 1961, 720 - 722

[6] Jacquez, J. A.: Compartmental Analysis in Biology and Medicine, University of Michigan Press, 1985

[7] Jones, R. W.: Principles of Biological Regulation, Academic Press, New York, 1973

[8] Kaplan, M., Brewer, N. R., Blair, W.H.: The Mouse in Biomedical Research, (Hrsg.: Foster, L. F., Small, J. D. Fox, H. G.), Academic Press, New York, 1983, 252 - 256

[9] Maybeck, P. S.: Stochastic Models, Estimation, and Control, Volume 1, Academic Press, San Diego, 1979

[10] Metzger H.: Simulation von Organ-Mikrozirkulationssystemen - stationärer Gas- und Metabolitaustausch in drei-dimensionalen Kapillarsystemen, in: Schneider, B., Ranft, U.: Simulationsmethoden in der Medizin und Biologie, Springer, Heidelberg, 1978, 184 - 208

[11] Sato, A., Fujiwara, Y.: Elimination of Inhaled Benzene and Toluene in Man, in: Jpn. J. Ind. Health 14, 1972, 224 - 225

[12] Sato, A., Nakajima,T.: Partition Coefficients of some Aromatic Hydrocarbons and Ketones in Water, Blood and Oil, in: Brit. J. Ind. Med. 36, 1979, 232 - 234

[13] Schmid-Schönbein, H., Grunau, G., Bräuer, H.: Exempla Hämorheologica, Albert-Roussel Pharma GmbH, Wiesbaden, 1980, 18 - 23

[14] Silbernagl, S., Despopoulos, A.,: dtv-Atlas der Physiologie, Deutscher Taschenbuchverlag/Georg Thieme Verlag, Stuttgart, 1983

[15] Perl, W.: Discussion of Nitrous Oxide, in: Uptake and Distribution of Anesthetic Agents (Hrsg. Papper, E.M. und Kitz, R. J., McGraw-Hill New York), 1963, 250 - 257

[16] Ranft, U.: Ein digitales Simulationsmodell des Herz-Kreislaufsystems, in: Schneider, B., Ranft, U.: Simulationsmethoden in der Medizin und Biologie, Springer, Heidelberg, 1978, 209 - 223

[17] Wosilait , W. D., Luecke, R. H., Ryan, M. P.: Numerical Simulation of the Infusion of Adriamycin Using a Physiological Flow Model, in: Witten, M.: Mathematical Models in Medicine, Pergamon Press, Oxford, 1988, 861 - 868

Eine neue Beschreibung der Dynamik der basalen Insulinsekretion

D. Overkamp, W. Renn, B. Jakober, M. Eggstein

Medizinische Universitätsklinik, Abt. IV
7400 Tübingen

Einleitung

Insulin ist das zentrale Hormon der Regulation des Glukosestoffwechsels. Es wird in den ß-Zellen der Langerhans'schen Inseln des Pankreas, die in großer Zahl über das gesamte Pankreas verteilt sind, gebildet. Von dort gelangt Insulin über die Pfortader in den peripheren Blutkreislauf. Schon bei der ersten Passage durch die Leber, noch vor Erreichen der peripheren Zirkulation, wird ca. 50% des aus dem Pankreas freigesetzten Insulins abgebaut. Das in die Peripherie gelangte Insulin wird über Leber und Niere augeschieden.
Die Sekretion von Insulin wird vor allem durch die Glukosekonzentration des Bluts reguliert. Daneben übt Glucagon, gebildet in den alpha-Zellen der Langerhans'schen Inseln, einen sekretionssteigernden Effekt aus, während Somatostatin aus den delta-Zellen der Inseln die Sekretion von Insulin hemmt.

Von einer Reihe von Hormonen ist bekannt, daß ihre Sekretion nicht kontinuierlich, sondern pulsatil erfolgt (Weigle, 1987). Den sich daraus ergebenden Konzentrationsschwankungen wird auf der einen Seite für die Kodierung von Informationen Bedeutung zugemessen (Rapp et al. 1981), auf der anderen Seite wird vermutet, daß durch die sich ändernde Konzentration eine Downregulation der entsprechenden zellulären Rezeptoren verhindert wird und daß daraus eine Steigerung der Hormonwirkung in Bezug auf die Dosis resultiert (Bratusch-Marrain et al. 1987).

Seit einer Reihe von Jahren ist bekannt, daß auch die basalen Insulin-spiegel nicht zufällige Konzentrationsschwankungen aufweisen (Goodner et al. 1977).

Diese Konzentrationsschwankungen lassen sich sowohl in verschiedenen Tierspezies wie beim Menschen nachweisen (Lang et al. 1979). Die Insulinfreisetzung erfolgt auch aus dem mit einer konstanten Glukosekonzentration durchströmten, isolierten Hundepankreas pulsatil, so daß ein Rückkopplungseffekt zwischen Insulin und Glukose als Ursache der Konzentrationsschwankungen weitgehend ausscheidet (Stagner et al. 1980).

Zeitserienanalysen der basalen Insulinkonzentrationsschwankungen mit Hilfe von Autokorrelation und Fouriertransformation sind dahingehend interpretiert worden, daß diesen eine Periodizität zugrunde läge (Matthews et al. 1983). Seither wird nach dem für die Periodizität verantwortliche Schrittmacher gesucht (Stagner und Samols, 1985).
Wir zeigen hier, daß sich die basalen Insulinkonzentrationsschwankungen auch als Geräusch einer bestimmten Charakteristik auffassen lassen, und daß diesem Geräusch ein reiner Zufallsprozess zugrunde liegen kann.

Patienten und Methoden

Die Untersuchungen wurden an 6 stoffwechselgesungen Probanden durchgeführt. Nach einer Nüchternperiode von 10 Stunden wurde am Morgen aus einer arterialisierten Handrückenvene für 160 Minuten kontinuierlich Blut abgenommen und in 1-minütigem Abstand portioniert. Die Bestimmung der Insulinkonzentration erfolgte mit einem ELISA System, dessen Variationskoeffizient im relevanten Konzentrationsbereich bei 5% liegt. Die Meßdaten der einzelnen Probanden wurden Fourier transformiert. Anschließend wurde eine Regressionsanalyse der Leistungsspektren gegen die Frequenz durchgeführt.

In der Simulation der Konzentrationsverläufe wurde angenommen, daß die Freisetzung von Insulin aus den ß-Zellen in zeitdiskreter Zufallsfolge mit minütlichen Pulsen erfolgt. Die Wahl dieses Zeitrasters gründet sich auf Beachtung der Kreislaufzeit, die höherfrequente Konzentrationsschwankungen nicht zu übertragen vermag.
Die Pulse wurden als Input in ein Zweikompartimentsystem verwandt, das Verteilung und Abbau von Insulin beschreibt.
Das erste Kompartiment stellt den Insulinabbau in der Leber dar, das zweite die periphere Zirkulation.
Mittelwert und Standarabweichung der diskreten Pulse standen im Verhältnis 1/4, die Halbwertzeit im ersten Kompartiment war 0.6 min, die im zweiten 3,5 min.

Ergebnisse

Der Insulinverlauf eines Probanden ist in Abb.1 dargestellt.

Wie bei diesem fanden sich bei allen Probanden nicht zufällige Konzentrationsschwankungen der Seruminsulinkonzentration (Runs Test: $p<0.01$).

Die Simulation der Meßwerte durch das o.a. Model ergab einen in der Charakteristik ganz ähnlichen Konzentrationsverlauf (Abb.2)

Die Regressionsanalyse des Logarithmus der Leistung gegenüber dem Logarithmus der Frequenz ergab bei allen Probanden eine signifikant negativ lineare Korrelation:

Proband	Steigung	Korrelations-koeffizient	Signifikanz-niveau
1	-1.46	-0.82	<0.0001
2	-0.66	-0.51	<0.0001
3	-0.81	-0.58	<0.0001
4	-1.07	-0.67	<0.0001
5	-1.00	-0.63	<0.0001
6	-1.56	-0.81	<0.0001

Nach Mittelung der Leistungsspektren der einzelnen Probanden ergab sich die in Abb.3 wiedergegebene Regression, die der der simulierten Daten (Abb.4) gleicht.

Proband 6
Insulin (mE/l)
Time (min)

Abb. 1

Simulation
Insulin (mE/l)
Time (min)

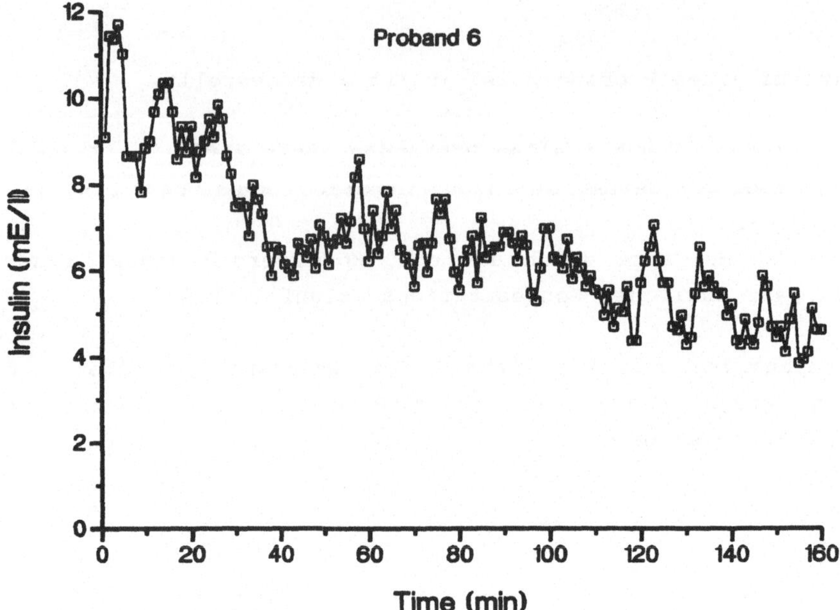

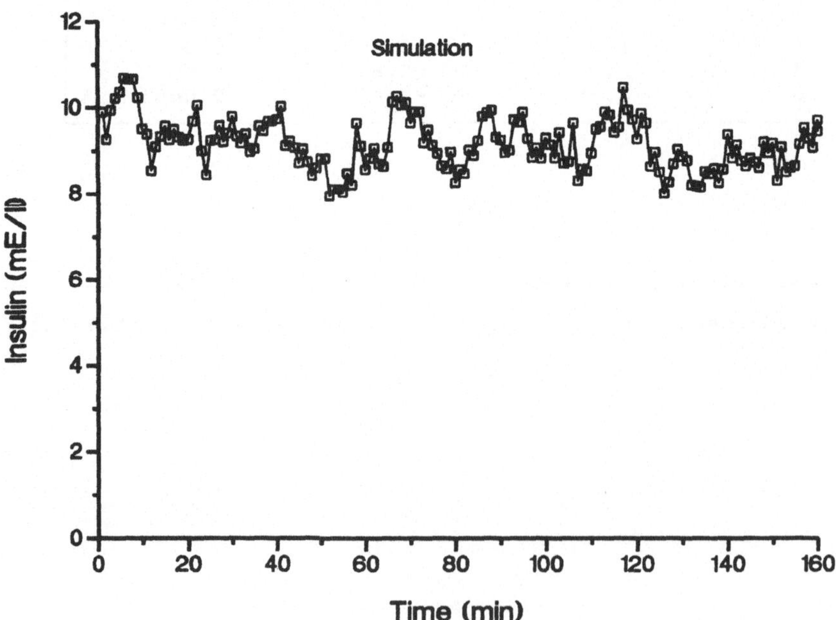

Abb. 2

log Power
Y = B(0) + B(1) * X
B(0) = -3.44
B(1) = -1.08
R² = .795
log Frequency

Abb. 3

log Power
Y = B(0) + B(1) * X
B(0) = -4.01
B(1) = -1.21
R² = .514
log Frequency

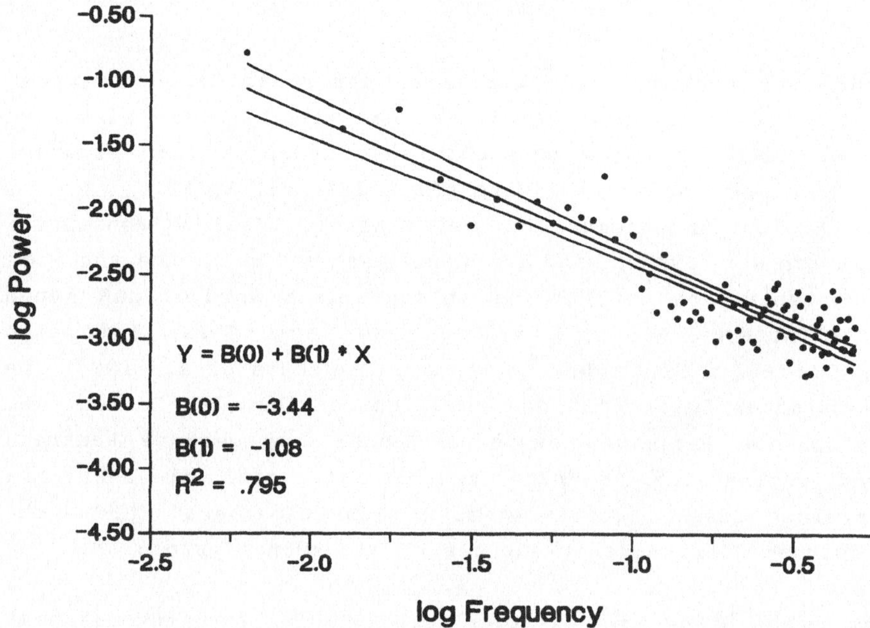

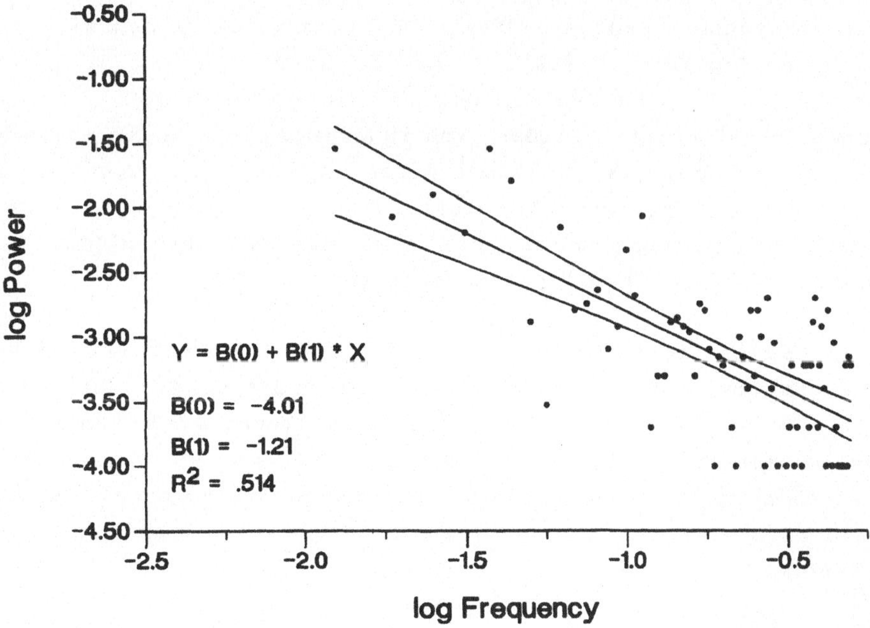

Abb. 4

Diskussion

Für eine Reihe von Hormonen läßt sich auch unter basalen Bedingungen eine pulsatile Sekretionscharakteristik nachweisen. Bei einzelnen Hormonen (z.B. LHRH) ist eine physiologische Bedeutung der Frequenz dieser Pulse nachgewiesen worden (Marshall und Kelch, 1986).
Auch die basale Insulinkonzentration weist nicht zufällige Konzentrationsschwankungen auf. Bisher sind die Ergebnisse von Zeitreihenanalysen der Insulinkonzentration dahingehend interpretiert worden, daß ihnen eine charakteristische Periodizität zugrunde läge. Diese läßt sich jedoch nicht bei allen Probanden nachweisen (Matthews et al. 1983). Da die Langerhans'schen Inseln mit den insulinproduzierenden Zellen ohne direkte anatomische Beziehung zueinander über das gesamte Pankreas verteilt sind, ergibt sich die Frage nach der Natur eines die Sekretion aus den einzelnen Inseln koordinierenden "Schrittmachers". Die Suche nach einem solchen Schrittmacher war bisher allerdings erfolglos.

Im Gegensatz zu den bisherigen Ansätzen haben wir den Insulinkonzentrationsverlauf als *Geräusch* interpretiert und nachgewiesen, daß sich bei allen Probanden eine signifikante Korrelation zwischen Amplitude des Leistungsspektrums und der Frequenz findet. Der funktionale Zusammenhang der Variablen ist dabei $1/f^{\beta}$. Für $0.5 < \beta < 1.5$ wird der Zeitverlauf als $1/f$ Geräusch bezeichnet. Prozesse dieser Charakteristik finden sich in der Natur sehr häufig (Voss, 1988).

Daten gleicher Charakteristik ließen sich innerhalb eines Zweikompartimentsystems mit bestimmten Zeitkonstanten dadurch erzeugen, daß Zufallszahlen als Input verwandt wurden. Diese Zeitkonstanten lagen dabei in der Größenordnung derer, die sich aus der physiologischen Halbwertzeit von Insulin im Organismus ergeben.

Schon früher ist gezeigt worden, daß sich der Konzentrationsverlauf einer Reihe von Hormonen, deren Sekretion in größeren Zeitabständen erfolgt (z.B. LH, GH, Cortisol), gut durch ein mathematisches Model der Sekretionsrate simulieren läßt, das die Summe von Gaußverteilungen mit unterschiedlichen Amplituden darstellt (Veldhuis und Johnson, 1988). Diese Sekretionsrate wurde wie bei uns als Input in ein Kompartimentsystem betrachtet.

Wir folgern, daß der Insulinkonzentrationsverlauf im Serum von Normalpersonen adäquat als $1/f$-Geräusch beschrieben werden kann. Nach dem

Ergebnis der Modelversuche ist es nicht notwendig, eine Synchronisation der Insulinfreisetzung aus den Langerhans'schen Inseln des Pankreas zu fordern. Ob der Insulinsekretion im Pankreas tatsächlich lediglich ein Zufallsprozess zugrunde liegt, oder ob ein nichtlineares Rückkopplungssystem mit möglicherweise chaotischem Verhalten operativ ist, läßt sich aufgrund der vorhandenen Daten nicht entscheiden.

Literatur

1. Bratusch-Marrain, P.R.; M. Komjati und W. Waldhäusl. Pulsatile insulin delivery: physiology and clinical implications. **Diabetic Medicine** 4: 197-200 (1987)

2. Goodner, C.J.; B.C. Walike; D.J. Koerker; J.E. Ensinck; A.C. Brown; E.W. Chideckel; J. Palmer und L. Kalnasy. Insulin, glucagon, and glucose exhibit synchronous sustained oscillations in fasting monkeys. **Science** 195: 177-179 (1977)

3. Lang, D.A.; D.R. Matthews; P.J. Peto und R.C. Turner. Cyclic oscillations of basal plasma glucose and insulin concentrations in human beings. **N.Engl.J.Med.** 301: 1023-1027 (1979)

4. Marshall, J.C. und R.P. Kelch. Gonadotropin-releasing hormone: role of pulsatile secretion in the regulation of reproduction. **N.Engl.J.Med.** 315: 1459-1468 (1986)

5. Matthews, D.R.; D.A. Lang; M.A. Burnett und R.C. Turner. Control of pulsatile insulin secretion in man. **Diabetologia** 24: 231-237 (1983)

6. Rapp, P.E.; A.I. Mees und C.T. Sparrow. Frequency encoded biochemical regulation is more accurate than amplitude dependent control. **J.Theor.Biol.** 90: 531-544 (1981)

7. Stagner, J.I.; E. Samols und G.C. Weir. Sustained oscillations of insulin, glucagon, and somatostatin from the isolated canine pancreas during exposure to a constant glucose concentration. **J.Clin.Invest.** 65: 939-942 (1980)

8. Stagner, J.I. und E. Samols. Pertubation of insulin oscillations by nerve blockade in the in vitro canine pancreas. **Am.J.Physiol.** 248: E516-E521 (1985)

9. Veldhuis, J.D. und M.L. Johnson. A novel general biophysical model for simulating episodic endocrine gland signaling. **Am.J.Physiol.** 255: E749-E759 (1988)

10. Voss, R.F. Fractals in nature: From characterization to simulation. In: The Science of Fractal Images, hrsg. von Peitgen, H.O. und D. Saupe. New York Berlin: Springer Verlag, 1988, p. 21-70.

11. Weigle, D.S. Pulsatile Secretion of fuel-regulatory hormones. **Diabetes** 36: 764-775 (1987)

Auswertung von ^{13}C-Glukose Tracerexperimenten zur
Bestimmung der Glukoseproduktion der Leber

W.Renn, D.Overkamp, A.Pickert, M.Eggstein
Medizinische Universitätsklinik Tübingen, Abt. IV.

Problemstellung

Ziel der dargestellten Arbeit ist eine Verbesserung der Differentialdiagnose von Typ II Diabetikern. Dies sind Diabetiker, die zwar eine
Insulinsekretion besitzen, jedoch erhöhte Blutzucker Werte aufweisen,
da die Insulinwirkung reduziert ist. Die Wirkung des Insulins besteht
bekanntlich darin, daß durch dieses Hormon die Glukose aus dem
Blutkreislauf in die Muskel- und Fettzellen des Körpers transportiert
wird. Daneben gibt es aber noch eine zweite Wirkung des Insulins, die
darin besteht, daß die Glukoseproduktion der Leber bei Überangebot von
Glukose im Körper, d.h. nach Nahrungsaufnahme, drastisch reduziert
wird. Bei der Differentialdiagnose des Typ II Diabetes wäre es deshalb
wesentlich feststellen zu können, welche der beiden Insulinwirkungen
gestört ist. Mit den herkömmlichen Verfahren der klinischen Diagnostik, kann diese Frage nicht beantwortet werden, da die Leberproduktion
von Glukose nicht gemessen werden kann. In zahlreichen wissenschaft-
lichen Untersuchungen wurde die Berechnung der Leberproduktion
versucht, wobei radioaktiv markierte Glukose als Tracer verwendet
wurde. Diese Methode ist jedoch praktisch nicht anwenbar, da die
radioaktive Belastung den Patienten nicht zugemutet werden kann. Um
das Problem der radioaktiven Belastung zu umgehen, wird bei unseren
Versuchen Glukosetracer eingesetzt, der mit den stabilen ^{13}C-Atomen
markiert ist. Dieser Tracer kann mit Hilfe eines Massenspektrometers
mit sehr hoher Genauigkeit gemessen werden. Außerdem ist die Auswertung der bisherigen Untersuchungen unter Verwendung der sogenannten
Steele Equation unbefriedigend, da man damit wiedersprüchliche
Ergenisse erhält, was sich im Auftreten von - physiologisch unsinnigen
- negativen Produktionsraten äußert. Die Problematik dieser Methode
ist in [1] dargestellt. Die Auswertung unserer Daten geschieht mit
Hilfe einer neu formulierten Dekonvolutionsmethode, die in der Lage
ist auch bei Bolusinjektion des Tracers , physiologisch vernünftige
Ergebnisse zu liefern. Diese Methode benötigt am wenigsten Tracersubstanz, wurde aber bisher bei der Untersuchung des Glukosemetabolismus wegen der mathematischen Schwierigkeiten nicht angewendet.

Mathematische Auswertung

Da diese Untersuchungen sehr aufwendig und teuer sind, wurde zunächst die mathematische Auswertung konzipiert und durch Simulationsrechnungen validiert, um ein optimales Versuchsdesign zu haben. Das verwendete mathematische Modell wurde von Cobelli et al. [2] zur Auswertung von radioaktiven Glukose Tracerexperimenten entwickelt.

Faßt man die Bereiche des Körpers, in denen die Zellen ohne Insulin Glukose aufnehmen können zu einem ersten insulinunabhängigem Kompartiment zusammen und beschreibt die Muskel- und Fettzellen durch ein zweites insulinabhängiges Kompartiment, so kann man für die Beschreibung des Glukosemetabolismus ein Zwei-Kompartiment-System verwenden. Das Plasmakompartment, in das man den Tracer injiziert und in dem die Glucose und der Tracer gemessen werden, wird dabei mit dem ersten Kompartiment zusammengefaßt. Der formaler Aufbau mit den entsprechenden Transportraten ist in Abb. 1 dargestellt.

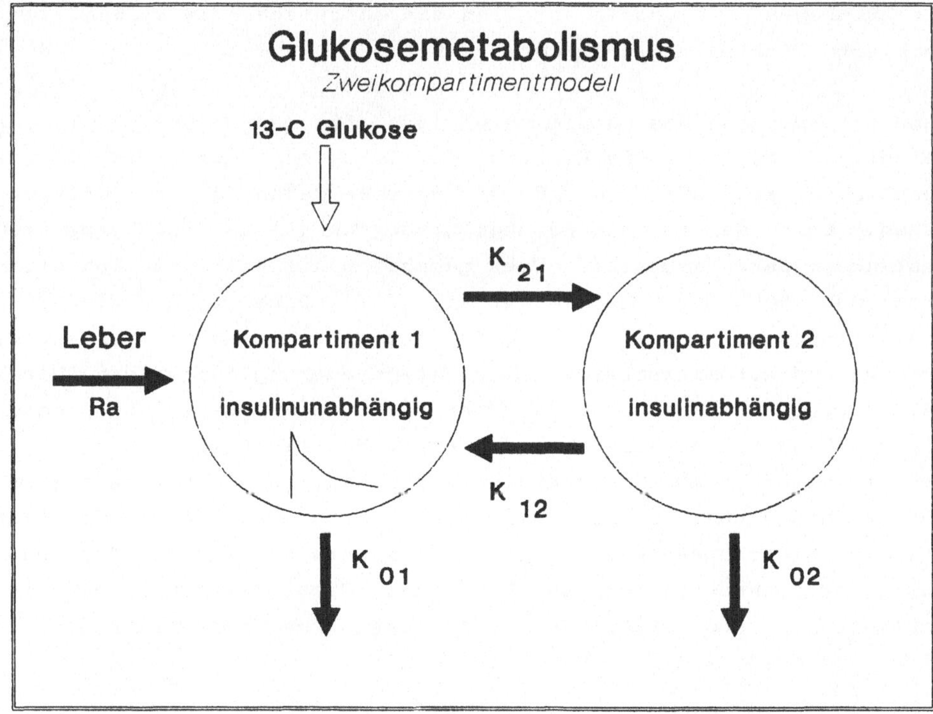

Abb 1. : Vereinfachtes Modellschema für den Glukosemetabolismus.

Die dazugehörigen Bilanzgleichungen haben folgende Gestalt:

$$dG_1/dt = -k_{11}G_1 + k_{12}G_2 + R_a + Dosis/V \qquad (1)$$
$$dG_2/dt = -k_{21}G_1 + k_{22}G_2 \qquad (2)$$

$$dg_1/dt = -k_{11}g_1 + k_{12}g_2 + dosis/V \qquad (3)$$
$$dg_2/dt = -k_{21}g_1 + k_{22}g_2 \qquad (4)$$

mit den Abkürzungen

$$k_{11} := k_{01} + k_{21} \quad und \quad k_{11} := k_{02} + k_{12} \qquad (5)$$

G_1 ist die Glukosekonzentration im ersten Kompartiment, G_2 ist die Glukosemenge im zweiten Kompartiment geteilt durch das Volumen V des ersten Kompartimentes. R_a (rate of appearance) stellt den Zufluß an Glukose aus der Leber, die gesuchte Leberproduktion dar. Die Menge an intravenös injizierter unmarkierter Glukose, die zu Testzwecken den Glukosespiegel im Plasma erhöht, wird mit Dosis bezeichnet. Die klein geschriebenen Symbole stellen die entsprechenden Größen für die markierte Glukose - den Tracer - dar.

Zunächst müssen die Gleichungen (3,4) für den meßbaren Tracer g_1 gelöst werden, um die Kinetik der Glukose, d.h. die Reaktionskonstanten k_{01} bis k_{02}, für die weitere Rechnung zur Verfügung zu haben. Dazu werden die Lösungskurven von (3) an die experimentell bestimmten Daten angepaßt. Dies geschieht mit Hilfe des Optimierungsprogramms NONLIN von Metzler et al. [3]. Dabei werden die Differentialgleichungen (3,4) mit Hilfe eines Runge-Kutta Verfahrens 4. Ordnung und automatischer Schrittweitensteuerung integriert. Die Minimalisierung der Abweichung zwischen Rechnung und Experiment geschieht mit dem Gauß-Newton Verfahren und einer Modifikation nach Hartlee. Das Programm liefert die vollständige Statistik, d.h. Vertrauensbereiche und Standardabweichung der geschätzten Parameter, Summe der quadratischen Abweichungen etc.. Ein wesentliches Problem bei der Schätzung der Reaktionskonstanten besteht in der Nichtidentifizierbarkeit der Parameter [1] und kann zu völlig falschen Ergebnissen führen, wenn es nicht beachtet wird.

Aus medizinischen wie auch aus systemtheoretischen Gründen sind zwei Zustände gesondert zu betrachten, der Basalzustand mit der Plasmaglukose im steady state und der Zustand nach Stimulation in dem der Plasmaglukosespiegel durch geeignete Zufuhr von Glukose erhöht wird.

1. Basalzustand (steady state)
Dieser Zustand ist für die unmarkierte Glukose G gekennzeichnet durch folgende Bedingungen:

$$\text{Dosis} = 0, \quad dG_1/dt = 0, \quad dG_2/dt = 0 \qquad\qquad (6)$$

Damit lassen sich aus (1,2) die unbekannte Glukosemenge im Kompartiment 2 und die Glukoseproduktion der Leber im steady state sofort berechnen

$$G_2^{ss} = G_1^{ss}\exp(-k_{22}t) \quad \text{und} \quad R_a^{ss} = (k_{11}k_{22} - k_{12}k_{21})G_1^{ss}/k_{22} \qquad (7)$$

Für die Zugabe der markierten Glukose hat man die Möglichkeiten, den Tracer als Infusion , als kurzzeitigen Bolus oder als eine Kombination aus Bolus und Infusion zu verabreichen. Um eine geignete Wahl zu treffen, wurden diese drei Möglichkeiten mit Hilfe von Simulationsrechnungen durchgespielt, wobei Reaktionskonstanten aus der Literatur [2] verwendet wurden. Unsere Wahl fiel auf die Bolusinjektion, da bei der Infusion die Versuchsdauer zu lang wäre, da es etwa sechs Stunden dauern würde bis der Tracer im steady state ist. Die Kombination aus Bolus und Infusion andererseits hätte den Nachteil, daß dabei zuviel der teuren Tracersubstanz verbraucht würde. Um das Identifizierungsproblem der Parameter zu umgehen wurde im Basalzustand angenommen, daß die Elimination aus dem insulinabhängigen Kompartiment k_{02} vernachlässigt werden kann, dadurch werden die restlichen Reaktionskonstanten eindeutig identifizierbar. Damit kann dann auch die basale Glukoseproduktion der Leber (7) berechnet werden.

2. Zustand nach Stimulation (Nonsteady state)
Bei der Berechnung der Glukosekinetik aus den Tracerdaten im nonsteady state muß berücksichtigt werden, daß die Elimination aus dem insulinabhängigen Kompartiment eine wesentliche Rolle spielt, da das Insulin nach Stimulation mit Glukose stark ansteigt, d.h. $k_{02} > 0$. Damit die Parameter aber weiterhin identifizierbar bleiben, nehmen wir an, daß die Rückflußrate k_{12} mit der im basalen Zustand identisch ist.

Durch Simulationsrechnungen für den Zustand nach Stimulation konnte gezeigt werden, daß der Fehler der Schätzung für R_a aus den experimentellen Daten am kleinsten ist, wenn die Stimulation mit unmarkierter Glukose im Zeitverlauf identisch ist mit der Zufuhr des Tracers. Deshalb wurde auch hier eine Bolusinjektion gewählt.

Die Berechnung von R_a aus dem Gleichungssystem (1,2) ist ein typisches Dekonvolutionsproblem und wurde unter anderem von Eaton et al. [4] für den Fall eines identifizierbaren 2-Kompartiment-Systems behandelt. Das Problem mit nichtidentifizierbaren Parametern und mehreren Kompartimenten läßt sich ebenfalls lösen, wie im Folgenden gezeigt wird.

Geht man von den Absolutwerten für die Plasmaglukose G über zu den Differenzen vom basalen Plasmaspiegel, die definiert sind durch :

$$D_1 := G_1 - G_1^{ss} \quad \text{und} \quad D_2 := (G_2 - G_2^{ss})/k_{21} \; ,$$

so erhält man, das neue Differentialgleichungssystem

$$dD_1/dt = - k_{11} D_1 + k_{12} k_{21} D_2 + R_a - R_a^{ss} \tag{8}$$
$$dD_2/dt = - \quad D_1 + k_{22} D_2 \tag{9}$$

mit den Anfangsbedingungen $D_1(0) = \text{Dosis}/V$ und $D_2(0) = 0$

Aus Gleichung (9) ergibt sich die Glukosedifferenz im zweiten Kompartiment nach Integration

$$D_2(t) = \int_0^t \exp(-k_{22}(t-u)) D_1(u) du \tag{10}$$

und aus (8) erhält man dann den zeitlichen Verlauf der Glukoseproduktion der Leber zu

$$R_a(t) = R_a^{ss} + dD_1(t)/dt + k_{11} D_1(t) - k_{12} k_{21} D_2(t) \tag{11}$$

Diese Gleichung enthält nur noch Parameterkombinationen, die immer identifizierbar sind, sodaß Fehler, die durch die Identifizierbarkeitsannahmen bei der Bestimmung der Glukosekinetik aus den Tracerdaten hierdurch kompensiert werden. Eine analoge Rechung kann man auch für mehr als zwei Kompartimente durchführen, ohne daß nichtidentifizierbare Parameterkombinationen auftreten.

Bei der numerischen Auswertung solcher Dekonvolutionsgleichungen werden üblicherweise Spline-Funktionen verwendet, um zwischen den Meßdaten zu interpolieren. Dieses Vorgehen ist jedoch nicht ohne Vorsicht anzuwenden, da systematische Meßfehler nicht erkannt werden. Wesentlich bessere Ergebnisse lassen sich erziehlen, wie in [5] gezeigt wurde, wenn man den Daten eine Modellvorstellung zugrunde legt, die durch mehrere Kompartimente beschrieben wird, und deren

Lösung an die Daten anpaßt. Wir verwenden deshalb den Ansatz

$$D_1 = \Sigma_i (A_i \exp(-\alpha_i t)) \quad \text{mit } i = 1, 3 \tag{12}$$

und bestimmen die darin enthaltenen Parameter wieder mit dem Optimierungsprogramm NONLIN [3]. Danach kann man die Differentiation und Integration analytisch durchführen und bekommt.

$$\frac{dD_1}{dt} = \Sigma_i (-\alpha_i A_i \exp(-\alpha_i t)) \tag{13}$$
$$D_2 = \Sigma_i (A_i (\exp(-k_{22} t) - \exp(-\alpha_i t))/(\alpha_i - k_{22})) \tag{14}$$

Setzt man diese Ergebnisse in (11) ein so erhält man die numerische Lösung für die Glukoseproduktion der Leber.

Experimetelle Methoden

Entsperechend der mathematischen Analyse des Problems erfolgt der Versuchsablauf in zwei Phasen. Zur Bestimmung der Glukoseproduktion der Leber im nüchternen Zustand (steady state) erfolgt eine venöse Bolusinjektion von ^{13}C-Glukose. Zur Bestimmung der Tracerkonzentration im Plasma wird aus einer Handrückenvene Blut kontinuierlich abgepumpt und mit Hilfe eines Fraktionssammlers in 1-Minuten-Fraktionen portioniert. Diese Versuchsphase wird über 150 Minuten durchgeführt. Direkt im Anschluß daran wird ein intravenöser Glukosetoleranztest durchgeführt. Dabei wird innerhalb von 2 Minuten 0,5 g Glukose/kg Körpergewicht, davon 5% ^{13}C-Glukose infundiert. Der Konzentrationsverlauf der markierten und nichtmarkierten Glukose im Plasma wird durch die oben beschriebene kontinuierliche Blutentnahme über weitere 150 Minuten verfolgt.

Ergebnisse

In Abb 2. sind die gemessenen und berechneten Werte sowohl für die unmarkierte Glukose im Plasma als auch die als Tracer verwendete ^{13}C-Glukose dargestellt. Aus optischen Gründen wurden nicht alle Meßwerte gezeigt, doch wurden charakteristische Außreißer belassen. Daran sieht man, daß die massenspekrometrisch erhobenen Tracerdaten aufgrund des komplexen Untersuchungsganges erhebliche Fehler aufweisen können. Dies hat zur Folge, daß ein 3-Kompartiment-Modell, wie es in [2] ebenfalls diskutiert wird, auf unsere Daten nicht angewendet werden konnte.

Das eigentliche Ergebnis unserer Arbeit ist in Abb. 3 aufgezeigt. In den ersten 150 Minuten befindet sich die Plasmaglukose im Basalzustand von 84 mg/dl. Dieser Wert wird durch die Zufuhr aus der Leber aufrecht

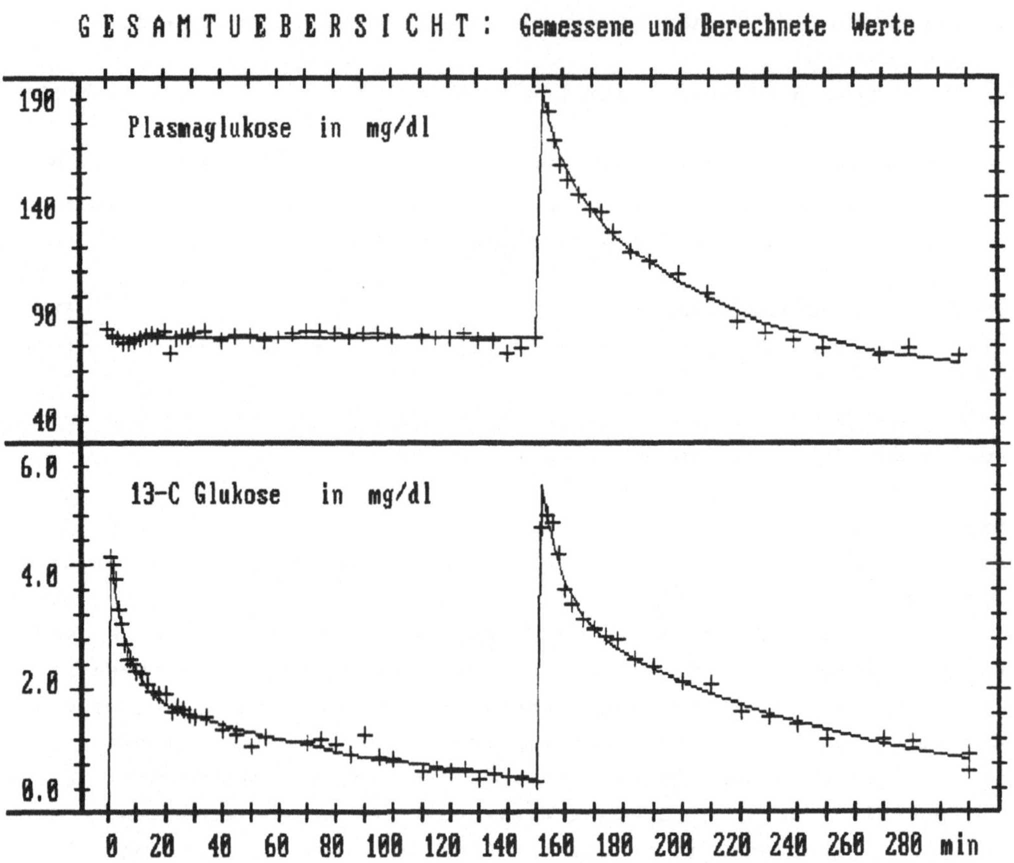

Abb 2. : Zusammenstellung der gemessenen und berechneten Werte

erhalten, die 2.52 mg/min und kg Körpergewicht (KG) beträgt. Nach dem Glukosestimulus bei 150 Minuten fällt die Leberproduktion stark ab und erreicht bei 170 Minuten ihr Minimum von 0.37 mg/min KG. Danach steigt die Glukose Produktion der Leber wieder an, hat aber bei 300 Minuten ihren Ausgangswert noch nicht erreicht. Dies ist der Grund dafür, daß der Blutzuckerspiegel ca. 100 Minuten nach einem intravenösen Glukose-toleranztest leicht unter den Ausgangswert absinkt, bei unserem Versuch bis auf 77 mg/dl und erst nach 180 Minuten normalisiert ist. Dieser Umstand wird gewöhnlich der peripheren Insulinwirkung zuge-schrieben. Im oberen Teil von Abb. 3 ist der an die Meßdaten angepaßte Verlauf der Plasmaglukose nochmals abgebildet. Im Vergleich dazu sind zwei Simulationsergebnisse aufgezeichnet. Die untere Kurve würde sich ergeben, wenn die Leberproduktion sofort auf Null absinken und nicht mehr ansteigen würde. Dies ergibt eine Differenz von 25 mg/dl bei 300

Minuten. Die obere Kurve simuliert den entgegengesetzten Fall, wenn
die Glukoseproduktion der Leber überhaupt nicht absinken würde, was zu
einer Enddifferenz von 40 mg/dl über dem Ausgangswert führt. Der erste
Fall wird in der medizinischen Praxis nur selten beobachtet, der
zweite Fall tritt bei Typ II Diabetikern sicher häufig auf, kann
jedoch mit herkömmlichen Untersuchungsmethoden nicht bestätigt werden.

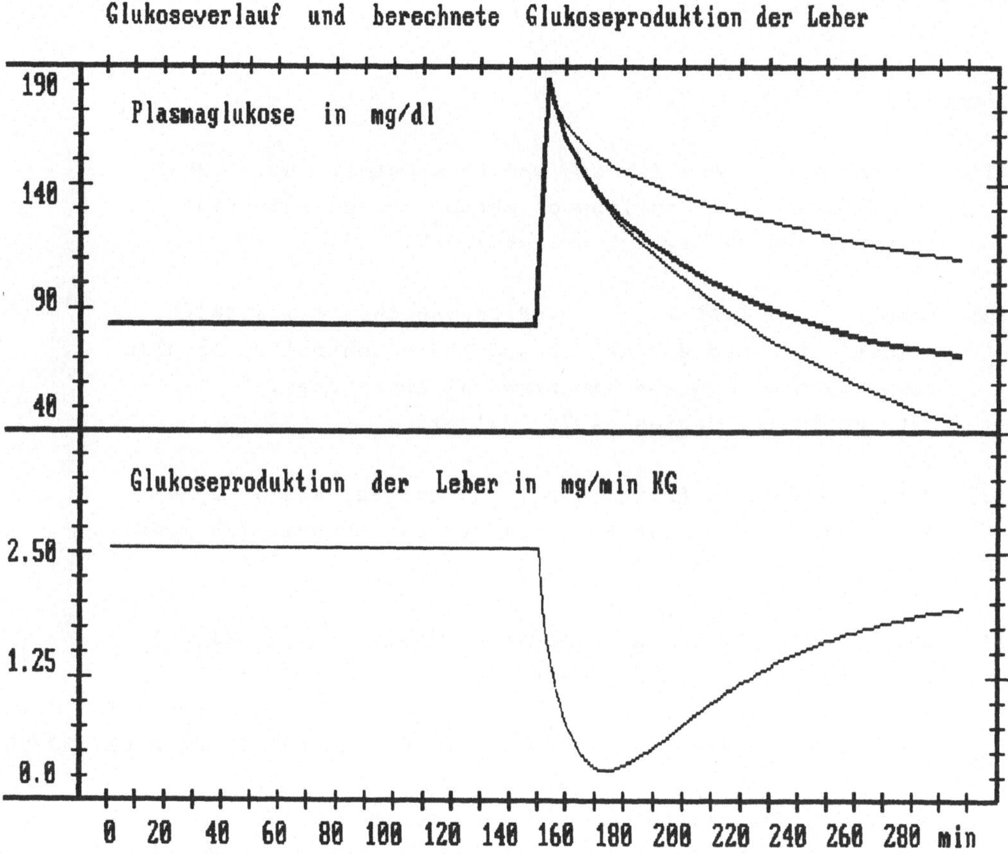

Abb 3. : Gegenüberstellung der berechneten Kurven für die Glukoseproduktion der Leber angegeben in mg pro Minute und kg Körpergewicht
(mg/min KG) im unteren Teil und die Konzentration von unmarkierter
Plasmaglukose im oberen Teil der Graphik. Über und unter dem
Glukoseverlauf nach Stimulation, sind mit dünnerer Strichstärke die
Ergebnisse von Simulationsrechnungen angegeben, bei denen angenommen
wurde, daß die Glukoseproduktion der Leber entweder total abfällt oder
konstant weiterläuft (obere Kurve im oberen Teil).

Zusammenfassung

Durch den Einsatz eines ungefährlichen Glukosetracers und eines durch Simulation validierten mathematischen Modelles, bei dem die Identifizierung der Parameter und der Einsatz auch analytischer Rechenverfahren die wesentlichen Merkmale sind, gelingt es verläßliche Daten über die Glukoseproduktion der Leber auch unter nonsteady state Bedinungen zu gewinnen. Damit könnte sich eine positive Auswirkung auf die Diagnostik und Therapie des Typ II Diabetes ergeben.

Literatur

[1]   Carson, E.R., Cobelli, C., and Finkelstein, L., (1983)
      The Mathematical Modeling of Metabolic and Endocrine
      Systems. John Wiley & Sons, New York.

[2]   Cobelli, C., Toffolo, G., and Ferrannini, F., (1984)
      A Model of Glukose Kinetics and Their Control by Insulin,
      Compartmental and Noncompartmental Approaches.
      Mathematical Biosciences 72, 291-315.

[3]   Metzler, C.M., Elfring, G.L., and McEwan, A.J., (1974)
      A Package of Computer Programs for pharmacokinetik modeling
      Biometrics, 30, 562-563.

[4]   Eaton, R.P., Allen, R.C., Schade, D.S.,   (1980)
      Erickson, K.M., and Standefer, J.,
      Prehepatic Insulin Production in Man: Kinetic Analysis Using
      Peripheral C-Peptide Behavior. J Clin Endocrinol Metab 51,520-528

[5]   Renn, W., Maulbetsch, R. and Eggstein,M. (1989)
      Digital and Analogous Presentation of Laboratory Results in
      Dependence of Time. Clinical Biochemistry, Vol 2, 81-121.
      Walter de Gruyter & Co, Berlin - New York.

PHYSIOLOGISCH ORIENTIERTE MULTIKOMPARTIMENT-MODELLIERUNG ARBEITSTOXIKOLOGISCHER PROBLEME

G.Wuenscher
Arbeitshygienisches Zentrum der Chemischen Industrie
4220 Leuna, PSF 31

EINLEITUNG

Noch vor wenigen Jahren wurden pharmakokinetische oder toxikokinetische Probleme (Pharmaka und Schadstoffe) methodologisch und methodisch einheitlich innerhalb des klassischen Konzeptes der Kompartiment-Modelle behandelt. Nun wird mit der physiologisch orientierten Modellierung (physiologically based pharmacokinetic modeling PB-PK) (/1-8/) eine Linie verfolgt, die ihren Ausgangspunkt in der Pharmakologie (Anästhesiologie) hat, aber eben besonders auf Anforderungen bei der Modellierung (arbeits-)toxikologisch interessierender Biosysteme (Säugetierorganismus als Vesuchstier, Mensch) antwortet. Es wird erwartet, auf diese Weise die Interpretation von Daten vor allem der Toxizitätstestung (Dosis-Wirkungs-Beziehung) und des Biomonitoring (Beziehng zwischen äußerer Exposition und innerer Belastung/Dosis) zu verbessern. Der auf diese Weise mögliche Zugang zur 'inneren Exposition' (target tissue dose) als Ausage über die Konzentration an Schadstoff oder dessen Biotransformationsprodukt in einem bestimmten, nicht mehr nur phänomenologisch als Kompartiment beschriebenen Organ oder Indikatormedium (Blut, Urin) muß als eine wesentliche Voraussetzung dafür angesehen werden.

Dabei ist auch eine Modellumgebung von Bedeutung, die Einflüsse auf das eigentliche toxikokinetische Transportsystem zu erfassen versucht, die über die äußere Umgebung und innere Zustände vermittelt werden. Diese Modellumgebung soll verschiedene Extrapolationen ermöglichen, um die Gültigkeit der unter bestimmten experimentellen Bedingungen erworbenen Daten erweitern und auf andere Bedingungen übertragen zu helfen.

Die Varietät der Bedingungen bezieht sich dabei auf

- die exponierte Spezies (Maus, Ratte, Hamster, ... - Mensch),
- expositionsmodifizierende Faktoren (z.B. körperliche Aktivität),
- Expositionsmuster, Expositionsniveaus (chronisch, akut) und
- Aufnahmewege (inhalativ, oral, perkutan, ...).

Das Kompartimentmodell der klassischen Pharmakokinetik (meist 2 Kompartimente; im linearen Fall Massenbilanz durch System gewöhnlicher Differentialgleichungen mit superponierbaren Lösungen beschrieben) kann diesen Anforderungen nur unvollkommen gerecht werden. Die aus den statistischen Daten-Anpassungs-Prozeduren (curve fitting) abgeleiteten Parameter und Kenngrößen (AUC/area under the curve, MRT/mean residence time) haben nur eine phänomenologische Bedeutung. Es wird insgesamt weniger zum Verstehen der dahinterstehenden physiologischen und biochemischen Grundprozesse beigetragen. Bei diesen statistisch orientierten Techniken wird ohnehin nicht genügend beachtet (/9/), daß

- das Verhalten des Systems nur für den Beobachtungsbereich gilt,
- aus der Anzahl der Exponential-Terme nicht auf die Zahl der Kompartimente des Realsystems geschlossen werden kann,
- der gleiche Datensatz u.U. durch unterschiedliche Gleichungssysteme beschrieben werden kann und
- Alternativmodelle, die dem gleichen Datensatz genügen nicht zu unterscheiden sind.

Modelle die dem Verstehen dieser Grundvorgänge dienen sollen haben einen etwas anderen Anspruch. Solche Modelle sollen

- Fragen aufwerfen,
- bei Experimenten Orientierung sein,
- helfen, vorhandene Konzepte, Hypothesen ,theoretisches Wissen und Daten bezüglich des Realsystems zu integrieren.

Ein so verstandenes KONZEPTUALES MODELL, das in hohem Maße Objektwissen integriert, kann in starkem Maße experimentelle Arbeit und Modellierung verbinden. So wird die Akzeptanz des Modellierens unter den Toxikologen erhöht. Der Flankierung experimenteller Arbeit durch Modelle kommt große Bedeutung zu, Modellierung ist dann auch weit mehr als ein Problem der Anpassung experimenteller Daten.

MODELLSTRUKTUR

Der Säugetier-Organismus ist strukturell vom Atmungs- und Kreislauf-system bestimmt (wobei gerade in der Industrietoxikologie dem Atmungs-system als Aufnahmeweg besondere Bedeutung zukommt). Das Kreislauf-system verbindet einzelne Kompartimente, die hinsichtlich der anato-misch-physiologischen Struktur und nicht nur phänomenologisch als En-titäten aufgefaßt werden können.

Die Zirkulation des Blutes, die einzelnen Kompartimente verbindend, ist die bestimmende Struktur des Modells. Systemtheoretisch ist die Rezirkulation des Fremdstoffes als positive Rückkopplung anzusehen. Wegen dieser Rückkopplung erweist es sich als recht schwer aus der Messung, z.B. der Konzentration eines Schadstoffes im zirkulierenden Blut, auf die Systemstruktur zu schließen, wenn die beobachtete Erscheinung durch die Rezikulation bestimmt wird (/1Ø/). Die entspre-chenden Transportfunktionen für HERZ-LUNGE-HERZ und für den KÖRPER bzw. die Kompartimente beziehen sich iterativ auf einander, so daß implizite Funktionen z.B. für die Konzentrationen im Blut entstehen, die bei der Systemanalyse Entfaltungsoperationen nötig machen.

Aber eine Systemsynthese ist möglich. Es läßt sich eine Modellstruktur entwerfen, die die Rezirkulation bei der Verteilung des Fremdstoffes im Körper berücksichtigt. In der Regel werden bis zu acht Kompartimente berücksichtigt, die sich hinsichtlich des Stofftransportes als zu unterscheidende Einheiten erweisen, weil diese sich entweder in den organ-spezifischen Parametern (VOLUMINA, DURCHBLUTUNGs-Raten) oder den substanz-spezifischen Parametern (VERTEILUNGS-KOEFFIZIENTEN) unter-scheiden. Diese drei Größen bestimmen den Stofftransport zwischen Herz-Lunge-Herz und den Kompartimenten nach der Aufnahme während der Rezirkulation. Die Transportraten werden bestimmt durch den Konzentra-tionsausgleich im pro Zeiteinheit durch das Gewebe fließenden Blut-teilvolumen. Im abfließenden Blutelement stellt sich eine durch die

Verteilungskoeffizienten bestimmte Konzentration ein. Die Annahme, daß in vivo der Konzentrationsausgleich im Vergleich zur Kontaktzeit (Flußrate !) schnell genug erfolgt, scheint hinreichend gesichert. Die entsprechenden Daten zu Gewebevolumina und gewebespezifischer Durchblutung sind verfügbar. Diese Größen lassen sich sogar aus Individuum- und spezies-ezogenen Grundgrößen wie Körpermaßen ableiten. Die Prinzipien der Allometrie sind hier anwendbar. Aber auch Daten, die den Stofftransport regieren, sind unter Berücksichtigung von Prinzipien der Thermodynamik (Löslichkeit) aus in-vitro-Messungen ableitbar. Die Validität dieses Modellierungsansatzes ist unter verschiedensten Bedingungen im Vergleich von Experiment und Modell nachgewiesen worden (/1,4,5/). Dabei ist vielfach beobachtet worden, daß nach abgeschlossener Verifizierung der Modellstruktur das Modell in Simulationsläufen das experimentell festgestellte Verhalten reproduziert, ohne daß die üblichen Prozeduren der Datenanpassung in Anspruch genommen werden. Das wäre ohnehin nur im begrenzten Umfang möglich, weil die Parameter einer objektbezogenen Plausibilität genügen müssen (physiologische Bereiche !). Systematische Abweichungen würden eher – im Sinne der Wechselbeziehung zwischen experimenteller Arbeit und Objektwissen/Modellierung – eine strukturelle Veränderung des Modells (Erhöhung des Informationsgehaltes) auslösen müssen.

MÖGLICHKEITEN EINES MODELL- UND DATENBANKSYSTEMS

Ein in dieser Hinsicht validiertes Modell ist zusammen mit der physiologischen Umgebung als BASISMODELL einer MODELL-UND DATENBANK denkbar. Dieses Basismodell muß dabei in einem leistungsfähigen Simulationssystem integriert sein. Hier hat sich das System SONCHES als hilfreich erwiesen , auf dessen Grundlage ein toxikokinetisches Modell- und Datenbanksystem (TOMODABA) aufgebaut wird (/11 - 15/). Das System integriert experimentelle und modellbezogenen Daten:

- Zeitreihen, Einzelwerte, statistische Maßzahlen im REALEXPERIMENT
  gemessener Konzentrationen etc.,
- physiologische Daten bezüglich der zu modellierenden BIOSYSTEME
  (Versuchstierarten, Mensch),
- Stoffbezogene Daten (Verteilungskoeffizenten von ca 4Ø SUBSTANZEN,
  die vor allem als Lösungsmittel in der Arbeitswelt vorkommen) und
- Angaben zu den äußeren Bedingungen, die aus der UMGEBUNG des Modells
  das zu simulierende Experiment bestimmen (zeitlicher Verlauf des
  Expositionsniveaus bzw. des Niveaus der physischen Aktivität etc.).

Durch Zuordnung der in verschiedenster Weise kombinierbaren Daten bezüglich des Biosystems, der Substanz und der Umgebung zum Basismodell
(Festlegung eines SZENARIOs in Vorbereitung eines Simulationslaufes)
kann eine Vielfalt von Experimenten simuliert werden, die der Vielzahl
denkbarer Realexperimente angemessen sein soll. Sachlich zusammenge-
hörende Daten werden in einem Makroservice, die zugehörigen Simula-
tionsresultate und Meßreihen in einem Szenarioservice verwaltet. Im
Rahmen einer Resultatnachverarbeitung werden die Modell-Experiment-
Vergleiche realisiert.

Der Nutzen eines solchen toxikokinetischen Modell- und Datenbanksystems besteht darin, experimentelle Ergebnisse vergleichen zu können,
die unter verschiedensten Bedingungen erhalten worden sind. Die Kon-
sistenz eines Informationspools zu einem bestimmten toxikologischen
Problem - meist sind Beobachtungen über mehrere Spezies hinweg zu
vereinen - kann wesentlich besser überprüft werden, wenn von der durch
physiologische und toxikokinetische Grundprozesse bestimmten Phänome-
nologie abstrahiert werden kann. Das betrifft vor allem Interspezies-
Extrapolationen im Zusammenhang mit der Bewertung von Dosis-Wirkungs-
Beziehungen. Hier ist man oft gezwungen, die Dosis (als Expositions-
surrogat) nachträglich durch allometrische Beziehungen auf andere
Spezies zu übertragen. Dabei wird nicht berücksichtigt, daß eigentlich
die Grundprozesse dieser Skalierung bedürfen. Diese wirken in der
durch das physiologisch orientierte Modell beschriebenen Struktur in
dynamischer Weise zusammen, so daß die Dosis zu extrapolieren, keines-
falls als arithmetische Operation denkbar ist. In diesem Sinne ist
eine solche TOMODABA natürlich als Werkzeug in der Hand des expe-

rimentierenden Toxikologen anzusehen. Die Möglichkeiten innerhalb des Simulationssystems sowohl objektbezogene Daten als auch Modellstrukturen verwalten bzw. entwickeln und für den Prozeß der Modellierung (Weiterentwicklung des Modells als Konsequenz eines Modell-Experiment-Vergleich) bereithalten zu können, unterstreicht diesen Werkzeugcharakter. Ein aktueller Aspekt der Modellierung der hier behandelten Biosysteme sind MONTE-CARLO-SIMULATIONEN. Die inter- und intraindividuelle Variabilität gehört zu den Grunderscheinungen der biologischen Systeme und es liegt nahe, dieses Verhalten zu simulieren. Insbesondere das Simulieren gruppen-bezogener Daten ist wichtig, kann doch so der Zusammenhang zu den sehr oft als statistischen Maßzahlen vorliegenden experimentellen Daten hergestellt werden. Hier gibt es durch die Arbeit einer schweizer Arbeitsgruppe vielversprechende Ergebnisse (/7/). Das Simulationssystem SONCHES unterstützt Monte-Carlo-Simulationen sehr weitreichend, entsprechende Simulationsexperimente finden eine Grenze dadurch, daß wenig Angaben über den das Biosystem charakterisierenden Verbund von Verteilungen der Modellparameter verfügbar sind. Das Problem der Variabilität verwendeter Daten läßt sich aber auch außerhalb der stochastischen Modellierung durch Sensitivitäts- und Verhaltensanalyse (SONCHES-Optionen) behandeln.

Ein wichtiger Aspekt einer TOMODABA besteht in der Möglichkeit der Kombination des Basismodells mit Modellen, die wie z.B. im Falle der Metabolisierung weiterreichende toxikologische Phänomene bschreiben. Es hat sich gezeigt /(12)/, daß die Prozesse der Metabolisierung nicht analysierbar sind, wenn die toxikokinetischen Prozesse der Verteilung bzw. des Antransportes im metabolisierenden Organ unbewertet bleiben. Das ist insbesondere im Zusammenhang mit Sättigungsphänomenen konkurrierender Metabolisierungswege bedeutsam. In Abhängigkeit von Expositionsmustern und der dadurch beeinflußten Toxikokinetik können verstärkt hochtoxische Metaboliten gebildet werden, die als kanzerogenwirkende Substanzen ein stark diskutierter Aspekt der modernen Toxiko-

logie darstellen. Hier ist verstärkt die Anwendung der beschriebenen
Modellierungsansätze zu beobachten.

LITERATUR

/1/ FISHEROVA-BERGEROVA,V.: 'Toxicokinetics of organic sol-
    vents.' Scand. J. Work. Environ. Health 11 (1985):suppl.1,7-21.
/2/ ANDERSEN,M.E.: 'Tissue Dosimetry, Physiologically-Based Pharmaco-
    kinetic Modeling, and Cancer Risk Assessment'
    Cell Biology and Toxicology 5(1989)4, 405-415
/3/ CLEWELL,H.J.; ANDERSEN,M.E.: 'Dosimetric Modells:Physiologically
    Based Pharmacokinetics' Health Physics 57(1989) Sup.1, 129-137
/4/ ANDERSEN,M.E.; MacNAUGHTON,M.G.; CLEWELL,H.J.; PAUSTENBACH,D.J.:
    'Adjusting Exposure Limits for Long and Short Exposure Periods
    Using a Physiological Pharmacokietic Model'
    Am. Ind. Hyg. Assoc. J. 48/4 (1987) 335-343)
/5/ GARGAS,M.L.; ANDERSEN,M.E.: 'Physiologically Based Approaches
    for Examining the Pharmacokinetics of Inhaled Vapors'
    In: Toxicology of the Lung; D.E. Gardner et al.(Eds);
    New York: Raven Press 1989
/6/ OPDAM,J.J.G.; SMOLDERS,J.F.J.: 'Alveolar sampling and
    fast kinetics of tetrachloroethene in man. II. Fast
    kinetics', Brit. J. Ind. Med. 44 (1987), 26-34
/7/ DROZ,P.O.; WU,M.M.; CUMBERLAND,W.G.; BERODE,M.:
    'Variability in biological monitoring of solvent exposure.
    I. Development of a population physiological model.'
    Brit. J. Ind. Med. 46(1989), 447-460
/8/ LUTZ,R.J.; DEDRICK,R.L.: 'Physiological Pharmacokinetics: Rele-
    vance to Human Risk Assessment' In: New Approaches in Toxicity
    and Their Application in Human Risk Assessment;
    A.P. Li (Ed.) New York:Raven Press 1985
/9/ RESCIGNO,A.; BECK,J.S.: 'Use and Abuse of Models'
    J. Pharmacokinetics Biopharmaceutics 15(1987) 327-344
/10/Weiss,M.: 'Steady-State Distribution Volume in Physiologic Multi-
    organ Systems'
    Biopharmaceutics Drug Disposition 4(1983) 151-156
/11/WENZEL,V.; MATTHAEUS,E.; FLECHSIG,M.: 'Generic Modelling in
    SONCHES.' In: Proceedings of the International Symposium
    held in Berlin (GDR) Sep. 1988, Sydow, A. et al (Eds)
    Berlin: Akademie-Verlag 1988, part I., 129-124
/12/FLECHSIG,M.; MATTHAEUS,E.; WENZEL,V.: Simulation Environ-
    ment in SONCHES, ditto. part II., 360-364
/13/WUENSCHER,G.; WENZEL,V.; FLECHSIG,M.; MATTHAEUS,E.: 'Simula-
    tion System SONCHES based toxicokinetic Modelling as a
    Tool in Biological Monitoring', ditto. part II., 332-335
/14/WUENSCHER,G.; KERSTING,H.; HEBERER,H.; WESTMEIER,I.;
    WENZEL,V.; FLECHSIG,M.; MATHAEUS,E.: 'Simulation system
    'SONCHES' based toxicokinetic model and data bank as tool
    in biological monitoring and risk assessment'. Science Total
    Environm., special issue 1989 (in press).
/15/WUENSCHER,G.; WENZEL,V.; FLECHSIG,M.; MATTHAEUS,E.: 'Toxico-
    kinetic Modeling by Simulationsystem 'SONCHES'. What should
    and can such a System do ?'
    Syst.Anal.Model.Simul. 7(1990)6; in press

BIOLOGIE

Analysis of Low-dimensional Complex Processes in Epidemiology

F. Drepper
Theoretical Ecology Working Group
Research Center Jülich, D 5170 Jülich, FRG

A new method of time series analysis based on measures of information for conditional one step predictions seems to be particularly suited to disentangle the role of chance and low dimensional unstable "deterministic" dynamics in generating the unpredictability in the behaviour of individual and collective biological systems. For deterministic motion the uncertainty of a conditional prediction becomes independent of its accuracy, if the precision of the prediction is coupled to the one of the knowledge of the past. The discriminating power of the analysis can further be increased by avoiding global averages.

Empirically obtained as well as model based time series of measles incidence patterns are used as examples to demonstrate the usefulness of the unpredictability profile as a new instrument of nonlinear time series analysis.

The behaviour of individual as well as of collective biological systems shows large fluctuations which are characterized by a complex mixture of regularities and irregularities. The basic common feature of these systems with organismic complexity is the fact that they are all open multi-component-systems which are maintained under far from equilibrium conditions and whose individual elements have strong cooperative interactions. The strong coupling between the numerous elements leads to a drastic reduction of the dimension of the dynamical rules to which the behaviour of multi-element organisms obey. The dynamics of their macroscopic state may be described by comparatively few degrees of freedom and is dominated by the attractors of the low dimensional dynamics. The question in how many cases the dimensionality is low enough to be verified empirically has to remain open at the moment.

Under far from equilibrium conditions open systems may have nonequilibrium attractors. The coexistence of different types of attractors can only occur in nonlinear dynamical systems. For more than two degrees of freedom the probability increases that the behaviour is influenced by hyperbolic fixed points. This may lead to chaotic attractors, characterized by sensitive dependence on initial conditions (bad predictability) and by fractal geometry.

There have been numerous studies of chaotic nonlinear *model* systems (e.g. May 1976). They could show that the apparent behavioural complexity of living systems is not in contradiction to low dimensional state space descriptions. On the other hand there have been so far very few successful attempts to prove the existence of low dimensional deterministic descriptions from *empirical* observations of living systems with complex nonequilibrium behaviour. The incidence patterns of measles in some large North American Cities have played a key role to show the relevance of low dimensional unstable determinism to explain unpredictable behaviour in biological systems (Schaffer 1987). Using a standard epidemiological model (Dietz 1976) Schaffer showed that measles incidence patterns in New York City are relatively well described by parameter values in the chaotic regime.

Whereas classical methods of time series analysis like Fourier spectra, histograms or (linear) ARMA models (Box and Jenkins 1976) are not able to distinguish between low dimensional nonlinear and high dimensional random unpredictability new graphical and information theoretic methods have proven to be useful in the context of nonequilibrium statistical physics in particular in turbulence research (Brandstätter at al.).

The main aim of the paper will be to show that the unpredictability profile (Drepper 1988) is comparatively well suited to identify "deterministic" (regular) uncertainty production in empirical data even for relatively short observation times. The Measles incidence patterns in England and Wales as well as in New York City will serve as examples. In the case of childhood epidemics the interesting question is not whether there is a determinisms at all, but the question whether the deviations from a strict periodic behaviour can partly be explained by instabilities of the underlying determinism.

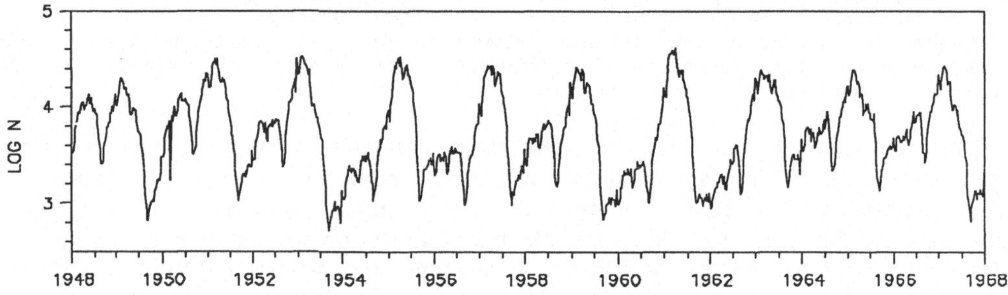

Fig. 1: Log_{10} of weekly cases of measles incidences in England and Wales in the years from 1948 to 1968 (Source: B. Grenfell, Manchester)

Figure 1 shows the logarithm of the weekly cases of measles incidences for England and Wales in the years from 1948 to 1968. The end of the time interval corresponds to the introduction of mass vaccinations which changed the dynamical pattern considerably. The pre-vaccination phase is dominated by a two year cycle as has been confirmed by Anderson, Grenfell and May with the help of Fourier analysis. A large epidemics is always followed by a small one. The sequence of amplitudes of low epidemics however seems to be quite unpredictable.

Parallel to the analysis of the empirical data we want to make use of the fact that measles incidence patterns can be understood relatively well in terms of a simple model. Following ideas of Dietz the host population is divided into four classes: the susceptibles S, the exposed E which are not jet infective, the infectives I and the recovered R which are permanently immunized in the case of measles.

$$\dot{S} \quad = \quad n \quad - \beta\, SI \tag{1a}$$

$$\dot{E} \quad = \quad \beta\, SI - aE \; + \; wE_0 \tag{1b}$$

$$\dot{I} \quad = \quad aE \; - \; gI \; + \; wI_0 \tag{1c}$$

$$\dot{R} \quad = \quad gI \; - \; n \; - \; w(E_0 + I_0) \tag{1d}$$

$$\beta = \beta_0 (1 + \beta_1 \cos(2\pi t)) \qquad\qquad (2a)$$
$$n = m\, \delta(\, t-[t] - 0.7) \qquad\qquad (2b)$$

Here t denotes the time measured in years and t-[t] the noninteger fraction of it. We notice the nonlinearity of the infection rate in form of the product of the two state variables S*I. As has been pointed out by London and Yorke it is essential for the nonequilibrium behaviour of the measles system that the contact parameter β has a yearly variation (2a). Following an idea of Schenzle we also take into account that the new susceptibles are not continuously born into the kindergarten or the school but enter them once per year at the start of the school year (2b). In addition to this we add small constant imports of exposed and infectives (wE_0, wI_0). E_0 and I_0 represent the equilibrium densities of exposed and infectives in the endemic equilibrium for $\beta = \beta_0$ and n = m. Using plausible parameter values: n = 1/50, a = 36.5, g = 100, β_0= 1800 (Dietz, Schenzle, Schaffer) and β_1= 0.24, w = .05 we get a model behaviour belonging to a periodic attractor with a four year period. The natural way to represent the dynamics is a phase diagram showing the joined dynamics of as many variables as is necessary to characterize the state of the system in a unique way. In our case it turns out that 3 variables are sufficient. In view to the possibility of a comparison with the empirical data we choose 3 delayed values of the logarithm of the number of infectives. The 3-dimensional phase diagram is shown in fig. 2a.

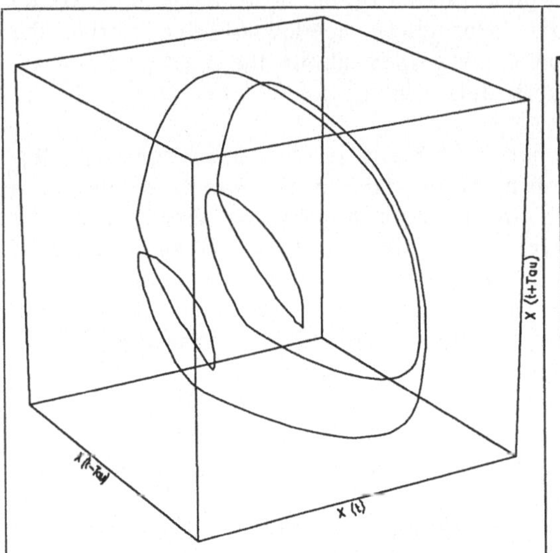

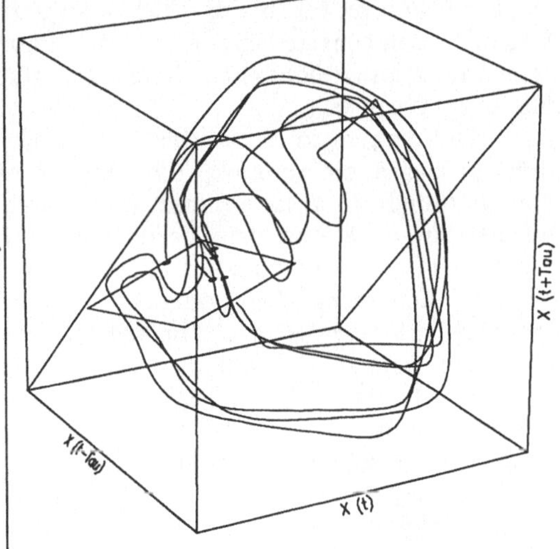

Fig. 4a: 3-dimensional reconstruction of the period 4 attractor of model 1a - 2b in an artificial state space spanned by delayed coordinates.

Fig. 4b: 3-dimensional reconstruction of the attractor corresponding to fig. 1. In both cases the delay time (Tau) is 16 weeks.

Applying the same embedding procedure (and a gentle spline smoothing) to the empirical data from fig.1 we get the attractor of fig. 2b. For better visibility of the similarity we have plotted only those trajectories which pass the indicated Poincaré window and reenter it approximately two years later. That means 4 out of 10 two year cycles have been left out. We notice that the explicative value of the relatively simple model is not bad. In particular we are able to recognize the kinks in fig. 2a at the end of the school year also in the empirical data in fig. 2b.

We further notice that the short term predictability of the empirically observed incidence pattern changes dramatically during a typical cycle. There are phases with high predictability resulting from a stable deterministic dynamical rule as well as areas with high unpredictability. To see whether this unpredictability is exclusively due to random noise (caused e.g. by spatial heterogeneity and migration) or also due to unstable determinism, it is useful to switch to formal information theoretic methods.

In systems with continuously varying state variables, information quantities depend on the level of accuracy with which the states are being measured or observed. The closer one looks, the more information is needed to characterize a state or a sequence of states. In these systems the interest concentrates on the scaling behaviour of information quantities for increased resolution of observation. Of particular interest for the present purpose is the information content of conditional short term predictions, where the conditioning refers to a knowledge of the recent history having the same accuracy as is required for the prediction. Analog to the dimensional analysis introduced by Grassberger and Procaccia - order 2 Renyi informations are being used.

The characteristic feature of low dimensional *deterministic* dynamics turns out to be the fact that for these systems the information content of conditional short term predictions is independent of the resolution of observation. When increasing simultaneously the accuracy of the prediction as well as of the knowledge of the recent history the additional information needed to characterize the future is compensated by the better knowledge gained about the past. This characteristic scaling behaviour holds for stable determinism as well as for an unstable one (Drepper).

The method to determine the information content of the conditional predictions is based on counting the number of pairs of trajectories, which remain in a neighborhood of a given size (accuracy) for a given number of time steps. The uncertainty of a one step prediction is related to the decrease of the number of pairs, when the number of time steps is increased by one. (Termonia, Grassberger, Drepper) In contrast to the graphical analysis the information theorecitical one is not limited to three dimensions. In all examples to be presented here a six dimensional embedding has been used.

To illustrate the interpretation of the unpredictability profile in particular the distinction of random vs. deterministic sources of unpredictability we use an extension of model (1) in the direction of a stochastic differential equation. We convert the deterministic imports of exposed and infectives into Poisson type random processes. In view to a comparison with the empirical data we choose a (macroscopic) noise level with a 14% relative standard deviation (relative to the equilibrium value). This corresponds to an effective total population size of 500.000 as origin of the stochastic imports. In addition to the Poisson type noise in the differential equation we further introduce observational type multiplicative white Gaussian noise with a 10% standard deviation.

The system under study has relatively long living (metastable) aperiodic transients. Therefor relatively low levels of noise in the differential equation keep the dynamics away from reaching the periodic attractor. The complex behaviour of nonlinear dynamical systems with low levels of noise can be imagined as a diffusion process which jumps between either weakly stable or unstable periodic orbits. It is typical for these systems that the probability for jumps varies systematically on the attractor.

We restrict the analysis on the unpredictability in the neighborhood of the weakly stable period four orbit, which is the most densely populated part of the attractor. It can be defined as the collection of trajectories which pass an appropriate Poincaré window and reenter it four years later. (Comp. fig. 2)

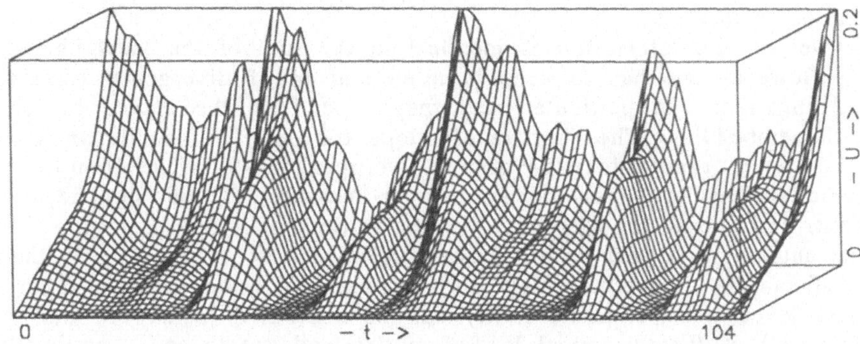

Fig. 3a shows the unpredictability profile of the stochastically perturbed weakly stable period four orbit of the SEIR model of fig. 2a. The backward direction indicates the dependence on the observational accuracy.

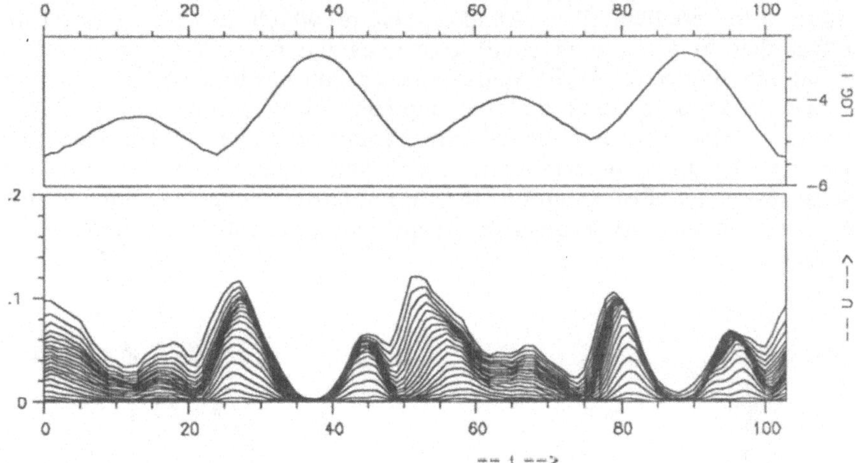

Fig. 3b shows the average number of infectives for the trajectory bundle defining the period four cycle.
Fig. 3c shows a cut through the unpredictability profile of fig. 3a. The horizontal axis gives the time elapsed since the beginning of the cycle measured in units of double weeks.

Fig. 3a shows the unpredictability along the period four orbit. The vertical axis represents the information content of a conditional one step prediction measured in bits. The horizontal axis shows from left to right the time phase within the four year cycle given by the time elapsed since January 1. of a year with a very large epidemics followed by a year with a very small epidemics. The backward direction in fig. 3a refers to the resolution of observation measured as the maximal distance between adjacent trajectories for which they are treated as undistinguishable. We see that there are time phases characterized by a horizontal slope in the direction of increased resolution. The horizontal slope extends over an intermediate range. For lower resolutions the determinism is not jet resolved. For higher resolutions the noise becomes visible as an increase of the conditional unpredictability. When the horizontal slope· occurs on the level zero, this corresponds to a phase with stable determinism. When the slope is horizontal on a finite level, this corresponds to an unstable deterministic phase with local divergence of trajectories.

The level of unpredictability is identical to the sum of the local characteristic exponents. However one has to keep in mind that *local* divergence rates are not topological invariants. In particular they may depend on the choice of the delay used for the embedding. The (horizontal) slope for deterministic dynamics on the other hand is a topological invariant (Drepper 1990). The length of the deterministic scaling area can be seen as a measure of the significance of a deterministic description as opposed to a stochastic one.

In the autumn of a year with a large epidemics (t = 44 and t = 95) there is a high unpredictability - whether next year's epidemic will be very small or on an intermediate level. This unpredictability can be attributed clearly to the instability (nonlinearity) of the dynamical law of motion and not to an exceptionally high momentary level of the noise source. Though less clearly the same applies to the situation at the end of a year preceding a large epidemic (t = 27 and t = 79), when the "decision" is taken, whether next years epidemic will again be small or a very large one. The large epidemics belong to stable deterministic phases with a high short term predictability. Another feature which can be inferred from fig. 3 is the fact that in years with small epidemics the noise has a relatively stronger influence on the dynamics as in years with an intermediate level epidemics.

We have to keep in mind that the unpredictability profile of fig. 3 belongs to a time series obtained from a model with parameter values which lead to a stable limit cycle in the fully deterministic case. The unpredictability profile however reveals that even in this situation the introduction of noise into a differential equation may give rise to a positive Lyapunov exponent or at least to a locally unstable behaviour.

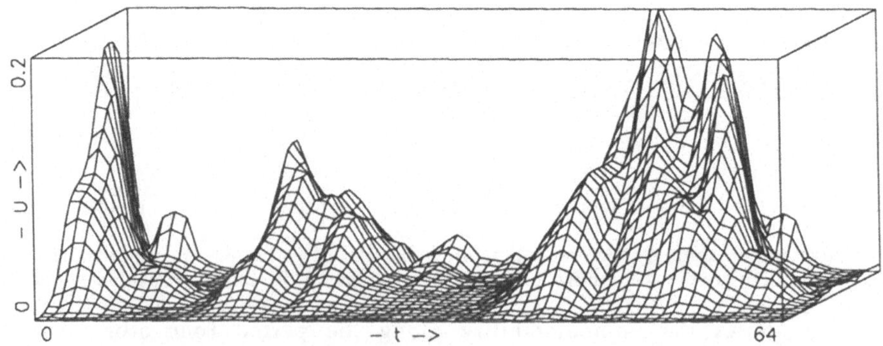

Fig. 4a shows the unpredictability along the period two orbit defined as the average behaviour on the 3-dim. attractor reconstructed from the empirical data for England and Wales as shown in fig. 2b. The backward direction indicates the dependence on the observational accuracy.

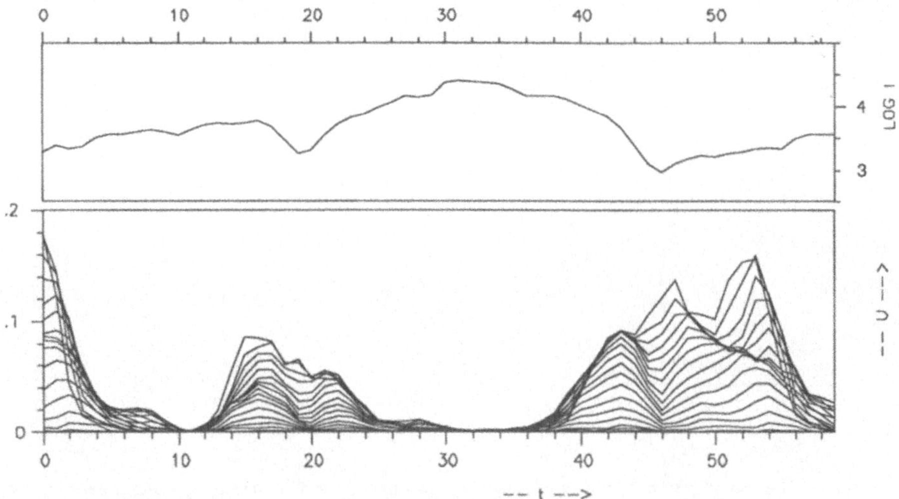

Fig. 4b shows the time series plot of one cycle of the period two orbit.
Fig. 4c shows a cut through the unpredictability profile of fig. 4a. The horizontal axis gives the time elapsed since the beginning of the two year cycle given in units of double weeks.

Fig. 4 shows the corresponding unpredictability profile for the empirical data of fig. 1. In the case of England and Wales the measles epidemics are phase locked to a two year cycle. To improve the statistics we restrict the analysis to a distinction of phases within a two year cycle. $t = 0$ corresponds to January 1. of a year with an even number. As in fig. 3 we notice that the spike of unpredictability in the autumn after a large epidemics ($t = 43$) can be explained by a local instability of the underlying determinism. The peak of unpredictability before a large epidemics is much less pronounced because of the exact phase locking.

For comparison we also include the unpredictability profile of the monthly cases of measles incidences in New York City in the time interval from 1928 to 1968. Comparing fig. 5 with fig. 4 we have to keep in mind that the uncertainty production is measured per time step, which is a month in the case of the New York data and two weeks in the case of fig. 4. The peaks of unpredictability at the

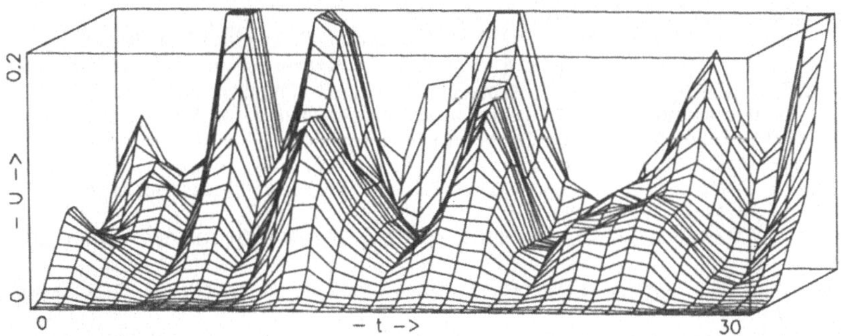

Fig. 5a shows the unpredictability along the unstable period two orbit of the 3-dimensional attractor reconstructed from the measles incidence pattern for New York City. (Source: Yorke and London 1973)

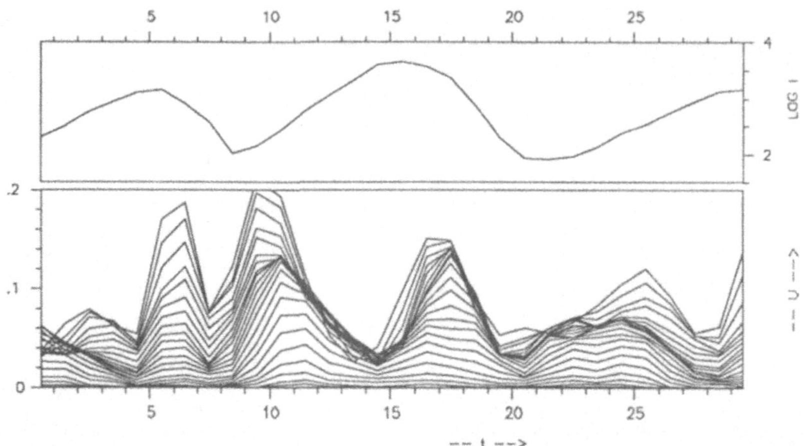

Fig. 5b shows the average behaviour for the trajectory bundle defining the period two cycle. Fig. 5c shows a cut through the unpredictability profile of fig. 5a. The horizontal axis gives the time elapsed since the beginning of the cycle given in units of months.

end of a large epidemics (t = 17) occurs earlier in the year as compared to England and Wales. The peaks at the beginning of a large (t = 10) as well as of a small (t = 24) epidemic are much more pronounced than in the case of fig 4.

Generally speaking one can say that the Unpredictability Profile is able to reveal phases of behaviour, which have a chance to be explained by a deterministic dynamical rule - irrespective whether the corresponding determinism is stable or not. Its usefulness is not only limited to the design phase of a model but may also show up in the validation phase. For an unstable deterministic system the simple comparison of a model trajectory with the empirical history is no longer a good indicator for the validity of a model. However the comparison of the unpredictability profiles is a much better indicator.

Anderson R. M., Grenfell B. T. and May R. M., J. Hyg. Camb., 93 (1984) 587-608
Box G. E. and Jenkins G. M., Time Series Analysis Forecasting and Control, Holden Day 1976
Brandstätter A. et al., Phys. Rev. Lett. 51 (1983) 1442
Dietz K., The Incidence of Infectious Diseases under the Influence of Seasonal Fluctuations, Lecture Notes Biomath. 11 (1976) 1-15
Drepper F. R., Unstable Determinism in the Information-Production-Profile of an Epidemio logical Time Series, in Wolff W. et al. (ed.s), Ecodynamics, proceedings of a workshop held at Jülich, FRG, Oct. 1987, Research Reports in Physics, Springer 1988
Grassberger P, Estimating the Fractal Dimensions and Entropies of Strange Attractors, in Holden A. V. (ed.), Manchester University Press 1986
Grassberger P. and Procaccia I., Phys. Rev. 28a (1983) 2591
London W. P. and Yorke J. A., Recurrent Outbreaks of Measles, Chickenpocks and Mumps I, Am. J. Epidemiol. 98 (1973) 453-468
May R. M., Simple Mathematical Models with very Complicated Dynamics, Nature 261 (1976) 459-467
Schaffer W. M. et al., in Markus et al. (ed.s), From Chemical to Biological Organization, Springer 1987
Schenzle D., An Age-structured Model of Pre- and Post-Vaccination Measles Transmission, IMA Journal of Math. Applied in Medicine and Biology 1 (1984) 169-191
Termonia Y., Phys. Rev. A 29 (1984) 2427
Yorke J. A. and London W. P. Recurrent Outbreaks of Measles, Chickenpocks and Mumps II, Am. J. Epidemiol. 98 (1973) 469-482

MATHEMATICAL MODEL OF PARAINFLUENZA VIRUS TYPE 1-INFECTIONS

Franke[1], H., Wichmann[2], H.E.

[1]Medizinische Klinik I, Universität Köln
[2] 'Arbeitssicherheit und Umweltmedizin', FB 14, Universität Wuppertal

Zusammenfassung: Die Stabilität endemischer Zustände bei Infektionen mit dem Parainfluenza Virus Typ 1 wird an einem mathematischen Modell untersucht. Bei jahreszeitlichen Schwankungen der Kontaktrate können Inzidenzmuster mit zweijährigem Zyklus auftreten, wenn die Dauer von Infektiosität und Immunität in einem bestimmten 'Parameterfenster' liegt.

Summary: The stability of endemic states due to parinfluenza virus type 1-infections is studied by a mathematical model. Taking into account seasonal periodicity in the contact rate, conditions are analysed for biannual incidence patterns, if the duration of infectiosity and immunity lies within a given 'parameter window'.

Parainfluenza viruses type 1 (PI1) play an important role in respiratory infections [Clarke 1973; Glezen et al 1984]. The occurrence patterns of the PI1 infection are endemic, but they can change easily to epidemic forms in autumn because of the seasonally growing infectiosity. Manifestations of the infection are found mainly among preschool children between 0.5 and 6 years of age, with a maximum at 2 years. A remarkable feature is the biannual periodicity in the occurrence.

Up to six months after birth partial immunity exists caused by maternal antibodies. Older children can achieve temporary immunity after an infection. Not humoral antibodies from serum but antibodies in oropharynx seem to be responsible for protection against further infections. The exact mechanism of immunization, however, is yet unknown. Probably there exists a process of stepwise immunization and reinfections which is negatively correlated to the antibody titer increasing from one infection to the next.

In the following, the features of PI1 infections shall be analysed by a mathematical model considering susceptibles, infectives and resistants. For such SIR models many theoretical investigations exist [e.g. Bailey 1975; Dietz 1976; Dietz, Schenzle 1985]. For PI1 it is of specific interest under which conditions an endemic equilibrium state exists and which parameters allow a biannual periodicity of the infection waves which are observed in humans.

In a first step we use an age-independent SIR model and consider only the age group mostly involved, i.e. approximately between $T_1 = 0.5$ and $T_2 = 6$ years. The individuals of this age class can be assigned to 3 states (compartments):

x susceptibles, y infectives, z immunes (resistants)

We obtain (Fig. 1) for the influx in the compartment of susceptibles $A + \delta z$ (the maternal immune protection is assumed to end 6 months after birth, babies enter therefore the x compartment with influx A at this time, an additional influx δz comes from the compartment of immunes, z, i.e. from older children loosing their immunity after a duration of $1/\delta$). For the efflux we find $\lambda x + Bx$ (λx children leave the x compartment by aging after a mean transit time of $1/\lambda = T_2 - T_1$, $Bx = \beta yx$ had contact with y infectives and become infectives themselves, β contact rate). The differential equation for the susceptibles is then given by

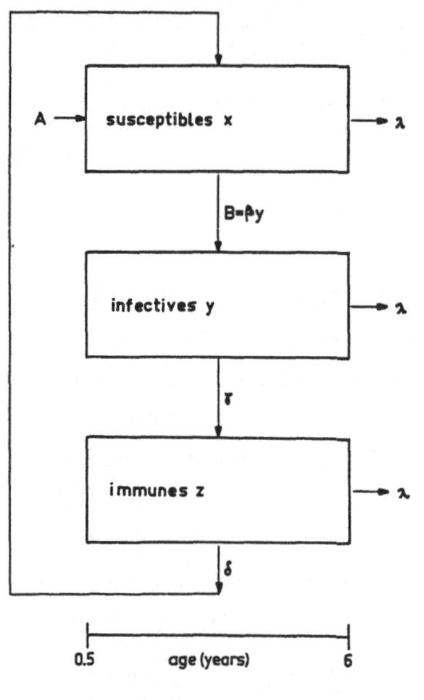

Fig. 1 Age-independent SIR model. The age class reaches from 0.5 to 6 years

$$dx/dt = A - \lambda x + \delta z - Bx .$$

Corresponding equations are obtained for the infectives and immunes ($1/\gamma$ duration of infectiosity)

$$dy/dt = -\lambda y + Bx - \gamma y$$
$$dz/dt = -\lambda z + \gamma y - \delta z ,$$

and for the total number $n = x + y + z$ of individuals

$$dn/dt = A - \lambda n.$$

Assuming a stable ($dn/dt = 0$) and normalized ($n = 1$) population one obtains $A = \lambda = 1/(T_2 - T_1)$. This enables us to eliminate one variable ($x = 1 - y - z$, $dx/dt = -dy/dt - dz/dt$)

$$dy/dt = (\beta - \gamma - \lambda)y - \beta y(y+z)$$
$$dz/dt = \gamma y - (\delta + \lambda)z .$$

Equilibrium states ($dx/dt = dy/dt = dz/dt = 0$) are therefore determined by the equations

$$0 = (\beta - \gamma - \lambda)y - \beta y(y+z)$$
$$0 = \gamma y - (\delta + \lambda)z .$$

The obvious solution $P_0 = (1, 0, 0)$ is called 'endemic nullpoint' (only susceptibles present), the second solution $P_E = (1 - y_E - z_E, y_E, z_E)$ with

$$y_E = (\delta + \lambda)/(\gamma + \delta + \lambda) \cdot (1 - (\gamma + \lambda)/\beta)$$
$$z_E = \gamma/(\delta + \lambda)y_E$$

describes an endemic equilibrium point. An endemic solution exists only if the contact rate β is strong enough, i.e. $\beta > \gamma + \lambda$. By determination of the Jacobian eigenvalues one can show [Wichmann, Köppen 1978; Wichmann 1979] that P_E is asymptotically stable, P_0 instable. In the case $\beta \leq \gamma + \lambda$ the nullpoint P_0 is asymptotically stable, resp. stable.

For the occurrence of a stable endemic situation, we conclude, a certain critical strength of the contact rate must be exceeded, otherwise the system runs back to the nullpoint state without infectives and immunes.

To study deviations from the endemic equilibrium point P_E we assume seasonal fluctuations of the infectiosity by a time-dependent contact rate $\beta(t) = \beta_0 \cdot a_P(t)$. The annual periodic function a_P is provided with a pronounced peak in the autumn and may be normalized, i.e. the annual mean value of the contact rate equals β_0. A simulation with $\beta_0 =$ 5/month, duration of infectiosity $1/\gamma$ $=0.5$ months, duration of immunity $1/\delta = 6$ months is shown in Figure 2. After an initial phase of about 2-3 years the system oscillates with a precise annual periodicity. The peaks in the number of infectives coincide with the October peaks of the contact rate (step function) while the peaks of the immunes follow with a delay of about one month.

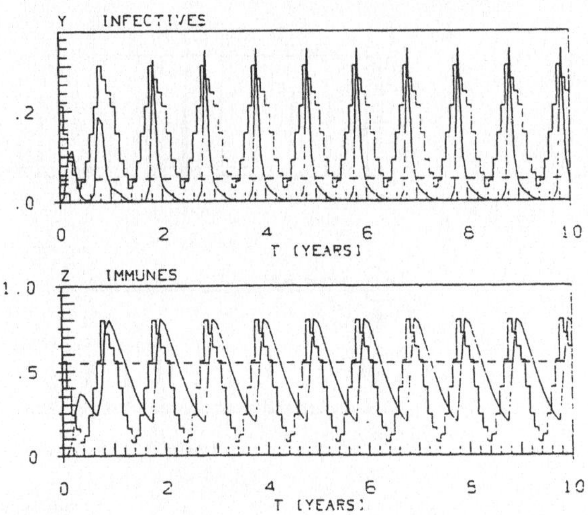

Fig. 2 Annual oscillations of the system induced by the time-dependent contact rate $\beta(t)$ with $\beta_0 = 5$/month, duration of infectiosity $1/\gamma$ $= 15$ days, duration of immunity $1/\delta = 6$ months, transit time $1/\lambda = 5.5 \cdot 12$ months, initial values $y_0 = 0.0001$, $z_0 = 0$. The step function gives the used annual periodic function $a_P(t)$. The endemic equilibrium values y_E, z_E are marked by dashed lines

If the duration of infectiosity $1/\gamma$ is increased to 30 days (keeping the other parameters constant) the system converges faster to the oscillatory equilibrium state, but the qualitative behaviour does not change (Fig. 3a). A decrease of $1/\gamma$ diminishes y because the compartment of infectives (Fig. 1) empties now more rapidly and the influx rate βy is reduced additionally. The system becomes more sensitive and instable, i.e. a longer time is necessary to come into a stable state of precise annual periodicity (Fig. 3b).

For $1/\gamma = 8$ days a stable biannual pattern occurs (Figs. 3c-d). The smaller peaks in the number of infectives are characterized by a delay of 1-2 months compared with the contact rate. Due to the longer interval from a big peak the y compartment empties to a higher degree. The

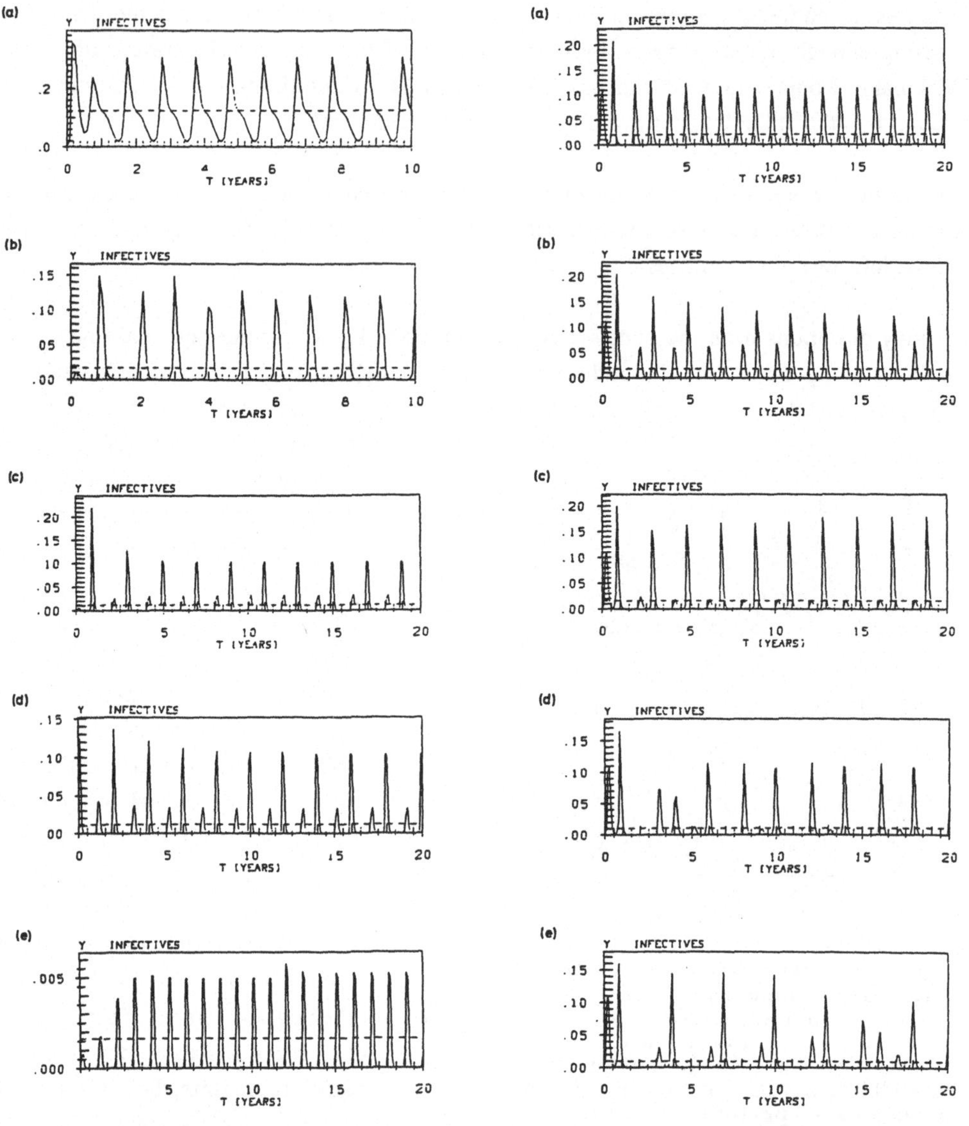

Fig. 3 Oscillating behaviour of infectives for different duration of infectiosity $1/\gamma$ (other parameters as in Fig. 2): **(a)** 30 days; **(b)** 9 days; **(c)** 8 days; **(d)** 8 days, initial values $y_0 = 0.1$, $z_0 = 0$; **(e)** 6 days

Fig. 4 Oscillating behaviour of infectives for different duration of immunity $1/\delta$ (other parameters as in Fig. 2): **(a)** 18 months; **(b)** 21 months; **(c)** 24 months; **(d)** 48 months; **(e)** 54 months

seasonal influx wave $\beta(t)y(t)$ has therefore a delayed onset and the β peak is less effective. During the next months the y compartment empties to a minor extend, the influx wave develops earlier with the autumn peak of the contact rate and a more powerful y peak is built up. In this way biannual stationary oscillations seem to be consistent with the model.

The simulations in Figure 3c and 3d differ only by the initial values. The results are identical biannual patterns with a phase shift of 1 year. In the case of a further reduction of $1/\gamma$ to 6 days the difference of the highly emptied y compartment may become too small to produce a biannual pattern. Only rather tiny annual peaks develop now (Fig. 3e).

A corresponding analysis is shown in Fig. 4 for constant duration of infectiosity and a stepwise change of the duration of immunity $1/\delta$. A short duration $1/\delta$ empties the z compartment and enlarges the x and subsequently the y compartment. This gives an oscillation which is soon stable and follows perfectly the seasonal fluctuations of the contact rate. In contrast, for a long duration of immunity the number of immunes z remains at a high level. The number of infectives y reaches therefore only lower values and in a corresponding way stable biannual oscillations are again possible for $1/\delta$ greater than 20 months (Fig. 4b-c). At a duration of 48 months biannual states may become instable (Fig. 4d), but at 54 months or above triannual patterns seem to develop (Fig. 4e).

From Figs. 3 and 4 one learns that there is a 'parameter window' for the crucial model parameters γ and δ where stable biannual oscillations occur. For the following simulations we chose $1/\gamma = 13$ days and $1/\delta = 12$ months, which are within this parameter window. They are consistent with the limited epidemiological knowledge on PI1 infections, where the duration of infectivity, $1/\gamma$, is about 1-2 weeks and the duration of immunity, $1/\delta$, is several months [Denny 1987]. However, different parameters may lead to similar results, and at present the identification of the 'true' parameter values is not possible.

To describe the complete chain of spread of PI1 infections, also adults have to be considered. Therefore, a more refined age-stratified model is constructed (Fig. 5). The most important first 6 years are divided into 12 equal age classes, older individuals up to 75 years are assigned to a further age class ($T_i = i \cdot 0.5$ years for $i = 0,..,12$, $T_{13} = 75$ years). Introducing a contact matrix β_{ij} with $B_i = \sum_j \beta_{ij} y_j$ corresponding ordinary differential equations can be derived as before. Newborns are immune by maternal antibodies. Entering the age class 2 (6-12 months) they are assumed to be-

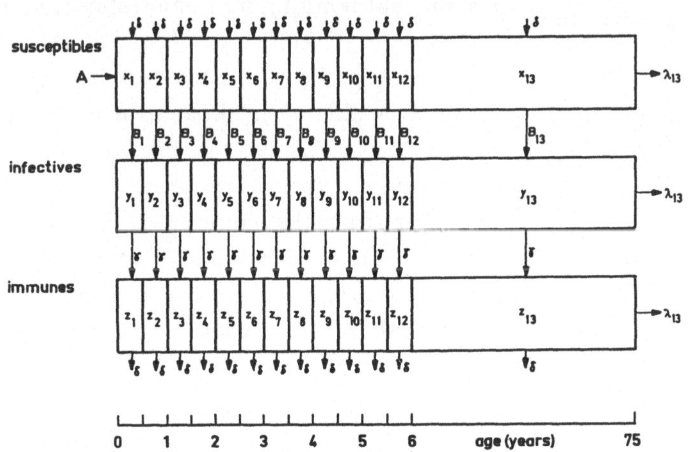

Fig. 5 Age-dependent SIR model with 13 age classes. Each of the first 12 age classes has a transit time of 0.5 years

come susceptible. We take therefore the birth influx A into the susceptible compartment x_1 and assume for the contact matrix $\beta_{1i} = \beta_{i1} = 0$ $(i = 1,..,13)$.

In our simulation we assume a symmetric contact matrix. Because of maternal antibodies the contact with the first age class should not result in an infection, $\beta^0_{1i} = \beta^0_{i1} = 0$ $(i = 1,..,13)$. Contact rates of the second and third age class (with higher classes) may be rather small but with increasing tendency, we took $\beta^0_{2i} = \beta^0_{i2} = 2/\text{month}$ $(i = 2,..,13)$ and $\beta^0_{3i} = \beta^0_{i3} = 9/\text{month}$ $(i = 3,..,12)$. Contact rates of the fourth and higher classes were assumed to be on a high and constant level $\beta^0_{ij} = \beta^0_{ji} = 31/\text{month}$ $(i,j = 4,..12)$. For the contacts with adults we chose again lower values $\beta^0_{13i} = \beta^0_{i13} = 3/\text{month}$ $(i = 3,..13)$.

The contact matrix has been chosen according to qualitative knowledge about increasing contact with age and 'tuned' using data on virus antibody titers in children [Döller, Rebmann unpublished]. These data (Fig. 6) show the percentage of children with positve PI1 antibody titer from a random sample of about 1000 hospitalized children between 0 and 6 years of age.

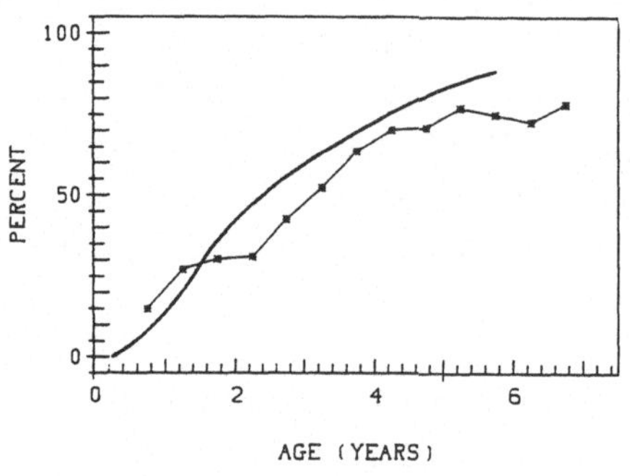

Fig. 6 Age distribution of the percentage of positive parainfluenza virus type 1-antibodies in blood of a random sample of about 1000 hospitalized children (stars) [Döller, Rebmann unpublished] and model result for individuals who were at least for one time in the infectious state (full line)

The model curve represents the corresponding percentage of individuals who were at least for one time in the infectious state.

Discussion

Although parainfluenza type 1 (PI1) virus infections are common and especially important in small children, only qualitative information on the epidemiological characteristics is available: (1) Infection occurs at all age groups exept in newborns, but the manifestation is different and has its maximum at preschool age. (2) For the first 6 months the newborns are (partially) protected by PI1 antibodies from the mother. (3) There is an annual fluctuation with many infectives in autumn and winter, due to better spread of the viruses during cold and wet weather. (4) On top of this, a biannual pattern of the infection is observed. (5) The duration of the infectivity is about 1-2 weeks. (6) Persons are immunized for several months after the infection.

Including these informations in a mathematical model of the SIR type, it can be shown that they are consistent with each other and can be interpreted from a quite simple concept. A minimum contact rate is necessary to have a stable endemic situation. If the contact rate is too low, compared with the duration of infectivity and an aging parameter, only a stable state with susceptibles is possible. Furthermore, there exists a 'parameter window' for the duration of infectivity and immunity, which allows for stable biannual patterns as observed epidemiologically for PI1 infections. However, without additional knowledge the 'true' parameters cannot be identified from this consideration. In addition, from the age-dependent increase of PI1 antibody titer more detailed information on the contact rates can be derived. However, the contact rate may be underestimated if not all persons who have been infected also show a positve titer.

One application of the model, namely on croup syndrome which is closely related to PI1 infections is described in [Wichmann, Franke 1990].

References

Bailey, NTJ: The Mathematical Theory of Infectious Diseases and its Applications.
Griffin (2nd edn) London (1975)

Clarke SKR: Parainfluenza virus infections. Postgrad Med J 49 (1973) 792-797

Denny FW: Privat communication (1987)

Dietz K: The incidence of infectious diseases under the influence of seasonal fluctuations.
Lecture Notes in Biomathematics. Springer NY (1976)

Dietz K, Schenzle D: Mathematical models for infectious disease statistics.
Celebration Statist. Springer NY (1985) 168-204

Döller G, Rebmann H: Antibody titers of respiratory viruses in hospitalized children.
(Unpublished results)

Glezen WP, Loda FA, Denny FW: Parainfluenza viruses. In: Evans AS: Virial Infections
of Humans. Wiley London (1984)

Wichmann HE: Asymptotic behaviour and stability in four models of veneral disease.
J Math Biol 8 (1979) 365-373

Wichmann HE, Franke H: Croup syndrome and parainfluenza virus type 1-infections
- a model analysis. (1990) This volume

Wichmann HE, Köppen L: Stability of nonlinear systems. EDV Medizin Biologie 9 (1978)
118-123

CROUP SYNDROME AND PARAINFLUENZA VIRUS TYPE 1-INFECTIONS - A MODEL ANALYSIS

Wichmann[1], H.E., Franke[2], H.

[1] 'Arbeitssicherheit und Umweltmedizin', FB 14, Universität Wuppertal
[2] Medizinische Klinik I, Universität Köln

Zusammenfassung: Pseudokrupp ist eine Atemwegserkrankung bei kleinen Kindern, die durch Viren, ungünstige Wetterbedingungen, aber auch durch Reizstoffe in der Atemluft verursacht werden kann. Infektionen mit dem Parainfluenza 1-Viren spielen dabei eine wichtige Rolle. Der Zusammenhang zwischen einer Infektion mit diesem Virus und Pseudokrupp wird an einem verallgemeinerten altersabhängigen SIR Modell untersucht. Unter der Annahme, daß das Risiko eines Krupp-Anfalls für ein infiziertes Kind von dem Durchmesser der Trachea abhängt, ergeben sich aus der Simulation Bedingungen, unter denen das zweijährige periodische Auftreten und die Altersverteilung der Erkrankung, die epidemiologisch beobachtet werden, reproduziert werden können.

Summary: Croup syndrome is a respiratory disease in small children which can be caused by different factors as viruses, weather conditions and irritants in the air. One of the most important cause are parainfluenza virus type 1-infections. Their influence on croup attacks is analysed by a generalized age-dependent SIR model. In the model, the chance that an infected child has a croup attack depends on the diameter of the trachea. The simulation identifies conditions, for which the biannual periodicity and the age distribution of the disease, which are epidemiologically observed, can be reproduced.

There are various viruses (parainfluenza, influenza, RS, adeno, rhino and other types) which are known as possible causes of acute lower respiratory tract infections in preschool children [Loda et al 1968; Glezen et al 1971; Foy et al 1973; Glezen et al 1973; Henderson et al 1979; Murphy et al 1981]. An important role play the parainfluenza viruses. They can be identified indirectly by the increase of the antibody/antigen titer, or by direct isolation in cultured material from the oropharynx.

The virus most frequently associated to croup syndrome (laryngotracheobronchitis with dyspnea, barking, cough, inspiratory stridor, hoarseness etc.) is parainfluenza virus type 1 (PI1, Fig. 1) with a high incidence rate in autumn, when wet and cold weather provides good conditions for the spread of infections [Buchan et al 1974; Denny et al 1983; Denny et al 1986]. Infections with parainfluenza virus type 2 also show a close relation to croup syndrome and also have their maximum in autumn. However, they are less frequently found than PI1. In contrast, parainfluenza viruses type 3 are mainly found in winter or spring. They may be responsible for an increase of croup admissions to hospitals and pediatricians during this time of the year (Fig. 1). In addition, other viruses which are mainly related to other respiratory diseases, may sometimes also lead to croup attacks.

In the following, the consideration is restricted to PI1 infections and their role in the etiology of croup. Both, PI1 isolations and croup admissions show not only a peak in autumn but also a biannual periodicity. Therefore, it shall be investigated how the interpretation of these stable oscillations found with a mathematical model for PI1 epidemics [Franke, Wichmann 1990] can be applied on croup syndrome. In addition, it is of interest to see how the age distribution of the disease is related to this phenomenon.

Pathophysiologically parainfluenza infections lead to a swelling of mucosa in larynx (Fig. 2) which may cause due to the small anatomical size of larynx and trachea in small children to respiratory problems. The age for the occurrence of croup syndrome is limited to about six years, and the age distribution reveals a maximum at 2 years.

To simulate the occurrence of croup by PI1 infections, the age-dependent SIR (susceptibles, infectives, resistants) model described in [Franke, Wichmann 1990] will be used. Here consideration is restricted to the infectives, y_i in the age group of 0.5 to 6 years.

Viral infections occur at any age, but they become apparent as croup only among young children. Therefore we introduce apparence factors α_i which take the growth of the respiratory

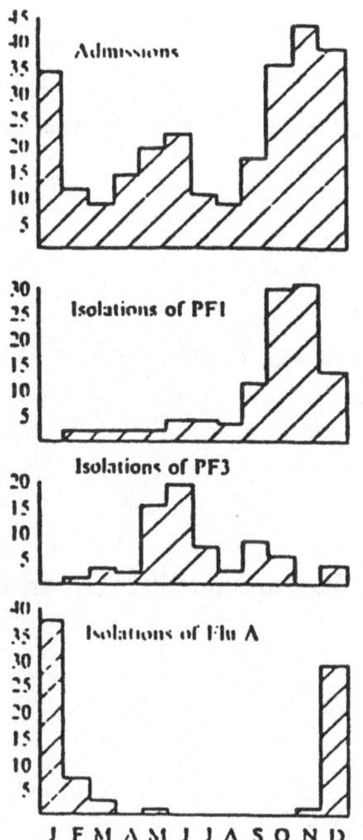

Fig. 1 Monthly distribution of croup admissions and isolations of parainfluenza viruses type 1 and 3 and influenza A viruses in Glasgow 1966-72 from [Buchan et al 1974]

system of the child into account. By definition y_i is the number of infectives in the age class i, the number of individuals who show croup syndrome may then be given by $\alpha_i y_i$. An infection causes swelling of mucosa in larynx and the subglottic tract. It can reduce the respiratory

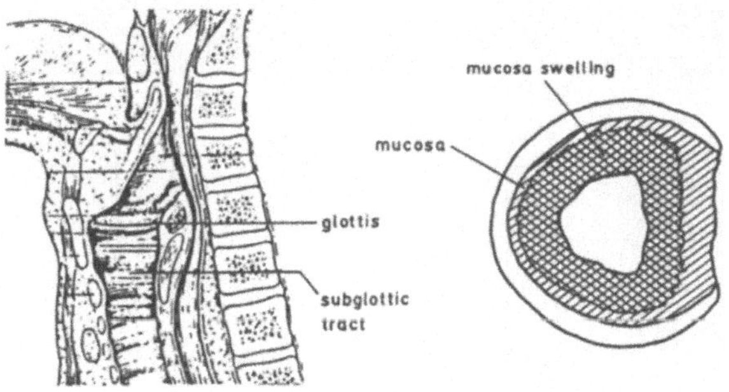

Fig. 2 Swelling of mucosa in the larynx and trachea is responsible for respiratory symptoms in croup syndrome (schematically)

volume very effectively in that early state of anatomical development. According to the law of Hagen and Poiseuille the volume is proportional to the fourth power of the diameter D of the trachea. We assume therefore a $1/D^4$ proportionality for the apparence function α and a linear growth by age T for D and obtain

$$\alpha(T) = \alpha_0/(1+aT)^4 .$$

From anatomical data for newborn (D≈5-7mm) and for the age T=10 years (D≈8-11mm) one can estimate a≈0.06/year. Fig. 3a shows the apparence function $\alpha(T)$ with $\alpha_0 = 1$.Choosing a plausible structure for the contact matrix (increasing rates for increasing age classes) we receive the age distribution of PI1 infections

$$y_i^{inf} = (\int_{t_o}^{t_o+1} y_i(t)dt) / t_1$$

($t_1 = 1$ year) in Fig. 3b. With $\alpha_i = \alpha((T_{i-1}+T_i)/2)$ it follows then for the age distribution of croup syndrome

$$y_i^{age} = \alpha_i\, y_i^{inf}$$

a shape (Fig. 3c) which looks quite similar to the observed data. The annual distribution is calculated by

$$y_i^{ann} = \sum_{i=1}^{12} \alpha_i \, (\int_{t_o+(j-1)t_1}^{t_o+jt_1} y_i(t)dt) / t_1$$

and a cumulated distribution can be defined for odd and even years:

$$y^{cum}(t) = \sum_{i=1}^{12} \alpha_i \sum_j y_i(t+jt_1), \text{ with } j = \{^{1,3,5,...}_{0,2,4,...}$$

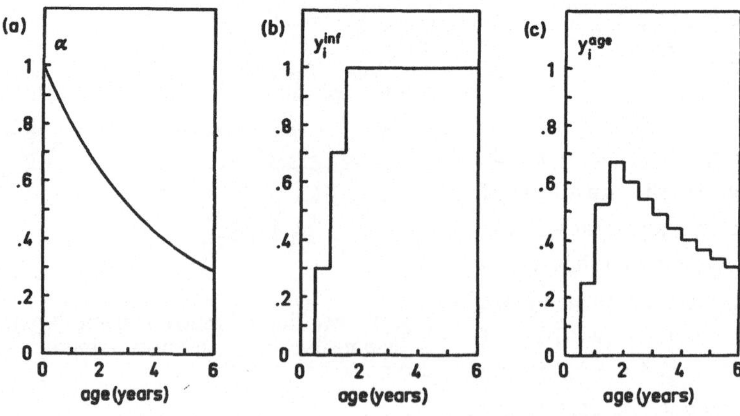

Fig. 3 Model description: (a) croup apparence function $\alpha(T)$; (b) age distribution of infectives y_i^{inf}; (c) age distribution of croup syndrome $y_i^{age} = \alpha_i y_i^{inf}$ (arbitrary units)

The results of the simulation are summarized in Figs. 4-6. The biannual periodicity becomes particularly evident from the annual distribution y_i^{ann} (Fig. 4b), which is compared with annual mean values of croup admissions in Cologne (Fig. 4a). A fairly good reproduction of the differences between odd and even years is achieved. The same holds true for the cumulated monthly distribution y^{cum} (Fig. 5). The position and height of the peaks for odd (—) and

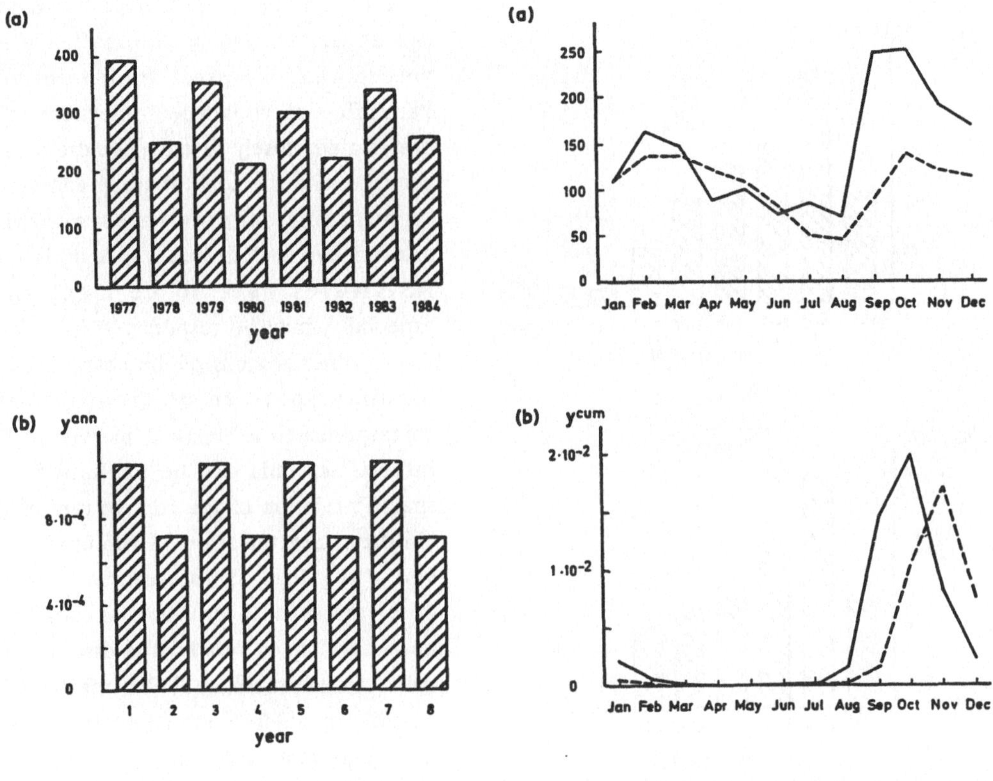

Fig. 4 Annual distribution of croup: (a) hospital admissions in Cologne 1977-84 [Wichmann et al 1990]; (b) y^{ann} from model simulation

Fig. 5 Cumulated monthly distribution of croup: (a) admissions in Cologne 1977-84 [Wichmann et al 1990] for odd (—) and even (---) years; (b) y_{odd}^{cum} (—) and y_{even}^{cum} (---) from model simulation

even (---) years correspond roughly to the data. It has to be kept in mind, that the analysis presented here is restricted only on PI1-related croup admissions in autumn. Croup attacks during other seasons of the year as seen in Fig. 1 may be caused by other viruses. Finally, the age distribution for croup syndrome agrees with the simulated distribution y_i^{age} (Fig. 6). The maximum at 1-2 years and the subsequent decline seem correctly approximated by the apparence function $\alpha(T)$.

Discussion

Croup syndrome is a respiratory disease in small children which can be caused by different factors as virus infections, wet, cold and foggy weather or irritants of the airways like cigarette smoke or air pollution. Despite this complex etiology ot the disease, the most important part of the time patterns of croup attacks, measured by admissions to the hospitals or visits in the offices of pediatricians, can be understood by the mathematical model analysis presented here: (1) Parainfluenza virus type 1-infections (PI1) are related closely to croup syndrome and coincide with the peak of croup admissions in autumn; other viruses of which parainfluenza 3, influenza A or B, and RS-viruses have to be mentioned especially, may be responsible for attacks during other seasons of the year. (2) The biannual pattern of croup attacks corresponds to a biannual pattern of PI1 infections. This can be explained by a specific relation of the contact rate of the children, the duration of infectivity and the duration of immunity. Similar observations of 2- or 3-annual oscillations are known from other virus diseases like measles and mumps. (3) PI1 infections are possible in children and adults and depend on frequency and type of contacts. In contrast, croup syndrome is restricted to small children. The most important reason for this is the anatomy of larynx and trachea: in small children, the lumen of both is small, and a swelling of the mucosa may lead to dyspnea. Taking the growth of the respiratory tract with age into account, it is possible to understand the age distribution observed for croup syndrome.

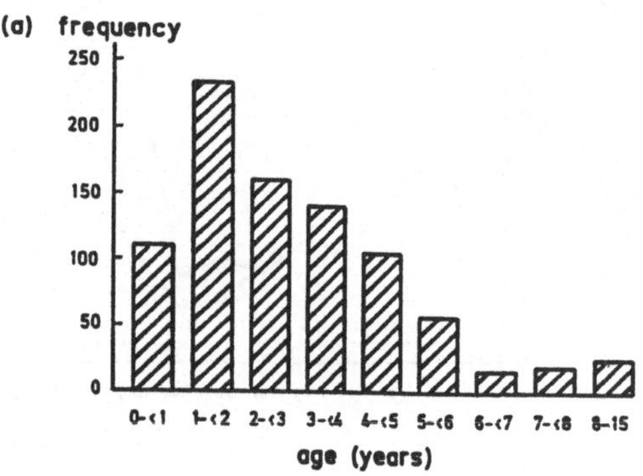
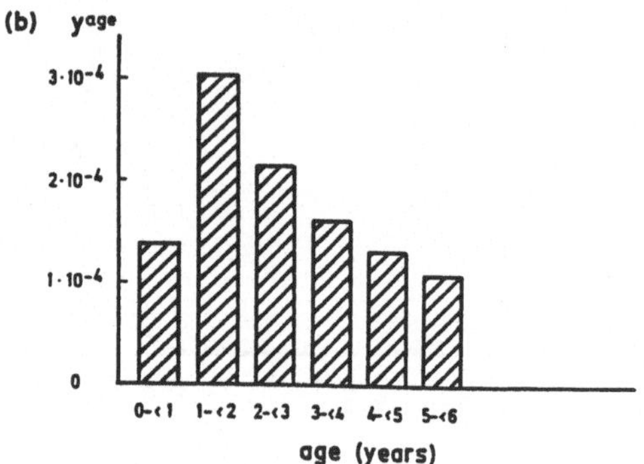

Fig. 6 Age distribution of croup: (a) frequency of croup attacks in Cologne 1977-84 [Wichmann et al 1990]; (b) y^{age} from model simulation

However, despite the reproduction of several features of the disease and PI1 infections by the proposed model, some questions are left open. So the identification of the most relevant epidemiologic parameters (duration of infectiosity and immunity) is not yet solved. Other etiologic factors for the disease seem to be too complex to be incorporated in a mathematical model. They have to be analysed by different tools [see e.g. Wichmann et al. 1990].

Acknowledgement

We are obligued to Drs. Dietz, Döller, Rebmann and Klöckner for carefully reading the manuscript.

References

Buchan KA, Marten KW, Kennedy DH: Aetiology and epidemiology of viral croup in Glasgow, 1966-72. J Hyg Camc 73 (1974) 143-150

Denny FW, Clyde WA Jr: Acute lower respiratory tract infections in nonhospitalized children. J Pediatr 108 (1986) 635-646

Denny FW, Murphy TF, Clyde WA Jr et al: Croup: An 11-years study in a pediatric practice. Pediatrics 71 (1983) 871-876

Foy HM, Cooney MK, Maletzky AJ et al: Incidence and etiology of pneumonia, croup and bronchiolitis in preschool children belonging to a prepaid medical care group over a four-year period. Am J Epidemiol 97 (1973) 80-92

Franke H, Wichmann HE: Mathematical model of parainfluenza virus type 1-infections. (1990) This volume

Glezen WP, Denny FW: Epidemiology of acute lower respiratory disease in children. N Engl J med 288 (1973) 498-505

Glezen WP, Loda FA, Clyde WA Jr et al: Epidemiologic patterns of acute lower respiratory disease in children in a pediatric group practice. J Pediatr 78 (1971) 397-406

Henderson FW, Clyde WA Jr, Collier AM et al: The etiologic and epidemiologic spectrum of bronchiolitis in pediatric practice. J Pediatr 95 (1979) 183-190

Loda FA, Clyde WA Jr, Glezen WP, Senior RJ, Sheaffer CI, Denny FW: Studies on the role of viruses, bacteria and M. pneumonia as causes of lower respiratory tract infections in children. J Pediatr 72 (1968) 161-176

Murphy TF, Henderson FW, Clyde WA Jr et al: Pneumonia: An eleven-year study in a pediatric practice.Am J Epidemiol 113 (1981) 12-21

Wichmann HE, Hübner R, Malin E, Schlipköter HW: Zusammenfassende Bewertung der 'Pseudokrupp-Studien' in Baden-Württemberg und Nordrhein-Westfalen. Projekt Europäisches Forschungsvorhaben zur Luftreinhaltung, Karlsruhe (1990)

SIMULATION VON CHEMISCHER KINETIK
MIT KISS AUF MIKRO-COMPUTERN

Björn A. Gottwald

(Fakultät für Biologie der Universität Freiburg)

Zusammenfassung

KISS (KInetisches Simulations-System) ist ein benutzer-
orientiertes interaktives digitales Simulations-System zur Modellierung chemischer Reaktions-Mechanismen. Die Eingabe erfolgt in der Form chemischer Gleichungen mit Geschwindigkeits-
konstanten für die Hin-Reaktion (und gegebenenfalls auch die
Rück-Reaktion) und ist im übrigen formatfrei. Die sich ergebenden Verlaufskurven für die Konzentrationen der reagierenden
Substanzen können ausgegeben werden als Tabellen, als Print-
Plot-Diagramme oder auf einem graphischen Bildschirm.

Summary

KISS (KInetics Simulation System) is a user-oriented interactive digital simulation system for modelling chemical reaction mechanisms. The input takes place in the form of chemical
equations with forward (and backward) reaction rate constants and is otherwise format free. The resulting progress
curves for the concentrations of the reacting substances can
be displayed as tables, as print-plot diagrams or on a graphical screen.

1. Einleitung

Eine früher von Gottwald (1981) publizierte Großrechner-Version von KISS war in der technisch-wissenschaftlichen Programmiersprache SIMULA geschrieben und daher in hohem Grade porta-
bel. Aus dem gleichen Grunde wurde PASCAL für die neue Mikrocomputer-Version von KISS gewählt, die auf IBM XTs, ATs und
damit kompatiblen Rechnern ohne Koprozessor lauffähig ist. Obwohl die Entwicklung mit dem leistungsfähigeren TURBO-PASCAL
3.0 durchgeführt wurde, wurde KISS so weit wie möglich in
Standard-PASCAL formuliert, damit es auch auf Großrechner

übertragen werden kann. Ein Struktogramm von KISS ist in Abbildung 1 dargestellt.

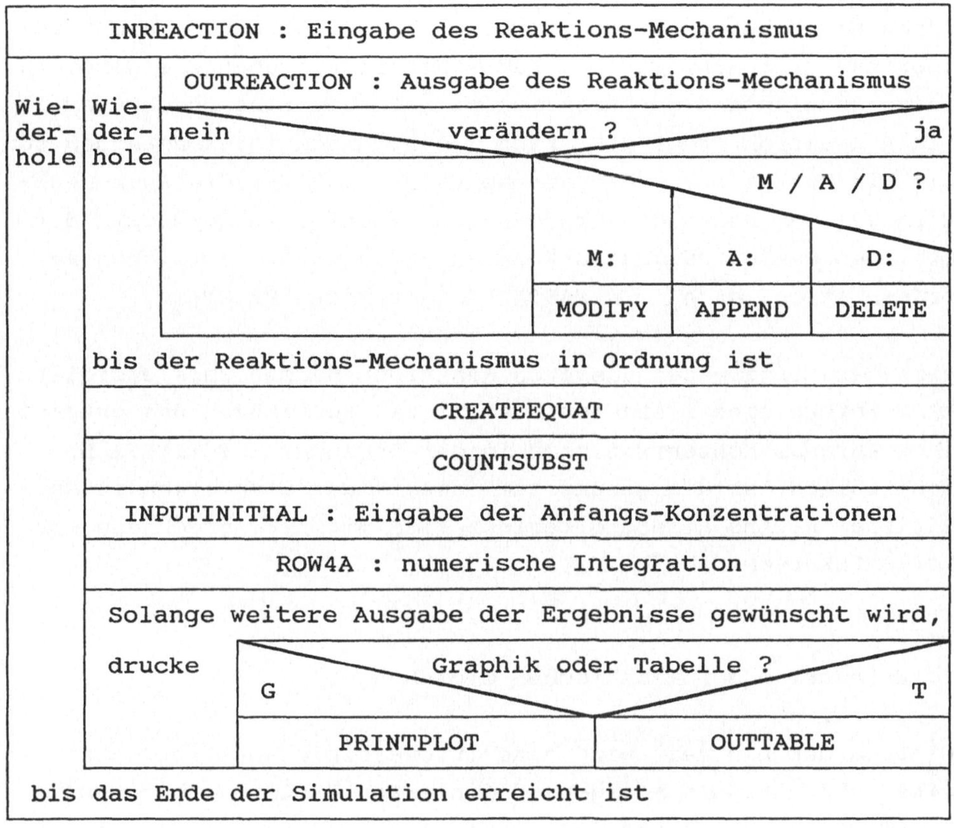

Abb.1 - Struktogramm von KISS

2. Dateneingabe

KISS verlangt als Eingabe vollständige chemische Reaktionen in der Form

$$a_1\ A_1 + a_2\ A_2\ \text{<-->}\ a_3\ A_3 + a_4\ A_4 \tag{1}$$

Hier bedeuten die a_i die stöchiometrischen Koeffizienten. Die A_i stehen für die Namen der an der jeweiligen Reaktion beteiligten Substanzen. Diese Namen können aus beliebigen auf der Tastatur verfügbaren Zeichen bestehen (Buchstaben, Ziffern und Sonderzeichen; jedoch nicht +, < und >) mit der Einschränkung, daß sie jeweils mit einem Buchstaben beginnen müssen. Entsprechend diesen Regeln prüft die Prozedur INREACTION die Syntax (aber nicht die Massen-Bilanz, weil die Substanznamen nicht auf chemische Symbole beschränkt sind) der Reaktion und gibt Hinweise bei fehlerhaften Eingaben wie bespielsweise unvollständigen Reaktionen. Nach der Eingabe der jeweiligen Reaktion fragt KISS nach Geschwindigkeits-Konstanten für die Hin-Reaktion (und gegebenenfalls auch für die Rück-Reaktion). Jede Reaktion wird dann gespeichert zusammen mit der bzw. den Geschwindigkeits-Konstanten in ihrer textlichen Repräsentation, d. h. als PASCAL-Zeichenkette REACTION.

Nach der Eingabe des gesamten Reaktions-Mechanismus erstellt KISS intern eine Liste der beteiligten Substanzen und erfragt ihre Anfangs-Konzentrationen in der Prozedur INPUTINITIAL. Schließlich ist die gewünschte Endzeit der Simulation zu spezifizieren. Anschließend beginnt KISS mit der Berechnung der Verlaufskurven.

3. Extraktion der kinetischen Gleichungen

Während der Laufzeit wird jede irreversible Reaktion durch eine Inkarnation des eigens definierten PASCAL-Verbundtyps EQUATION repräsentiert. Entsprechend wird jede reversible Reaktion durch zwei solche Verbunde repräsentiert. Obwohl es nicht notwendig ist, daß der Benutzer sich mit den Einzelheiten des PASCAL-Programmes beschäftigt, ist hervorzuheben, daß PASCAL zur Verwaltung von Datenstrukturen in Simulations-Systemen wie KISS besonders geeignet ist:

Alle Verbunde des Typs EQUATION (erzeugt durch die Prozedur CREATEEQUAT) sind jeweils charakterisiert durch

    1) die Anzahl der Substanzen auf der linken Seite NLEFT
    2) die Gesamt-Anzahl der Substanzen NTOTAL
    3) die stöchiometrischen Koeffizienten KOEFF [1..NTOTAL]
    4) die textliche Repräsentation der Namen der Substanzen
                                           SUBSTNAME [1..NTOTAL]

5) die internen Nummern der Substanzen SUBSTNR [1..NTOTAL]

6) die Geschwindigkeits-Konstante K

Darüberhinaus enthält KISS eine Prozedur RATE(YACT), die für jede Reaktion die jeweils aktuelle Reaktions-Geschwindigkeit R_i aus den aktuellen Konzentrationen YACT der beteiligten Substanzen berechnet gemäß Gleichungen wie (hier bedeutet ** die Potenzierung)

$$R_i = K * [A_1]**a_1 * [A_2]**a_2 \tag{2}$$

Aus diesen Reaktions-Geschwindigkeiten werden die zeitlichen Ableitungen der Konzentration für alle Substanzen N_j in dem vollständigen System der gekoppelten Reaktionen berechnet gemäß

$$d[N_j]/dt = \sum_{i=1}^{n} \pm a_j * R_i \tag{3}$$

Die erforderlichen Elemente der Jacobi-Matrix werden in entsprechender Weise berechnet.

4. Integrations-Algorithmus

Die Haupt-Aktivität während eines Laufs von KISS ist die numerische Lösung des Systems von Differentialgleichungen (3). Hierfür wird oft das numerische Integrations-Verfahren von Gear (1971) verwendet. KISS benutzt jedoch den schnelleren Integrations-Algorithmus ROW4A von Gottwald und Wanner (1981) :

ROW4A ist ein verallgemeinertes Runge-Kutta-Verfahren vierter Ordnung mit drei Funktions-Auswertungen pro Schritt und einem eingebetteten Verfahren dritter Ordnung zur Schrittweiten-Steuerung. Es verlangt bei jedem Schritt eine Berechnung der Jacobi-Matrix sowie die Lösung eines linearen Gleichungssystems der Ordnung n. Mit dem neuen Koeffizientensatz von Hairer und Wanner (1990) ist das Verfahren L-stabil und dadurch ein besonders geeigneter und robuster Integrations-Algorithmus für KISS.

5. Ausgabe der Ergebnisse

Während eines Laufs gibt KISS die berechneten Konzentrationen
der ersten vier Substanzen in einer Tabelle mit äquidistanten
Zeitschritten aus, die unabhängig von der intern benutzten
Integrations-Schrittweite sind. Da die Werte während der Berechnung gespeichert werden, können anschließend sowohl Tabellen als auch Verlaufskurven (Print-Plot-Diagramme oder Gra-
phiken, sofern eine entsprechender Bildschirm bzw. Drucker zur
Verfügung steht) für alle Substanzen ausgegeben werden.
Schließlich kann der Benutzer den Lauf mit dem gleichen oder
einem anderen Reaktions-Mechanismus und neuen Anfangs-Bedin-
gungen wiederholen.

6. Schlußbemerkung

Die Großrechner-Version von KISS ist in SIMULA auf der UNIVAC
1100/81 des Rechenzentrums der Universität Freiburg entwickelt
worden. Sie wurde mit verschiedenen Reaktions-Meachanismen von
Gottwald (1981) getestet, wobei auch versucht wurde, KISS mit
anderen publizierten Simulations-Systemen für chemische Kinetik zu vergleichen, insbesondere im Hinblick auf den Rechen-
zeit-Bedarf. Dies war jedoch schwierig, weil sie alle auf
verschiedenen Rechenanlagen implementiert waren.

Die Mikrocomputer-Version von KISS wurde mit einfachen Reaktions-Mechanismen wie dem Michaelis-Menten-Mechanismus getestet. Weitere Tests umfaßten den ziemlich steifen Mechanis-
mus von Robertson (1966) sowie eines der schwierigsten Probleme in der Simulation von chemischer Kinetik, den Oregona-
tor. Dabei handelt es sich um einen von Field und Noyes (1974)
zur Erklärung der oszillierenden Belousov-Zhabotinskii-Reak-
tion vorgeschlagenen Reaktions-Mechanismus. In allen Fällen
ergeben sich mit KISS die auch mit anderen Verfahren erhalte-
nen Verlaufskurven. Dies zeigt, daß KISS zur Simulation von
chemischer Kinetik auf Mikrocomputern besonders geeignet ist.

Eine Kopie der neuesten Version von KISS kann beim Autor unter der unten angegebenen Anschrift durch Einsendung einer formatierten Diskette (5¼" / 360 kB / Double Sided / Double Density) angefordert werden.

Literatur

R. J. Field and R. M. Noyes (1974) : "Oscillations in Chemical Systems. IV. Limit Cycle Behavior in a Model of a Real Chemical Reaction"
J. of Chem. Phys. 60, pp. 1877 - 1884.

C. W. Gear (1971) : "Numerical Initial Value Problems in Ordinary Differential Equations", Prentice Hall, Englewood Cliffs, N. J.

B. A. Gottwald (1981) : "KISS - a Digital Simulation System for Coupled Chemical Reactions"
Simulation 37:5, pp. 169 - 173.

B. A. Gottwald and G. Wanner (1981) : "A Reliable Rosenbrock Integrator for Stiff Differential Equations"
Computing 26, pp. 355 - 360.

E. Hairer and G. Wanner (1990) : "Solving Ordinary Differential Equations II. Stiff and Diferential Algebraic Problems", Springer-Verlag.

H. H. Robertson (1966) : "The Solution of a Set of Reaction Rate Equations" in "Numerical Analysis, An Introduction" (J. Walsh ed.), Academic Press, London, pp. 178 - 182.

Anschrift des Autors : Prof. Dr. Björn A. Gottwald
                        Fakultät für Biologie der
                        Albert-Ludwigs-Universität
                        Schänzlestrasse 1
                        D-7800 Freiburg im Breisgau

ALTERSKLASSEN-MODELLIERUNG MIT HILFE DES SIMULATIONSSYSTEMS SONCHES UND ANWENDUNG FÜR AGROÖKOSYSTEME

E. Matthäus

Zentralinstitut für Kybernetik und Informationsprozesse

Akademie der Wissenschaften zu Berlin

Zusammenfassung

Es wird die Modellierung von Entwicklungsprozessen mit Hilfe des Simulationssystems SONCHES diskutiert. Die fachgebietsspezifische Unterstützung für altersklassen-orientierte Modellierung wird an Insektenmodellen eines Agroökosystems Winterweizen erläutert.

1.Einführung

Die mathematische Beschreibung der Generationsfolge von Populationen erfordert immer dann detailliertere Betrachtungen, wenn die Asynchronität zwischen Geburt, Entwicklung und Sterben individuen-bezogen modelliert wird. Eine Einzelbeschreibung jedes Individuums führt zu einem hochdimensionalen Problem, wenn man eine Population realer Größe wirklich nachbilden will. Simulationstechnische Unterstützung ist bei diesen speicheraufwendigen Problemen sinnvoll.

Ein Kompromiß zwischen großer Individuenanzahl und mathematischer bzw. rechentechnischer Behandelbarkeit besteht in der Einführung von Zustandsklassen. Individuen, die sich um mehr als eine diskrete Zustandsänderung (z.B. Alter) voneinander unterscheiden, werden verschiedenen Klassen zugeordnet. Die Anzahl der Individuen pro Zustandsklasse ist eine wesentliche, die Populationsstruktur beschreibende, Zustandsvariable.

Die zeitliche Abfolge von Entwicklungsstadien und die unterschiedlichen Verweilzeiten in den Stadien (bei einer Insektenpopulation z.B. Eiablage, Reifung, Larvenentwicklung, Verpuppung, adultes Insekt) führt auch bei künstlicher Synchronität - durch Start des Experimentes mit einer Subpopulation gleichen Alters - zu einem Auseinanderlaufen (Dispersion) der ursprünglichen Impulsverteilung. In natürlichen und naturnahen Ökosystemen sind solche Experimentsituationen nicht gegeben. Für ein Verständnis der Populationsmächtigkeit in verschiedenen Stadien und zu verschiedenen Zeitpunkten ist die Dynamik der Altersklassenverteilung von essentieller Bedeutung.

In einem gemeinsamen Projekt mit Forschungsinstituten der Akademie der Landwirtschaftswissenschaften und der Martin-Luther- Universität Halle zur Modellierung eines Agroökosystems Winterweizen wurden Modelle von Getreideschädlingen (Getreideblattkäfer, Getreideblattlaus, Mehltau) als Voraussetzung für einen integrierten Pflanzenschutz mit Hilfe des Simulationssystem SONCHES erarbeitet (AUTORENKOLLEKTIV, 1986).

2. Entwicklungsmodellierung über diskrete Zustände

Für eine Modellierung der Populationsentwicklung bietet das fachgebietsorientierte Simulationssystems SONCHES (WENZEL, MATTHÄUS, FLECHSIG, 1990) mehrere Varianten. Eine Grobbeschreibung besteht in der Notation der verschiedenen Entwicklungsstadien und der Modellierung der Gesamtübergänge zwischen diesen diskreten Stadien.

Beispiel: Modell eines holometabolen Insekts (Getreideblattkäfer)

Alle Individuen innerhalb der Stadien sind ununterscheidbar. Bei einem Übergang wird nur die Gesamtrate beschrieben. Bei n Stadien (Compartments) mit einer Übergangsrate proportional zur Abundanz (Rate=1/τ) ergibt sich eine Gesamtverzögerung T_n= n·τ. Die SONCHES Notation lautet für das Beispiel, wobei das Präfix Z benutzt wird, um eine Übertrittsrate zwischen zwei Compartments zu deklarieren

Eiablage: ZEier= p_1· Weibchen

Larvenschlupf: ZLarven = p_2· Eier

Verpuppung: ZPuppen = p_3 · Larven

3. Entwicklungsmodellierung unter Nutzung von Zeitverzögerungen

Zur direkten Modellierung der Verweilzeit einer Subpopulation in einem Stadium verwendet man Zustandsvariable mit Zeitverzögerung. Im Simulationssystem SONCHES können beliebige diskrete Zeitverzögerungen direkt bei Argumentvariablen notiert werden. Die Dimensionierung des Modells ergibt sich aus der maximalen Verzögerungszeit (Entwicklungszeit). Zur Abbildung des Entwicklungszustandes (Ontogenese) nutzt man als Ersatzvariable meistens die Zeit oder Temperaturfunktionen. In einem Populationsmodell für Pelikane (Teil des Donaudelta-Ökosystemmodells, STANCIULESCU 1985) wird der Jahreszyklus der adulten Tiere mit Hilfe eines zeitverzögerten Prozesses zur Abbildung der Immigration in den Brutstandort wie folgt notiert.

1. ZADULTE: ADULTE(5),MONAT   Prozessdeklaration für Immigrationsrate
2. >ADULTE: ADULTE   Sterberate
3. YABFLUG: ADULTE,MONAT   Emigrationsrate

```
mit  1.  If (MONAT.eq.3)
            ZADULTE=ratio·ADULTE(5)
         else
            ZADULTE=0.
         endif

     2.  >ADULTE=mortality·ADULTE

     3.  if (MONAT.eq.11) then
            YABFLUG=ADULTE
         else
            YABFLUG=0.
         endif
```

1. Immigrationsrate proportional zur Populationsdichte 5 Monate fruher

2. einfache Sterberate

3. Emigration im November

Im Falle variabler, von Umweltgrößen abhängigen Entwicklungszeiten wird die maximale Verzögerungszeit des Argumentes notiert. Auf den Wert der Zustandsvariablen mit aktueller Verzögerung kann über eine SONCHES-Nutzerfunktion TIME_DELAY zugegriffen werden, die ein Pointerkonzept realisiert.

z.B. Temp_sum: temp

     ZLarven: Eier(10),Temp_sum

```
mit  Temp_sum=temp_sum+temp
     Entw_zeit=10*exp(-k*temp_sum)
     ZLarven=TIME_DELAY(Eier,Entw_zeit)
```

mit Hilfe einer Temperatursumme wird eine variable Entwicklungszeit berechnet, die als Argument in die Nutzerfunktion eingeht.

4. Entwicklungsmodellierung auf der Basis von Altersklassen

Bei variablen Entwicklungszeiten und Stadienwechsel in dem Sinn, daß alle Individuen mit einem Entwicklungswert $\theta > \theta_{th}$ in ein neues Entwicklungsstadium übergehen, ermöglicht SONCHES die Einzelbeschreibung der Kohorten als Altersklassen. Diese Notation erlaubt klassenbezogen die Entwicklung zu modellieren und den Übergang für alle Klassen unabhängig zu berechnen.

N_1	N_2						N_i						N_n
E_1	E_2						E_i						E_n

N_i Anzahl Individuen

E_i Entwicklungsgrad pro Altersklasse i

Wie in einem Paternoster werden die Altersklasseninhalte pro Zeittakt weitergeschoben. Es gibt keine Vermischung zwischen den Alterklassen. Ist die Verweilzeit pro Altersklasse, bzw der diskrete Altersunterschied zwischen den Klassen τ, so ergibt sich bei n Altersklassen eine Gesamtzverzögerung $T_n = n \cdot \tau$.

In SONCHES erfolgt der Übergang in die nächste Altersklasse nicht durch den Taktgeber getrieben, sondern abhängig von einem Immigrationsprozeß. Dieses ereignisorientierte Paternosterprinzip modelliert relative Altersklassen. Für eine objekt-orientierte Modellierung des Entwicklungsprozesses können alle vorgestellten Varianten gemischt werden. Nutzt man verschiedene Compartments, die unterscheidbare Entwicklungsstadien beschreiben und eine Altersklassenbeschreibung für den Reifungsprozeß innerhalb eines Compartments kann die Altersklassenstruktur beim Übergang zwischen Compartments verwischt werden, da aus beliebigen Altersklassen ein Übertritt erfolgen kann, jedoch alle "Neuankömmlinge" eines Zeittaktes in der ersten Altersklasse gesammelt werden.

Das absolute Alter wird in einer vom Simulationssystem bereitgestellten Argumentvariablen AGE gemerkt.

Das Paternoster-Prinzip ist eine numerische Realisierung des Leslie-Modells (LESLIE,1945), in dem Individuen jeder Altersklasse sterben oder neue Individuen der Altersklasse 0 erzeugen. Die Herleitung der kontinuierlichen oder der diskreten Variante beruht auf folgenden Annahmen:

Die Altersklassenverteilung kann sich ändern,

a) Individuen einer Alterklasse reifen oder sterben,

b) Individuen einer Alterklasse, die überleben, gehen in die nächste Altersklasse.

Sei $f(a,t)da$ die Anzahl der Individuen zur Zeit t mit dem Alter zwischen a und $a+da$ und $\gamma(a)dt$ die Sterberate der Individuen mit dem Alter a, dann folgt aus a) und b)

$$f(a,t+dt) = (1-\gamma(a-dt)dt) \cdot f(a-dt,t)$$

und für diskrete Änderungen

$$f(a,t+\tau) = P_{a-1} \cdot f(a-1,t) \qquad \text{mit } P_{a-1} = (1-\gamma(a-1)\tau)$$

wobei die Leslie-Matrix P_{a-1} den Anteil der Population im Alter $a-1$ beschreibt, der in das Alter a übergeht. Das Zeitintervall τ entspricht der Breite der Alterklasse.

In SONCHES notiert man einen klassenbezogenen Prozeß indem man die Zustandsvariable mit dem Suffix @ versieht. Die Wiederholung des Prozesses für alle Altersklassen erfolgt automatisch.

Folgende Notation ist ein Beispiel für einen klassenbezogenen temperatursummenabhängigen Geburtsprozess.

```
Temp_sum@: Temp                    mit Temp_sum@=Temp_sum@ + f(Temp)
ZWeibchen: Weibchen@,Temp_sum@        if (Temp_sum@.lt. 50) then
                                         ZWeibchen@=0.
                                      else
                                         ZWeibchen@=Weibchen@·Geburten
                                      endif
```

Diese Prozessnotation führt bei einer nur aus einem Weibchen bestehenden Beispielpopulation zu folgendem Abarbeitungsschema. Der Parameter "Geburten" sei in unserem Beispiel 2.

	Weibchen								Temp_sum						
t=1	1								5						
t=i	1								50						
t=i+1	2	1							5	55					
t=i+2	2	2	1						10	15	65				

Beispiel Getreidelaus

Bei der Modellierung der Getreidelaus als Teilmodell eines Winterweizenagroökosystems wurde dieses Altersklassenprinzip zur Beschreibung des Reifungsprozesses innerhalb verschiedener Entwicklungsstadien, die einen Generationszyklus bilden, genutzt. Das Strukturbild mit den schematischen Altersklassen ist in Fig. 1 dargestellt. Die Besonderheiten der Getreidelaus sind, daß Adulte Junglarven gebären und in Abhängigkeit der Ernährungssituation (Weizenontogenese, Anzahl Lause pro Ahre)

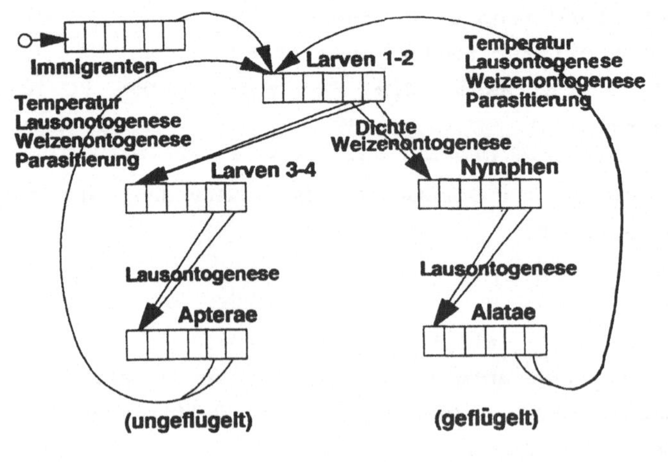

Fig.1 Compartments des Lausmodells

ungeflügelte oder geflügelte Adulte entstehen. Die wesentlichsten Einflußfaktoren sind an den Pfeilen vermerkt. Folgende Prozesse sind zu modellieren:

Compartment I (Immigranten)

ZI   : TSUM   (Immigration)   TSUM: Temperatursumme
ZL12@: I@,ONI@,TM,ON,PA2,XPR (Übertritt)   ONI: Ontogenese der Läuse
>I@   : PP,ON,I@,XPR,PA2,FU,ONI@   PA2: Parasitierungsgrad
ONI@ : TM,TIME   (Entwicklung)   FU: Verpilzungsgrad

Compartment L12   (Larvenstadium 1 und 2)

ZL34@ : L12@,ONL12@,ON,I,L12,L34,N,A1,A2   ONL12: Ontogenese der Larven
ZN@   : L12@/O,ONL12@   ON: Weizenontogenese
>L12@ : PP,ON,L12@,XPR   (Sterberate)   PP: Niederschlag
ONL12@: TM   XPR: Phloemsaftaufnahme
                                             TM: Temperatur

Compartment L34   (Larvenstadium 3 und 4)

ZA1@ : L34@,ONL34@   PL34: Parasitierungsgrad
>L34@ : PP,ON,L34@,XPR,PL34,FU   ONL34: Ontogenese der Larven
ONL34@: TM

Compartment N   (Nymphen)   analog zum Compartment L34

Compartment A1   (Apterae)

ZL12@ : A1@,ONA1@,TM,ON,PA1,XPR   PA1: Parasitierungsgrad
>A1@ : PP,ON,A1@,XPR,PA1,FU,ONA1@   ONA1: Ontogenese der Apterae
ONA1@ : TM

Compartment A2   (Alatae)   analog zum Compartment A1

Mit 20 Alterklassen pro Stadium wird durch das Simulationssystem ein Übergangsverhalten generiert, das einer Leslie-Matrix mit 120x120 zeitvariablen Elementen entspricht.

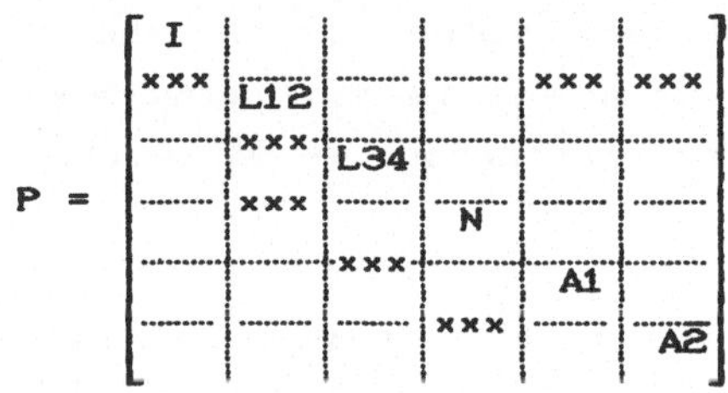

Die Nebendiagonalelemente z.B.

von   L12 : $[l^{L12}_{21},\dots,l^{L12}_{n,n-1}]$

sind wie folgt mit den SONCHES prozessen verknüpft:

(L12@- >L12@- ZL34@)/L12@

Die Fertilitätskomponenten (xxx) entsprechen·den Übertrittsraten Zc..c@, wobei c..c durch den jeweiligen Compartmentnamen zu ersetzen ist.

In Fig.2 sind für verschiedene Jahre, die sich qualitativ unterscheiden, Zeitverläufe der Populationsentwicklung dargestellt.

	Tempe-ratur	Nieder-schlag	Verpil-zung	Parasitierung L34	N	A1	A2
1980	niedrig	hoch	hoch	niedrig	niedrig	mittel	niedrig
1982	hoch	mittel	gering	hoch	hoch	niedrig	hoch
1984	mittel	niedrig	keine	mittel	niedrig	hoch	mittel

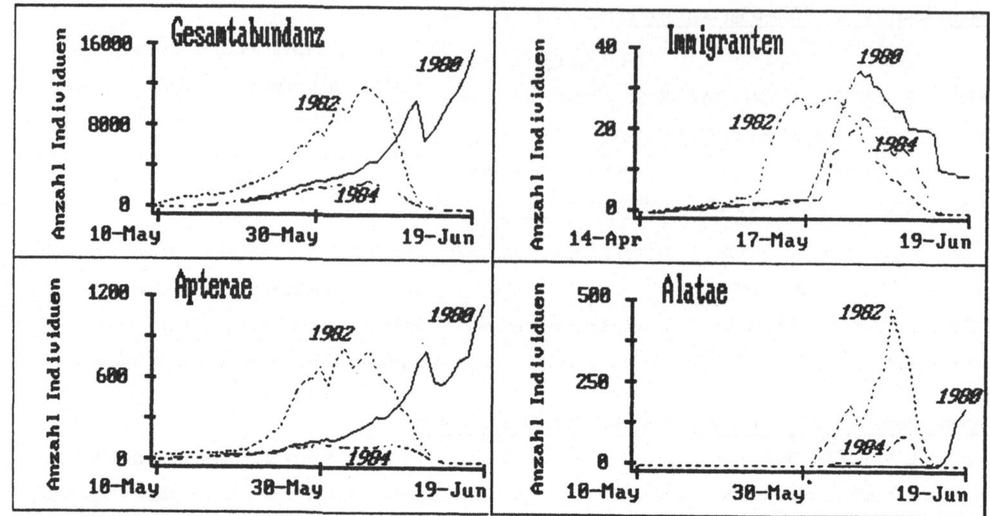

Fig. 2 Dynamik der Lausentwicklung für verschiedene Witterungen

Die Gesamtabundanz weist entsprechend der niedrigen Temperatur des Jahres 1980 eine verzögerte Entwicklung im Vergleich zu den anderen Jahren, aber aufgrund der auch verlangsamten Entwicklung der Wirtspflanze einen verlängerten Generationszyklus auf. Dies führt zu hohen Abundanzwerten. Ein Vergleich der geflügelten und ungeflügelten Weibchen verdeutlicht diese unterschiedliche Entwicklung. Die geflügelten Weibchen entstehen aufgrund des längeren Nahrungsangebotes 1980 erst sehr spät, zu diesem Zeitpunkt ist in den anderen Jahren die Lauspopulation schon zusammengebrochen.

Mit dem Simulationssystem kann auch die zeitliche Entwicklung der einzelnen Altersklassen studiert werden, solche Ergebnisse können unter natürlichen Bedingungen nicht gemessen werden, sind aber für den Experimentator als zusätzliche Informationsquelle wertvoll.

<u>Literatur</u>

Autorenkollektiv, 1986: Computer-Aided Modelling and Simulation of the Winter-Wheat Agroecosystem (AGROSIM_W) for Integrated Pest Management. Tagungsbericht., Akad. Landwirtsch.-Wiss. DDR, Berlin, Bd. 242

Leslie, P.H.,1945: On the use of matrices in certain population mathematics, Biometrika, 33, 183-212

Wenzel,V., E.Matthäus, M.Flechsig, 1990: One Decade of SONCHES, Syst. Anal. Model. Simul., 7 (to be published)

Stanciulescu, F., 1985: Principles of modelling and simulation large scale and complex systems. Application in ecology. In: A.Sydow, M.Thoma, R.Vichnevetsky (Eds): Systems Analysis and Simulation, Proceedings of the International Symposium, Berlin, Mathematische Forschung, Bd.27, Akademie Verlag, Berlin,133-138

MÖGLICHKEITEN UND GRENZEN DER MODELLIERUNG ÖKOLOGISCHER SYSTEME

MITTELS SONCHES

VOLKER WENZEL
Zentralinstitut fur Kybernetik und Informationsprozesse
der Akademie der Wissenschaften zu Berlin

ZUSAMMENFASSUNG

Als vor etwa 10 Jahren das Simulationssystem SONCHES mit einer
fachbezogenen Modellierungssprache konzipiert wurde, war in mehrerer
Hinsicht Neuland zu bestellen. Von besonderer Tragweite waren zwei
Aspekte: Entwurf eines Kalküls zur strukturierten, inkrementellen
Modellierung unter direktem Bezug auf die Objekte des unterstützten
Wissenszweiges Ökologie sowie auf deren typisches Verhalten und die
Einbettung dieser Sprache in ein integriertes Simulationssystem, das
alle Aktivitäten rund um ein Modell von dessen Aufstellung und
Eingabe über seine Verifikation und Validierung bis zur praktischen
Nutzung für Vorhersagen, Überprufung von Hypothesen und den Entwurf
von Steuerstrategien im Dialogbetrieb unterstützt.

In diesem Artikel liegt der Akzent auf dem ersten Aspekt, der
anhand der ökospezifischen Interpretierbarkeit von Flussprozessen
erläutert wird, wofür die Darstellung der Struktur des komplexen
Agroökosystems ein Beispiel gibt.

1. ALLGEMEINES ÜBER SONCHES

Hilfsmittel zur computergestützten Simulation wurden früher
charakterisiert durch die mathematische Form, in der die Modelle
vorliegen mussten: Differentialgleichungen, Differenzengleichungen,
Bedienungstheorie etc.

Die Untersuchung isolierter ökologischer Inhalte als
Fragestellungen der mathematischen Ökologie, die etwa durch einen
bestimmten Gleichungstyp wie Volterra-Lotka oder die
Leslie-Matrizen repräsentiert werden, kann man sich damit
sicher erleichtern. Für die Systemökologie, also das Studium realer
Ökosysteme sind sie jedoch nicht geeignet, da die fur den
Modellaufbau zunächst vorhandenen Kenntnisse sich um so
weniger in eine solche geschlossene mathematische Form bringen
lassen, je komplexer das System ist.

Das Simulationssystem SONCHES (Simulation Of Nonlinear Complex Hierarchic Ecological Systems [1,2]) bietet deshalb als fachgebietsorientiertes Werkzeug ein Sprachkalkül, das den Aufbau des Ökosystemmodells aus einzelnen Elementarprozessen selbst leistet. Letztere sind Kausalbeziehungen zwischen den Variablen und Raten unterschiedlichen Typs so, dass sie ökologisch inhaltlich interpretiert werden können. Ihrer so ermittelten Bedeutung gemäss werden sie dann in das Modell eingefügt, wobei kanonische Abläufe wie Bilanzierung von Raten automatisch ergänzt werden.

Die Raten sind als vollwertige Variablen überall referierbar und dazu durch Präfixe (X - Anspruch/Input, XX - unbefriedigter Anspruch, Y - Output, Z - Compartmentübertritt, < - Wachstum, > - Mortalität) vor den Namen der Objekte oder Pools gekennzeichnet.

Die wichtigsten Gesetze und Regeln für Ökosysteme sind dem Simulationssystem prozedural einbeschrieben und bewirken eine semantische Prüfung, die fachgebietsgerechte Ordnung des Modells wie seine recheneffiziente Verdichtung.

Die Elementarprozesse müssen bezogen auf ein passendes Eigenzeitintervall durch eine Berechnungsvorschrift realisiert werden, die sich auch nur auf Beobachtungen und Messungen dieser isolierten Kausalzusammenhänge stützt.

Wegen des für die Ökosystemelemente zugrundegelegten Automatenkonzepts kann das Modell a posteriori als verallgemeinertes Differenzengleichungssystem angesehen werden. Der Unterschied zu a priori gleichungsorientierten Simulatoren äussert sich eben darin, dass

- die Differenzengleichungen nicht der Ausgang sondern Ergebnis eines halbautomatisierten Modellierungsprozesses sind
- die Gleichungen in ihre Terme zerlegt verwaltet und strukturiert werden ; in dieser Struktur manifestiert sich eigentlich die ökospezifische Semantik
- die mathematische Realisierung der Gleichungsterme neben analytischen Ausdrücken ebenso Algorithmen oder Funktionstabellen sein dürfen.

Die Gesamtheit der etwa 20 inhaltlich unterschiedenen Prozesstypen, denen die Elementarprozesse zugeordnet werden, bildet zusammen mit den auf dieselben anzuwendenden Ordnungsprinzipien den abstrakten Ökosystemprototyp, in den nach dem Genotyp-Phänotyp-Prinzip jedes konkrete Modell einbeschrieben wird.

Das Anwendungsspektrum des Simulationssystems umfasst neben Agroökosystemen solche für aquatische und forstliche Systeme, ferner Populationen, Räuber-Beute-Systeme, primäre und sekundäre Sukzession, Epidemien, Modelle der Tierphysiologie, der Pharmakokinetik und Toxikokinetik bis hin zur Verhaltensbiologie.

Letztere Modelle sind gewöhnlich stark ereignisorientiert. Solche Systeme lassen sich jedoch nur dann mittels SONCHES modellieren, wenn sich die Ereignisabfolge mit einem vorgegebenen Zeittakt synchronisieren lässt, so dass imperative Ablaufsteuerung möglich bleibt. Das Auftreten von asynchronenen, zufälligen Ereigniszeitpunkten erfordert eine interrogative Ablaufsteuerung, die für ökologische Systeme nicht typisch ist und deshalb nicht einbezogen wurde.

2. WIE ZB. FLUSSPROZESSE DIFFERENZIERT ZU INTERPRETIEREN SIND

Es genügt schon, zwischen (i.a. abiotischen) Ressourcen R und (i.a. biotischen) Compartments C zu unterscheiden, um einen Fluss der Form

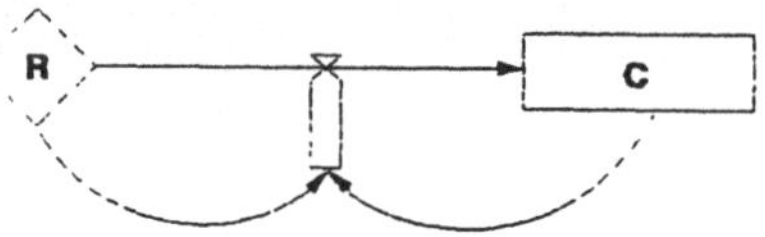

als *Nahrungsaufnahme* zu interpretieren. Dabei symbolisiert ein Pfeil auf das Ratensymbol den Anspruch von C, der dessen optimalen Lebensbedingungen entspricht, der andere aber die Limitierung durch den laufenden Füllungsstand von R.

Im dynamischen Modell sind nun die Einträge in den Pool R zu beschreiben und der genannte *Anspruch* (DE — demand)

$$XR : C,... \qquad\qquad * * \text{ MOD DE}$$

Der Simulator kann den linken Pfeil dann automatisch realisieren.

indem er *R* mit *XR* vergleicht und *XR* bei Limitierung zum möglichen Input für *C* reduziert. Dabei kann auch *Konkurrenz* (CT - competition) zwischen mehreren Compartments abgebildet werden.

Kann der Anspruch nicht erfüllt werden, so stellt der Simulator den unbefriedigt gebliebenen Anspruch *XXR* bereit, der zur naheliegenden Modellierung von Leistungsminderungen des Compartments weiterverwendet werden kann — oder aber zur Kompensation des Defizits durch dann gerade von *XXR* abhängigen Anspruch an eine zweite Ressource *RE*

$$XRE : XXR,... \qquad\qquad * * MOD\ CO$$

Dies wird als ein dem Modul CO - input coupling zuzuordnender Prozess erkannt und entsprechend eingeordnet.

Steht anstelle der Ressource ein zweites Compartment, dh. der Fluss ist von *C1* nach *C2* gerichtet und die Flussrate hängt nur von *C1* ab, so ist die Übertrittsrate als *Migration* oder *Ontogenese* interpretierbar. Durch Ergänzung eines Suffix (ZC2@) kann dieser Prozesstyp darüber hinaus zum Aufbau und zur Verwaltung einer Altersstruktur für das Zielcompartment benutzt werden (siehe [3]).

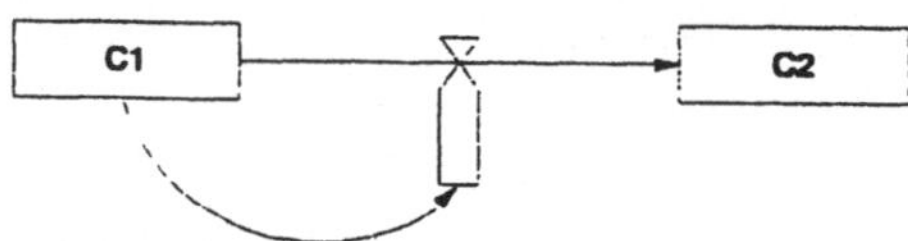

Von ganz besonderem Interesse für den Aufbau ökologischer Modelle ist der folgende Fall

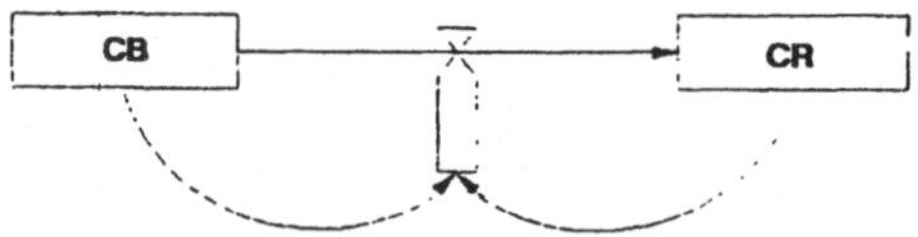

Dh. *CR* stellt einen Anspruch so, als sei *CB* eine Ressource. Im Unterschied zur passiven Limitierung der Ressource *R* bei der einfachen Nahrungsaufnahme ist *CB* aber ein biotisches Compartment, das sich dem Anspruch von *CR* entziehen oder ihm durch eigene Dynamik und Verhalten zumindest entgegenwirken kann. Es praktiziert also eine aktive von seinen Möglichkeiten (laufenden Zuständen) abhängige Limitierung für die Nahrungsaufnahme von *CR*.

Das geschilderte Phänomen ist gerade die *Trophie*. Um sie als freies

Spiel der Kräfte in vollem Umfange und bei grösstmöglicher Unterstützung durch das Simulationssystem abbilden zu können empfiehlt es sich, das Forrester-Raten-Symbol zu verlassen, den Fluss in zwei Teilflüsse zu dekomponieren und eine zu diesem Zwecke zu definierende trophische Ressource (hier TR) dazwischenzuschalten.

Die Modellierungsleistung des Ökologen besteht in der Angabe des Angebots durch die Beute

$$YTR : CB,... \qquad\qquad * * MOD TO$$

und des Anspruches durch den Räuber

$$XTR : CR,... \qquad\qquad * * MOD DE$$

Das Simulationssystem übernimmt

- die Summation der Trophischen Angebote $TR = \sum YTR(i)$, $i=1,n$
- die Summation der Trophischen Ansprüche $XTR = \sum XTR(j)$, $j=1,m$
- Bestimmung der tatsächlichen Flussraten unter Berücksichtigung von Limitierung und Konkurrenz
- Rückbilanzierung der ermittelten Flussraten bei CB_i
- richtige Anordnung aller Prozesse.

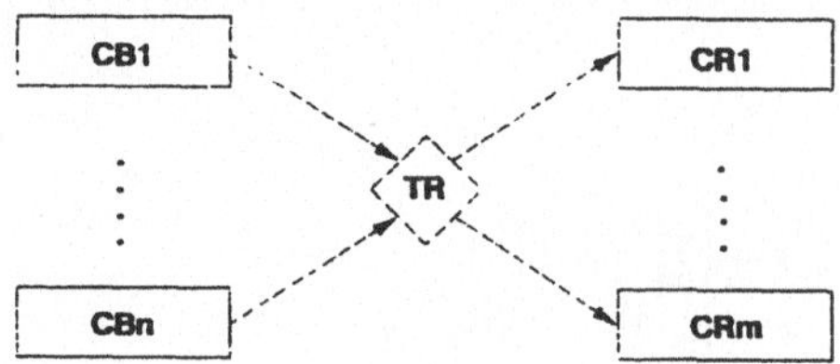

3. BEISPIEL EINES KOMPLEXEN AGROÖKOSYSTEMS

Es soll zeigen, wie durch Systematisierungsleistungen über ein objekt- und verhaltens-orientiertes Sprachkalkül auch sehr komplexe Systeme (mit 1000 und mehr Variablen) auf ihre inhaltliche Essenz gebracht werden können und überschaubar bleiben.

In der oben erläuterten Symbolik zeigt das unten folgende Schema das verallgemeinerte trophische Netz des Agroökosystems AGROSIM ([4]):

Die Nutzpflanze Winterweizen (Triticum) besteht aus drei Compartments: der Wurzelbiomasse R , dem generativen Teil G und dem oberirdischen Compartment A , das neben dem Assimilatepool als seiner Abundanzvariablen mit der grünen Biomasse GBM noch eine

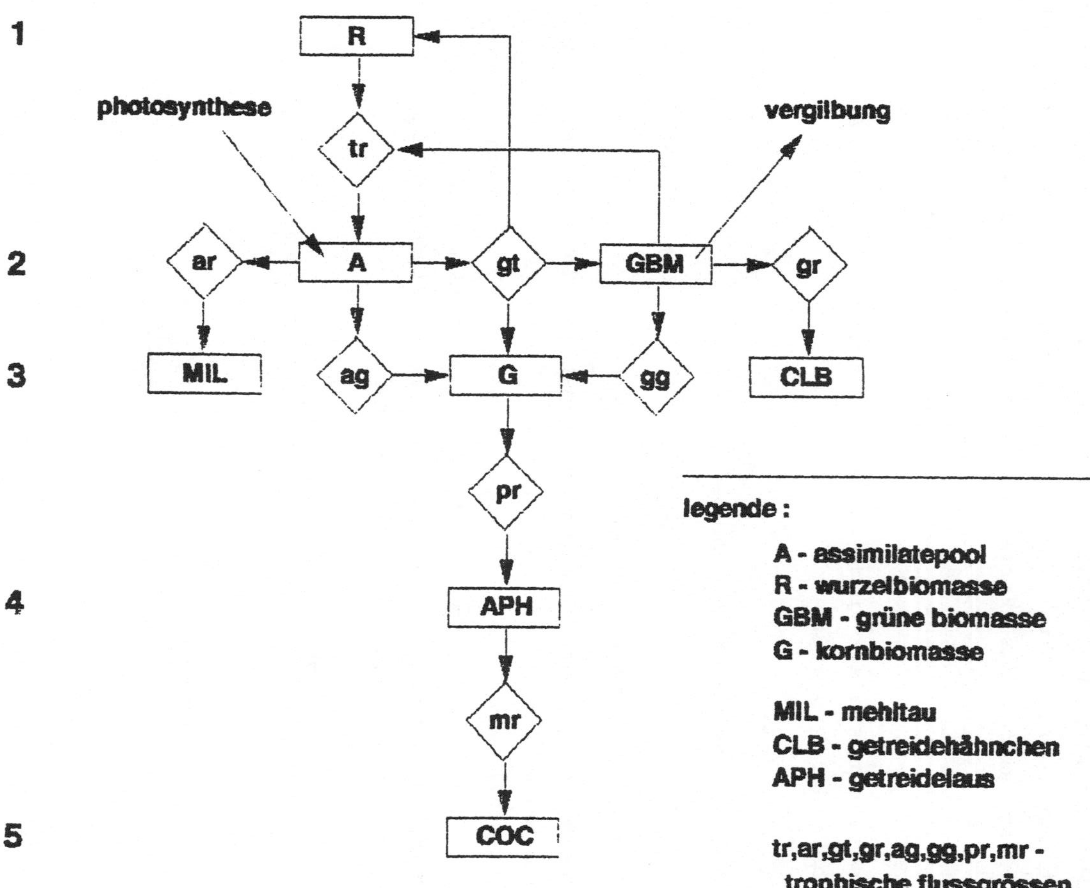

'TROPHISCHES NETZ' IN AGROSIM

Trophische Flüsse gibt es zu drei Schaderregern

- dem Mehltau (Erysiphe) *MIL* , der den Assimilatepool *A* dezimiert
- dem blattfressenden Getreidekäfer (Oulema lichenis) *CLB*
- der Phloemsaft saugenden Getreidelaus (Macrosiphum avenae) *APH* , von der ihrerseits ein trophischer Fluss zum natürlichen Feind Marienkäfer (Coccenelidae) *COC* modelliert ist, um auch biologische Bekämpfung simulieren zu können.

Ausser diesen 'echten' trophischen Flüssen sind über *TR* evtl. benötigte Assimilatumlagerungen aus der Wurzel und die Ansprüche aller drei Weizen-Compartments an das Resultat der Wachstumsatmung *GT* als Angebot/Anspruch-regulierte Flüsse. Ausserdem befriedigt das Compartment *G* seinen Bedarf nach dieser Ressource qua Input-Kopplung auch direkt aus *A* oder schliesslich durch Umlagerung aus *GBM*.

Es ergigt sich ein Modell mit 5 trophischen Schichten mit einem trophischen Zyklus innerhalb der zweiten.

Zur Vervollständigung des Fluss-Systems sind noch Photosynthese und Vergilbung als Quelle und Senke hinzuzufügen. Die Intensität der Abläufe wird durch das Taktmass einer biologischen Uhr in Gestalt einer Ontogenesefunktion bestimmt, die ihrerseits von den je herrschenden Witterungsbedingungen getrieben oder gebremst wird.

MIL, *CLB*, *APH* sind hochdimensionale Populationsmodelle, deren Struktur durch das Altersklassen-Konzept eine vergleichbar prägnante Darstellung erlaubt (siehe [3]).

REFERENZEN

[1] Knijnenburg,A.,Matthäus,E.,Wenzel,V.: Concept and Usage of the Interactive Simulation System for Ecosystems SONCHES. Ecol. Model. 26(1984),pp.51-76.

[2] Wenzel,V.: Problemlösungen bei Entwurf und Realisierung eines fachgebietsorientierten Simulationssystems (SONCHES). Diss., Berlin, 1987.

[3] Matthäus,E: Altersklassenmodellierung mit Hilfe des Simulationssystems SONCHES und Anwendung für Agroökosysteme, in diesem Band.

[4] Tagungsbericht der Akademie der Landwirtschaftswissenschafen, 242, Berlin, 1986.

ÖKOLOGIE

Interactive Modeling and Simulation of Environmental Systems on Workstations

Andreas Fischlin[1]

Abstract

Many systems dealt with in environmental sciences such as ecology or environmental biology could be easily modelled and efficiently simulated on personal computers or on workstations. Thanks to their graphical capabilities such computers make it possible to model systems interactively, e.g. supported by graphical structure editors, or allow for interactive simulation featuring sophisticated graphical output of the simulation results. However, in practice this potential remains often underexploited, since traditional, simulation software is mostly batch oriented, largely ignores computer science research, and offers rarely the functionality needed for a sensible interactive use. Instead of porting simulation software from main-frames onto workstations we propose new concepts based on Wymore and Zeigler's modeling theory, enhanced by some new interactive user oriented task concepts. This paper presents a scheme called RAMSES for the architecture of a modeling and simulation environment on a workstation particularly suited for the working with environmental systems. Furthermore it reports on some results which have been obtained by implementing portions of the RAMSES architecture, in particular an open and extensible modeling and simulation environment for the two classical model formalisms SM (Sequential Machine), DESS (Differential Equation System Specification) featuring modular modeling. Finally the modeling and simulation of a system from population ecology is presented as an example to illustrate and evaluate some of the concepts of RAMSES in ecological research.

1   Introduction

The demands for interactive modeling and simulation of environmental systems are old (HOLLING, 1964; DAVIDSON & CLYMER, 1966), whereas the possibility to satisfy them are rather new. Interactive modeling and simulation have never been the strength of the traditional main-frame simulation software, but today, with the wide-spread usage of personal computers and recently the more powerful workstations, the possibility to bring the computational power of the main-frames together with the interactivity and user-friendliness of workstations has become economical and is indeed most attractive.

However, straightforward realizations seem difficult, since an application of existing modeling and simulation techniques has first to overcome several obstacles: First the majority of simulation software has an architecture which is mainly batch oriented (CELLIER, 1975, 1979, 1982, 1984a; ANONYMOUS, 1988); secondly it largely ignores current computer science research (KREUTZER, 1986); and thirdly, maybe most importantly, aside from very few exceptions the software is not founded on a mathematically sound basis, i.e. it ignores the modeling and simulation theory developed in the last decades (ZEIGLER, 1976; WYMORE, 1984).

Interactivity is particularly attractive in the modeling of so-called ill-defined systems (INNIS, 1972; CELLIER & FISCHLIN, 1982; FISCHLIN & ULRICH, 1987). Typical for them is that essential portions of the mathematical properties of the studied system are poorly understood or even un-

---

[1] Current address: Systems Ecology Group, Institute of Terrestrial Ecology, Department of Environmental Sciences, Swiss Federal Institue of Technology Zürich (ETHZ), ETH-Zentrum, CH-8092 Zürich, Switzerland.

known. This is particularly true for environmental systems, which KARPLUS (1976) places in the middle of his model spectrum calling them grey-boxes. Compared with the black boxes on the one end of the spectrum their modeling appears attractive; on the other hand they lack the simplicity and clarity of the white boxes on the other end. A full list of advantages of interactive modeling in the area of environmental systems has been formulated elsewhere (FISCHLIN & ULRICH, 1987).

The attractive interactive modeling and simulation are, the much it poses difficulties, since it requires the development of dedicated simulation software. Interactive programs have a structure radically different from that of batch-oriented software (NIEVERGELT & WEYDERT, 1980; NIEVERGELT & VENTURA, 1984; FISCHLIN, 1986). The situation is furthermore complicated by the fact, that most simulation studies tend to develop very large computational demands, but sensible interactivity requires that response times remain within certain limits. Hence the design of interactive simulation software must cater to the two conflicting goals of batch and interactive simulations at once.

After having analyzed programming languages and programming styles in the context of simulation and studying current simulation techniques in much detail, KREUTZER (1986) concludes that existing simulation software largely ignores current computer science research. For instance the majority of simulation software has been and still is written with out-dated programming languages poorly fit for their purpose. In a recently published catalog (ANONYMOUS, 1988) from 191 world-wide listed simulation software packages 79% use Fortran, the rest either C or at best some object oriented extensions of C. In the bibliography of textbooks on simulation by KREUTZER (1986) 83% from the programming oriented books use Fortran or simulation languages which are Fortran precompilers. Precompilers separate the modeler only partially from the underlying implementation language. Often either the simulation language reflects the spirit and concepts of the used implementation language, e.g. naming rules for identifiers, or it forces the modeler to use it sooner or later, in cases he/she wishes to program an unavailable algorithm. CELLIER (1979; 1984b) claims that mainly due to the use of the programming language Fortran, the simulation software packages lack robustness. He argues that in order to increase the quality and reliability of the simulation software, one should use formally defined programming languages of type LL(1) to optimally support structured programming (BROOKS, 1979; WIRTH, 1985, 1986).

On workstations interactive modeling and simulation concepts must be simple, user-oriented, and have to support modern graphical user interfaces including items such as menus (pull-down or pop-up), windows, scrolling of window contents, and selection or dragging of graphical objects etc. The latter requires adequate programming support such as structured data types, dynamic memory allocation respectively deallocation (heap-technique), recursion etc. (GUTKNECHT, 1983; FISCHLIN, 1986; WIRTH, 1986). Some of the few programming languages which satisfy the requirement of formal definition and support a good programming practice are the procedural languages Modula-2 (WIRTH, 1985, 1986) and Ada, or Oberon as an example for a formally defined, object oriented language (WIRTH, 1988, 1989a, b).

Finally aside from very few rare cases (ÖREN, 1984; VANCSO, 1990) existing simulation software ignores the whole body of systems and modeling theory which has been developed during the last two decades (KLIR, 1979; ZEIGLER, 1976, 1979, 1984; WYMORE, 1984). Important for the modeling of environmental systems are the levels 1 (IORO) and 3 (System). On level 1 a system is defined by a time basis, the sets of the inputs, outputs, and input segments, plus the I/O relation relating outputs with input segments. On level 3 a system is similarly but in more detail defined by the additional set internal states, the state transition function mapping the product of inputs and internal states to internal states, and the output function mapping internal states to outputs. Furthermore the standard formalisms DESS (Differential Equation System Specification), DEVS (Discrete Event System Specification), and SQM (Sequential Machine) have been formulated to support frequently used classes of mathematical models. For a full summary see VANCSO et al. (1987).

From all this follows, that in order to meet todays requirements and to tap the potential of tomorrows computer technology, a new modeling and simulation software has to be developed. Therefore we have dared to begin with the development of an interactive modeling and simulation software called RAMSES[1] particularly designed for modern personal computers and workstations. Our approach is user-oriented, i.e. a conceptual frame-work derived from research activities, to support interactive modeling and simulation. The software should be particularly well suited for the modeling of ill-defined dynamic systems (INNIS, 1972; CELLIER & FISCHLIN, 1982) as often present in the environmental sciences (KARPLUS, 1976;). For the implementation we chose Modula-2 as the programming language and based the software on the *Dialog Machine* (FISCHLIN, 1986). The RAMSES software architecture is based on systems and modeling theory and it supports an object oriented working with such concepts. This paper focuses on RAMSES' session concept.

2 Interactive Modeling and Simulation with RAMSES

RAMSES provides software support for the activities typically followed by a user who constructs and simulates models of ill-defined systems by grouping them into the four sessions (Fig. 1):

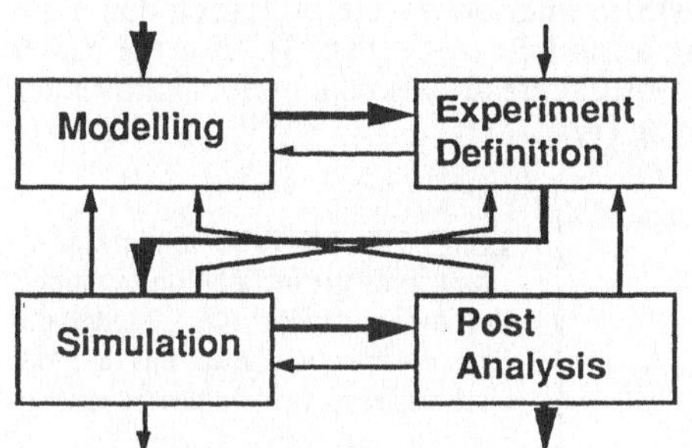

Fig. 1: State transition diagram depicting RAMSES' four task oriented sessions (states) and the possible user movements (transitions). 1) Modeling session resulting in the declaration of models, model objects and the formulation of model equations. 2) Experiment definition which consists of a definition of an experimental frame plus its association with a particular model definition. 3) Simulation session which produces model behavior. 4) Interactive postanalysis of previously computed simulation results.

- Modeling session: This activity serves the declaration of models, model objects, and the formulation of model equations. A mathematical model (not to be confounded with a simulation model) defines a certain mathematical structure but not necessarily a particular time domain, parameter and initial value sets.

- Experiment definition session: It consists of a specification of an experimental frame (ZEIGLER, 1976, 1979) plus its association with a particular mathematical model. Its result is at least an experimental frame or an experiment. The latter fully specifies a simulation model (not to be confounded with a mathematical model) which incorporates in addition to a mathematical structure also a particular time domain, parameter and initial value sets etc.

- Simulation session: Given a particular experiment has been defined and a simulation model exists, the simulation session is used to produce model behavior in time or space or both. Results can be saved for an analysis at a later point in time.

- Postanalysis session: Simulation results previously computed during a simulation session, can be analyzed interactively without having to recompute any model behavior.

Fig. 1 shows a state transition diagram of the meaningful and legal transitions among the RAMSES sessions. Note that iterative model development cycles are supported by various paths.

---

[1] Acronym for Research Aids for the Modelling and Simulation of Environmental Systems

For the user's convenience, sessions can be interrupted and resumed later in the same state they have been left any time. RAMSES has two user interfaces: The end-user interface and the client or programmer's interface. With only very few, but then intentionally introduced exceptions, any function offered by RAMSES can as well be executed via the end-user interface interactively or via the client interface by writing a program.

2.1 The Modeling Session

The modeling session serves the declaration and installation of models, their model objects (Fig 2), and the formulation of model equations.

A RAMSES model definition consists of at least one model and every model usually contains model objects such as state variables, expressions, model parameters, output and input variables for submodel coupling, and auxiliary variables. Every model object can also be declared as a monitorable variable. Not the output variables, which are reserved for the coupling of submodels, but only monitorable variables allow the user to display simulation results. In this respect RAMSES differs from the modeling theory by ZEIGLER (1976) or WYMORE (1984). RAMSES requires to associate with model objects certain real values such as derivatives respectively new values, parameters, initial, minimum, or maximum values. The RAMSES interfaces were built such that mandatory values must always be provided while declaring a model object (Fig. 3). For the easier recognition and identification of model objects by the end-user there are also the optional attributes: long textual description, short identifier, and the unit string (Fig 2) .

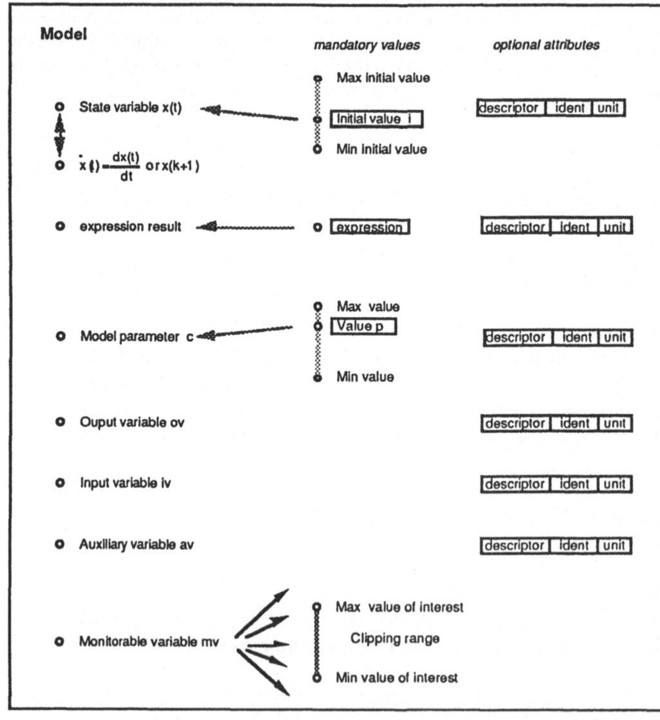

Fig. 2: A RAMSES model definition serves the installation of models and model objects (O): Model objects are state variables plus their derivatives resp. new values, model parameters, expressions, auxiliary, monitorable, input, and output variables.

The purpose of the model and model object declarations is to make their associated real variables known to RAMSES. During simulations RAMSES will then maintain these values: E.g. it uses the derivatives respectively new values (in case of discrete time difference equations) to compute and repeatedly update the values of the state variables (numerical integration). Otherwise RAMSES ignores these objects and their values. This offers the modeler the potential to use them freely, e.g. in structured data types, in a manner which is rather problem adapted than forced by the idiosyncrasies of the used simulation technique.

```
PROCEDURE DeclareSV    (m: Model; VAR s, ds: REAL; initial, minRange, maxRange: REAL;
                        descriptor, identifier, unit: ARRAY OF CHAR);
PROCEDURE RetrieveSV (m: Model; VAR s: REAL; VAR defaultInit, minCurInit, maxCurInit: REAL;
                    VAR descriptor, identifier, unit: ARRAY OF CHAR);
PROCEDURE ModifySV     (m: Model; VAR s: REAL; defaultInit, minCurInit, maxCurInit: REAL;
                        descriptor, identifier, unit: ARRAY OF CHAR);
PROCEDURE UndeclareSV(m: Model; VAR s: REAL);

PROCEDURE GetSV        (m: Model; VAR s: REAL; VAR curInit: REAL);
PROCEDURE SetSV        (m: Model; VAR s: REAL;     curInit: REAL);
```

Fig. 3: Excerpt from the client interface of RAMSES showing procedure declarations (they consist in Modula-2 of a heading only). The listed procedures provide all RAMSES functions needed to work with state variables.

There are three basic techniques by which modeling can be done within the RAMSES modeling session: 1) Via the client interface using the host <u>programming language</u> enriched with particular objects needed for modeling and simulation; 2) via an <u>interactive end-user interface</u> accessing individually by entry-forms the functions exported by the client interface in order to add (declare), modify, and remove (undeclare) system theoretical objects (model and model objects) from the model data base. 3) Via the <u>Editing</u> of a graphical representation, i.e. <u>relational digraphs</u> (Fig. 4), of a model system to support a more abstract view. Subsequently relations are specified by functions, i.e. declared as expressions, in a manner which resembles that used by the second technique.

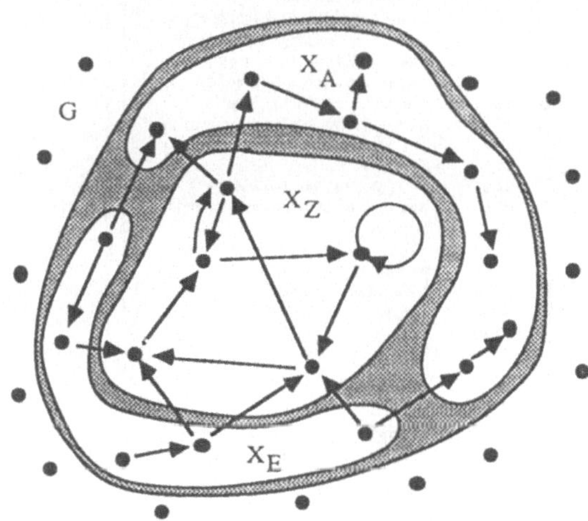

Fig. 4: Relational graph of a general system S represented as a 9-tuple of sets $S = (X_I, X_Z, X_O, R_I, R_{IZ}, R_Z, R_{ZO}, R_O, R_{IO})$. The universe of the system X is partitioned into three sets: Set of state variables X_Z, inputs X_I, and outputs X_O. Where: $X_Z = \{z \mid z$ from z reachable $\}$, $X_I = \{e \mid e \notin X_Z \land (\exists z : z \in X_Z \land z$ from e reachable$)\}$ $X_O = \{a \mid a \notin X_Z \land [(\exists z : z \in X_Z \land a$ from z reachable$) \lor (\exists e : e \in X_I \land a$ from e reachable$)] \}$. Form from ordered pairs within the system universe the following six structure sets: input structure $R_I = \{(e_1,e_2) \mid e_1,e_2 \in X_I \land (e_1,e_2) \in R_o\}$, input-state-coupling structure $R_{IZ} = \{(e,z) \mid e \in X_I \land z \in X_Z \land (e,z) \in R_o\}$, dynamic structure $R_Z = \{(z_1,z_2) \mid z_1,z_2 \in X_Z \land (z_1,z_2) \in R_o\}$, state-output-coupling structure $R_{ZO} = \{(z,a) \mid z \in X_Z, a \in X_O \land (z,a) \in R_o\}$, output structure $R_O = \{(a_1,a_2) \mid a_1,a_2 \in X_O \land (a_1,a_2) \in R_o\}$, and input-output-coupling structure $R_{IO} = \{(e,a) \mid e \in X_I \land a \in X_O \land (e,a) \in R_o\}$. For the unifications X_T resp. R_T must hold $X_T = X_I \cup X_Z \cup X_O \neq \emptyset$ and $R_T = R_I \cup R_{IZ} \cup R_Z \cup R_{ZO} \cup R_O \cup R_{IO} \neq \emptyset$.

The first technique is the most versatile one with the least restrictions; it even supports non-classical formalisma like cellular automata or recursive model equations etc. However it is less convenient, since it requires programming. The third is most attractive, since it allows to support hierarchical model structure editing: a node representing a subsystem can be collapsed or expanded into a separate window containing again the tuple of the subsystem or the supersystem.

Independent of the three ways the user chooses to work with, RAMSES applies always the same basic technique to manage models and model objects. They are installed or instantiated by calling declaration procedures, which allocate a memory block in the heap to store the object together with its associated values plus attributes. However, the actual objects like state variables or parameters remain fully in the scope of the user model definition. If the modeler uses the first modeling technique via the client interface, this scope corresponds exactly to the scope concept in Modula-2. For instance state variables may be part of any data structure and may be used in any type of statement sequences such as e.g. recursion etc.. Moreover, models and model objects can be removed (undeclared) any time from the model and model object base or be edited in any way. All these functions are realized according to the same principle. Fig. 3 shows an excerpt from the client interface for all procedures needed to manage the model objects of the type *State Variables*.

2.2 The Experiment Definition Session

It consists of the specification of an experimental frame (ZEIGLER, 1976, 1979) plus its association with a particular mathematical model. What results is a simulation model which contains no longer any missing concrete values necessary to fully specify e.g. an initial value problem. Furthermore a time domain or spatial domain for which the model behavior is of interest, and any other needed parameters such as integration method, maximum local error (absolute and relative) etc., have also been defined.

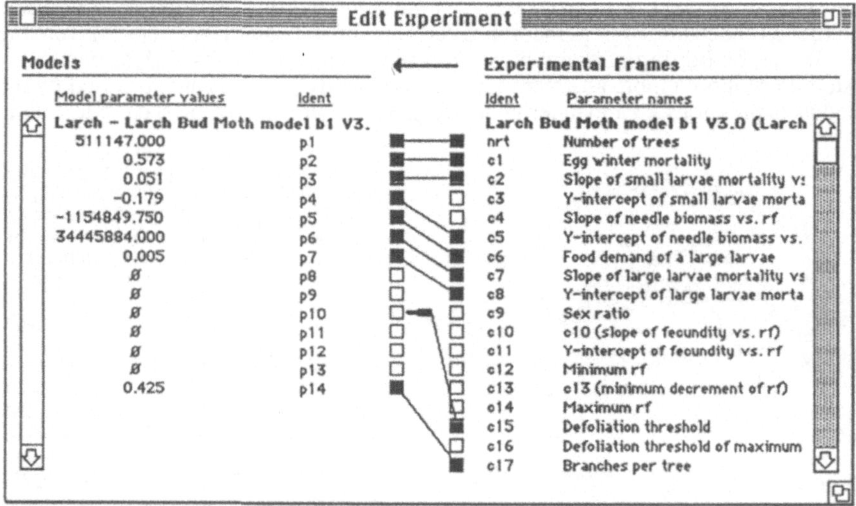

Fig. 5: Editing an experiment by combining a mathematical model with an experimental frame during an experiment definition session. A set of parameters defined during the modeling session have partly no defined values (ø). Specific values with a concrete meaning are contained in the experimental frame. The user can now assign them to the parameters of the mathematical equations via connecting items from the experimental frame with parameters in the mathematical model similar to establishing a plug connection. The direction of the top arrow can be toggled and determines the direction of the assignment.

This session is user oriented, since the thinking in terms of an experimental frame or a particular mathematical model is much easier than to think always in terms of the ever exploding numbers of combinations of the two (Fig. 5). For instance is it much easier to keep track of a number of data sets, each with its particular time domain, and the number of alternative models which could at least hypothetically be applied. If mathematical models can be freely combined with experimental frames, a technique which has been proposed e.g. by ÖREN (1982), the definition and combination of two members of these classes are easier to manage.

There remains the issue of compatibility between a mathematical model and an experimental frame to be resolved. In RAMSES a mathematical model and an experimental frame are called compatible if their order is the same and the dimensions of the model parameter, input, plus output vectors are the same. RAMSES accepts an experiment or a simulation model for a simulation session only if the combination of model and experimental frame are fully compatible (s.a. Fig. 1).

2.2 The Simulation Session

The purpose of a simulation session is to produce model behavior in time or space or both by solving a simulation model over a particular domain of the independent variables, normally time. This is called an elementary simulation run. A complex simulation experiment formed from several elementary runs is called a structured simulation run. RAMSES allows either to directly execute an elementary or a structured simulation run, each an arbitrary number of times (Fig. 6: k, n).

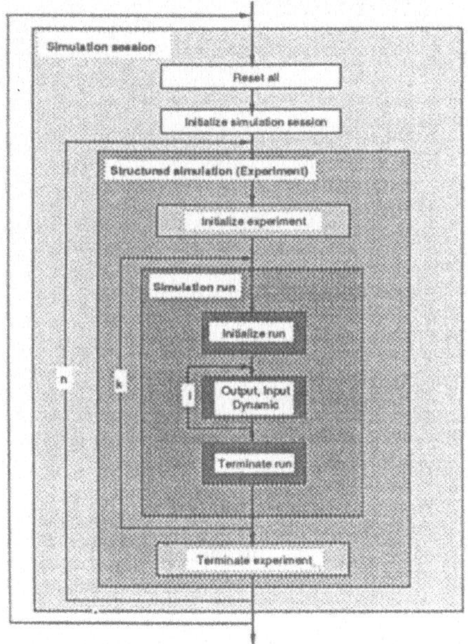

Fig. 6: Structure and flow chart of a RAMSES simulation session. The simulationist may execute directly an arbitrary number n of structured or of elementary simulation runs. A structured simulation («experiment») consists of a programmed number k of elementary simulation runs. Every simulation run consists of an *Initialize run*, dynamic (includes the sections *Output*, *Input*, plus *Dynamic* s.str.), and *Terminate run*. The dynamic section is executed according to the chosen time step and simulation time an arbitrary number of times i. Dark grey shaded areas: mandatory, i.e. must be defined by every model definition program; light grey shaded areas: optional and under full control of the modeler.

RAMSES automatically assigns the initial value i to the state variable x at the begin of every simulation run, and the value p is assigned to the model parameter c at the beginning of the simulation session or after any interactive change (Fig. 2). RAMSES maintains also the current values of state variables, parameters, and monitorable variables and remembers their initially specified values for eventual restoration (so-called resetting). During simulation experiments the unknown values, which the monitorable variable mv may obtain, are written on the stash file, tabulated in a table, or displayed in graphs (Fig. 7). The latter is only the case if the values fit within a particular range of interest as specified by the modeler; otherwise they will be clipped.

Often at the begin and at the end of an elementary simulation run particular actions must be taken, e.g. to initialize states or compute and record final results. The simulation environment of RAM-

SES provides facilities to install such procedures (Fig. 6: *Initialize run, Terminate run*). Furthermore in structured simulations, which typically execute several elementary runs, there is the possibility to initialize and terminate the whole experiment. This may be particularly useful in the case of a stochastic model, where e.g. means, variances and other data collections ought to be calculated from many runs (Fig. 6: *Initialize experiment, Terminate experiment*). Finally the whole simulation session may also require an initialization; this is provided in the simulation environment by allowing for the installation of an initialization procedure.

A basic purpose of interactive simulation is to allow for the easy monitoring of the simulation results by focussing on particular sensitive or otherwise interesting results in order to be able to temporarily halt or even interrupt the simulation. This is most useful in early stages of model development, where the user may want to abort a particular run quickly, once its main characteristics have become apparent.

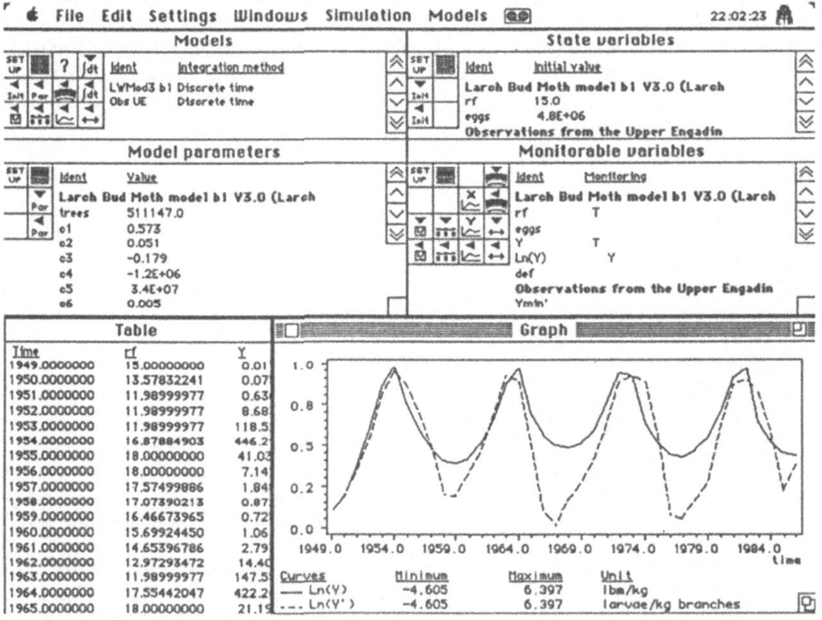

Fig. 7: Typical screen of a RAMSES simulation session. The model shown is the larch bud moth system described in the text, in the graph its simulated behavior is compared with observations.

For user convenience the simulation session of RAMSES allows also to change currently used parameter values, scaling values for monitorable variables, initial values etc. This avoids having to quit completely the simulation session for a run just testing a particular parameter value. This can be considered as working in the simulation session just with a scratch copy of the whole model and its associated values, whereby the values specified within the experiment definition session can any time be resumed by a reset. Note that the reverse is also possible, i.e. they can be copied to the model definition and be stored as an experimental frame. The arising need for selection of models or model objects, e.g. to change the value of a specific parameter, is provided by the so-called IO-windows. Fig. 7 shows a typical screen of a RAMSES simulation session with the menu bar plus its menus, four IO-windows, the table, and the graph window.

Since the user's monitoring is always restricted to a few variables, there arises the need, in particular for long lasting simulations, to save the results for a later inspection and exploration. RAM-

SES adopts the technique of the so-called stash file to temporarily save all potentially interesting results. The data are written according to a formally defined LL(1) syntax for easier scanning and parsing by the postanalysis program. This approach allows also to completely omit the control of the simulation by the end-user interface except for the initiation of a so-called structured simulation run (since it may consist of an arbitrary number of elementary runs it can be considered as a freely programmable, hence versatile form of batch-processing). For batch-processing the client interface can be used to program a structured simulation run, which may then even be executed on another machine, e.g. a super-computer, another currently unused workstation, a host serving as a simulation server, or a set of transputers within the workstation.

2.3 The Postanalysis Session

In the postanalysis session simulation results, previously computed during a simulation session, can be analyzed without having to recompute any model behavior. It serves the interactive exploration of model behavior even in cases where an interactive simulation of the results would last too long. Surprising results, e.g. in behavior of indicator variables, may be traced back to the temporal behavior of other, internal system variables, thus often allowing for a better understanding of the system mechanisms. During such an exploratory data analysis of the simulation results, the visualization and the interactive testing of ad-hoc formulated hypothesis play an important role.

In addition to the simulation results the stash file contains detailed information on the global parameters of the simulation environment, on the model and on its model objects which have been used to produce the data. This allows the postanalysis session to install in RAMSES a model and its objects exactly as they have existed during the simulation session, except that the postanalysis model (IORO level 1) is not used to produce the model behavior but to read the already computed behavior from the stash file and display it for the user by using the ordinary RAMSES monitoring mechanisms from the simulation environment. The latter encompasses graphical representations (scattergrams, line charts with optional error bars, 3-dimensional grid plots, contour maps etc.) and the tabular display of numerical values. This approach has not only the advantage of reducing the implementation work, but also to be easier to learn by the user, since every RAMSES user will be familiar with at least the simulation environment (Fig. 7). Whether he/she actually uses the simulation or the post-analysis session to analyze the behavior of a complex model will be of minor importance.

To grant a virtually unlimited data exchange between the simulation session and the postanalysis session, the secondary storage medium containing the stash file is used to store the simulation results. For an efficient access during the postanalysis session the organization of the stash file is crucial. We chose a formally defined LL(1) syntax, but otherwise it is a sequential text (ASCII) file. The latter is a compromise in terms of efficiency and the need for data exchange with other programs than just the postanalysis session software. For instance spread-sheet applications, statistical packages, or document processors can also open the stash file.

More details on the EBNF describing the stash file syntax, the mathematics, and the internal structure of the postanalysis session can be found in GYALISTRAS (1990).

3   An Application Example from Population Ecology

Population systems are used in many areas of the environmental sciences: for instance in ecotoxicological studies or in pest management. The larch bud moth system has been studied intensively for now over four decades and serves as a useful example to illustrate and evaluate RAMSES.

3.1 The population cycles of larch bud moth

Larch bud moth, *Zeiraphera diniana* GN. (*Lep., Tortricidae*), is a univoltine forest defoliating insect, which periodically attacks larch trees in the European Alps between 1700 to 2000 m a.s.l. (BALTENSWEILER & FISCHLIN, 1988). The mean of the cycle length is 9.2 years and the average amplitude amounts to 226.9 larvae/kg larch branches (FISCHLIN, 1982; BALTENSWEILER & FISCHLIN, 1988). The ecological mechanisms causing these population cycles are only partly understood and there exists a range of competing hypothesis postulated by numerous authors (FISCHLIN, 1980; BALTENSWEILER & FISCHLIN, 1988; CLARK *et al.*1967; WILSON, 1975; ANDERSON & MAY, 1982; MAY, 1981). We evaluated and tested all hypothesis first for their plausibility by comparing their assumptions and statements with all known ecological facts and data from the 40 year study and secondly their capability to predict quantitatively the observed system behavior. The latter has resulted in a family of mathematical models (FISCHLIN & BALTENSWEILER, 1979; FISCHLIN, 1982). Members of this model family have been implemented using experimental versions of RAMSES. The simulation session has been realized by embeding the simulation environment ModelWorks[1] (ULRICH, 1987; FISCHLIN *et al.*, 1990) in RAMSES.

3.2 Using RAMSES to model the Larch Bud Moth System

Several aspects of RAMSES, such as modular modelling, were important during the modeling and the simulation of the larch bud moth system. Members of the model family were modeled as submodels in form of separate modules (level 3 System) and measurements were implemented as parallel data-models (level 1 IORO) (Fig. 8). This allowed to compare simulated results with measured and observed time series (Fig. 7).

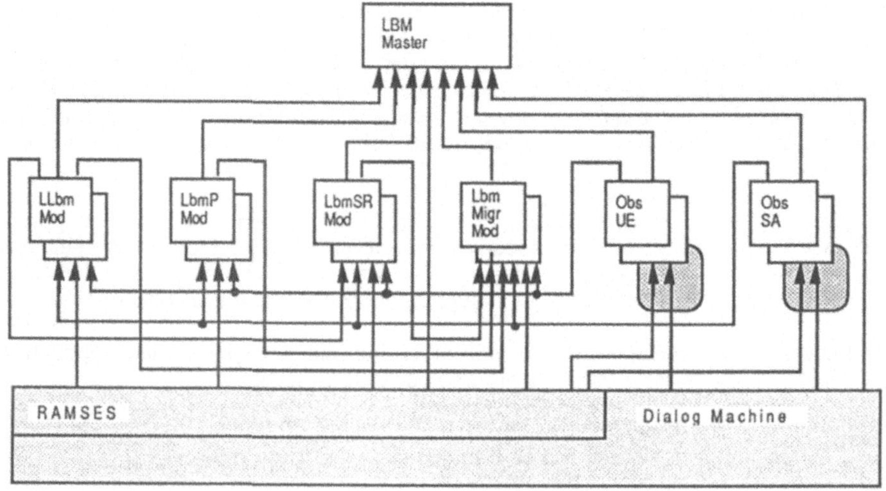

Fig. 8: Module structure of the model definition program implementing the family of population models of the larch bud moth system. ☐ (on top) - program module; ▤ - definition modules in front of implementation modules; ➤ - imports; ◯ - data files; ▭ (on bottom) - module libraries.

---

[1]ModelWorks may be used independently from other RAMSES tools, i.e. the modelling and postanalysis session.

The modeling was done with the first technique via the client interface. The master module, the program module *LBMMaster*, combines all modules to a model definition program (Fig. 8). Depending on the current user needs the corresponding submodels can be installed or deinstalled in the RAMSES model base. For instance note the following two submodels: The first submodel, module *LLbmMod*, describes the ecological interaction of the host plant larch (*Larix decidua* MILLER) with the herbivorous insect larch bud moth (*Z. diniana*). The second, parallel data-submodel, module *ObsUE*, mimics the real system by using field data from the Upper Engadine valley in Switzerland, which were sampled from 1949 till the present (BALTENSWEILER & FISCHLIN, 1988). At the begin of a simulation session the parallel data-submodel simply reads the observations stored in a data file into the memory and assigns the measured values during simulations to a monitorable variable, which the simulationist can compare with simulated results (Fig. 7).

4 Discussion

4.1 Strengths and Weaknesses of Interactive Modeling and Simulation

Modeling and simulation of ill-defined systems substantially benefits from interactivity, since iterative system structure identification and systems behavior analysis are typical for these systems, especially also in the field of environmental systems. Interactively connecting and disconnecting submodels, activating or deactivating models are powerful while navigating through a model family which is only partially explored. Systematic model combinations are only possible thanks to such an approach.

The more advanced a study becomes, the more modeling will focus just on a few subsystems or processes. The strengths of interactivity is during the model development phase. Once the structure of a model and its equations become relatively fixed, interactive modeling looses its attractivity and a rather batch-oriented simulation study becomes predominant. The two interfaces of RAMSES, the interactive end-user and the batch-oriented client interface, support these varying needs during the course of a study well.

Especially in the modeling of ill-defined systems there becomes another disadvantage of interactive modeling apparent. It gets easily difficult to keep track of all involved objects and to keep a good overview. This is particularly true for the second technique (s.a. modeling session) where the modeler uses no relational graph editor and works directly on the models and model objects. Therefore hierarchical modeling becomes a necessity; unfortunately RAMSES does currently not support real hierarchical modeling. The latter could be realized by an explicit subordination of submodels which would not only be reflected in the module structure but also would be recognized by RAMSES. Implementing hierarchical modeling such that it starts always from a single root model, would also offer the advantage of steering the user towards a top-down model design.

Disadvantages of interactive simulation become also particularly evident if computational needs are big. The more complex a model, the larger the conflict with the patience of the user, because.the simulationist can only watch a more and more limited selection of the results. For complex systems the interactive simulation degenerates too often to a monitoring of a few, uninteresting indicators, the actual results remaining hidden. Not only does this call for the separation of the simulation and the post-analysis as realized in RAMSES, but also offers the advantage of a transparent use of computing servers, such as a simulation server or a super computer. RAMSES' session concept appears to optimally support such solutions; for instance, except for restrictions inherent to the programming languages available on the Cray-XMP, did we encounter no fundamental difficulties when experimenting with a super-computer implementation of the RAMSES simulation environment for structured simulation runs. Except for the simulation session, which may be transferred to a simulation server, all other sessions remain on the workstation and can there profit from the available interactivity. This unites and allows to have the best of both worlds, the world of the batch-oriented hosts, strong in number crunching, and that of the workstations, strong in interactive use.

Some rather technical, implementation problems shall also be mentioned: Against common expectations, continuous consistency checking during interactive editing is often impossible, since users violate consistency while editing all the time: E.g. a first portion of an entry may already have been typed, a second not; a completely normal, but inconsistent situation, since some terms may already have been referenced, but are not yet present. Due to the difficulties to have a good overview when working with ill-defined systems, however, it would be crucial to have the computer perform consistency checks. Thanks to the state diagram of RAMSES' session concept (Fig. 1), a solution could be found, namely if the model definition and the experimental frame are not compatible, RAMSES rejects the transition from the modeling or experiment definition session into the simulation session. Inasmuch expert systems could really help to support the modeling process and consistency testing is currently not well understood. We have experimented with little expert systems, but only achieved disappointing results; just complete novices could really profit from the available advices.

Compared with previous implementations of the presented model family made with more cumbersome simulation tools (FISCHLIN & BALTENSWEILER, 1979) or commercially available simulation software (Marr, 1989) indicate that RAMSES is not only efficient and elegant, but also tends to support the user in such a way, that results are obtained in a more systematic manner. This has several reasons, but among the more important ones is certainly the choice of the programming language Modula-2.

Thanks to Modula-2, RAMSES supports elegantly a modular implementation of the members of a model family. Each submodel forms a self-contained unit, typically a Modula-2 module, yet they may exchange informations by IO-links. E.g. in the presented example relationships between submodels, such as the computation of initial states from observations, could be implemented with output to input coupling (Fig. 8). All model equations, no matter how complicated or of which form, could also be elegantly implemented; e.g. the migration model uses recursion to model spatial orientation and moth flying behavior (FISCHLIN, 1982).

The power of the programming language used to implement RAMSES has shown to be most suitable to achieve our originally set goals, because it supported modular and structured programming and helped to implement efficiently the software tools needed for the interactive exploration of model behavior. Moreover the RAMSES user profits as well, e.g. a sensitivity analysis could be implemented by programming a few lines of code and installing the executing procedure as a structured simulation run; the testing of pesticide applications or the heuristical design of a management scheme were also straightforward and easy to realize.

Much effort had to go into the design of a software system which is attractive for the beginner as well as open enough for the specialist who is interested in advanced techniques. This resulted in a basic symmetry between the end-user and the client interface. However, in case of conflict we opted rather for an optimally open system structure, than for the ease of use for the beginner, who is more likely to use just the end-user interface. These design goals are reflected e.g. in the maximal functionality of the client interface. Building on a solid model-based foundation and adding later front-end modules for easier use of the interactive end-user interface by beginners seemed to be an appropriate development strategy. It offered also right from the begin the sophisticated user optimal access to RAMSES. Offering both, the end-user and the client interfaces simultaneously has also the advantage that the user may smoothly transgress from major reliance on the end-user interface to the client interface by implementing step-wise more and more elaborate tasks, such as statistical analysis of data collection from many runs or parameter identification techniques etc. This way the risk to loose investments in complex model implementations can be kept minimal.

RAMSES demonstrated also that modeling and simulation software have to be tightly coupled. In particular interactive modeling and simulation require a common kernel managing dynamically models, model objects, and associated values in a model base accessible by both the modelling and the simulation sessions. The heap technique we adopted to achieve this behavior was surprisingly efficient: Compared with more traditional simulation software architectures the efficiency losses in computational speed were on the average as little as 10-20% (s.a. ULRICH, 1987).

An important disadvantage of RAMSES is of course the rather big software development effort. However, it appears to be unavoidable, since the new concepts required a substantial rewriting or at least restructuring of existing simulation software. For instance modular modeling required to implement integration algorithms anew. This is because the IO-links among submodels must be calculated before the numerical integration of differential equations. The user interface development is often underestimated and forms the dominant fraction of a rigorous, user-friendly interactive program. However, designing a good man-machine interface can be crucial for the final quality of a software. Only thanks to the *Dialog Machine* (FISCHLIN, 1986) could these efforts be minimized and the port of the simulation environment ModelWorks to another target machine, given a *Dialog Machine* was available, could be accomplished in only a few days labor.

Finally either the design as well as the use of the RAMSES tools were facilitated by the modelling theory. This is because the software architecture could be based on a model base kernel. It provides the installation (or deinstallation) of systems theoretically well understood objects, such as state variables or model parameters etc. Furthermore the user is assisted in or even guided to adopt a structured modelling approach. RAMSES recognizes and hence maintains values only if they conform to the mathematically defined concepts of the standard model formalisms SM, DESS, or DEVS. Yet this does not prevent the modeler from using rather unconventional formalisms such as recursion.

4.2 Perspectives

Since workstations and personal computers become increasingly popular and ever more powerful, interactive modeling and simulation appear to gain rapidly in importance for the analysis of ill-defined systems. This is particularly true in the field of environmental systems, where much remains to be done. Although heuristic approaches still dominate the field, and not all concepts are yet well understood, it seems that progress can be achieved along the described lines. In order to improve the health of our environment it is hoped that the modeling of environmental systems will not just remain a highly specialized activity reserved to a few specialists, but will actually contribute to the better understanding and solving of environmental problems.

5   References

ANDERSON, R.M. & MAY, R.M., 1980. *Infectious diseases and population cycles of forest insects.* Science, **210**: 658-661.

ANONYMOUS, 1988. *Catalog of simulation software.* Simulation **51** (4): 136-156.

BALTENSWEILER, W. & FISCHLIN, A., 1988. *The larch bud moth in the Alps.* In: Berryman, A.A. (ed.), *Dynamics of forest insect populations: patterns, causes, implications.* New York a.o.: Plenum Publishing Corporation, p. 331-351.

BROOKS, F.P. JR., 1979. *The mythical man-month - Essays on Software Engineering.* Reading, M. a.o.: Addison-Wesley Publ. Comp., 195pp.

CELLIER, F.E., 1975. *Continuous-systems simulation by use of digital computers: a state-of-the-art-survey and prospectives for development.* In: Hamza, H. (ed.) Proc. of the international Symposium SIMULATION'75, Zurich, Switzerland, June 1975, (To be ordered from:) Acta Press, P.O. Box 354, CH-8053 Zurich, p. 18-25.

CELLIER, F. E., 1979. *Combined Continuous/Discrete System Simulation by Use of Digital Computers: Techniques and Tools.* Diss. ETH No 6483.

CELLIER, F.E., 1984a. *Simulation software: Today and tomorrow.* Zeitschr. Schweiz. Ges. Automatik, **4**: 7-22.

CELLIER, F.E., 1984b. *How to enhance the robustness of simulation software.* In: Ören, T. I., Zeigler, B. P., Elzas, M. S.(eds), *Simulation and Model-Based Methodologies: An Integrative View*, 651pp., Springer, Berlin a.o., p. 519-537.

CELLIER, F.E. (ED.), 1982. *Progress in Modeling and Simulation.* Academic Press, London a.o., 466pp.

CELLIER, F.E. & FISCHLIN, A., 1982. *Computer-assisted modeling of ill-defined systems.* In: In: Trappl, R., Klir, G.J., Pichler& F.R. (eds.), *Progress in cybernetics and systems research. Vol. VIII: General systems methodology, mathematical systems theory, fuzzy sets.* Proc. of the 5th European Meeting on Cybernetics and Systems Research, University of Vienna, Austria, April 8-11, 1980, p. 417-429.

CLARK, L., GEIER, P., HUGHES, R. & MORRIS, R.F., 1967. *The ecology of insect populations in theory and practice.* London: Methuen, 232pp.

DAVIDSON, R.S. & CLYMER, A.B., 1966. *The desirability and applicability of simulating ecosystems.* Ann. N.Y. Acad. Sci., **128**: 790-794.

FISCHLIN, A., 1982. *Analyse eines Wald-Insekten-Systems: Der subalpine Lärchen-Arvenwald und der graue Lärchenwickler Zeiraphera diniana Gn. (Lep., Tortricidae)..* Diss. Eidg. Tech. Hochsch. Zürich No. 6977, 294pp.

FISCHLIN, A., 1986. *Simplifying the usage and the programming of modern workstations with Modula-2: The 'Dialog Machine'.* Internal report, Project-Centre IDA/CELTIA, Swiss Federal Institute of Technology Zürich (ETHZ), Zürich, Switzerland, 13pp.

FISCHLIN, A. & BALTENSWEILER, W., 1979. *Systems analysis of the larch bud moth system. Part I: the larch-larch bud moth relationship.* Mitt. Schweiz. Ent. Ges., **52**: 273-289.

FISCHLIN, A. & ULRICH, M., 1987. *Interaktive Simulation schlecht-definierter Systeme auf modernen Arbeitsplatzrechnern: die Modula-2 Simulationssoftware ModelWorks.* Proceedings, Treffen des GI/ASIM-Arbeitskreises 4.5.2.1 *Simulation in Biologie und Medizin*, February, 27-28, 1987, Vieweg, Braunschweig, p. 1-9.

FISCHLIN, A., ROTH, O., GYALISTRAS, D., ULRICH, M. & NEMECEK, T., 1990. *ModelWorks - An interactive simulation environment for personal computers and workstations.* Manual for Version 2.0. Systems Ecology Group, Internal Report 8, Swiss Federal Institute of Technology Zürich, Switzerland, 182pp.

GUTKNECHT, J., 1983. *System programming in Modula-2: mouse and bit-mapped display.* Internal report No. 56, Department of computer science, Swiss Federal Institute of Technology, Zürich, Switzerland, 58pp.

GYALISTRAS, D., 1990 (In prep.). *Interactive post-analysis of simulation results on a worksta- tion.* Systems Ecology Group, Swiss Federal Institute of Technology, Zürich, Switzerland.

HOLLING, C.S., 1964. *The analysis of complex population processes.* Can. Entomol., **96**: 335-347.

INNIS, G.S., 1972. *Simulation of ill-defined systems: Some problems and progress.* Simulation, **19**: 33-36.

KARPLUS, W.J., 1976. *The spectrum of mathematical modeling and system simulation.* In: Dekker, L. (ed.), Proc. of the 8th AICA Congress on Simulation of Systems, Delft, The Netherlands. North-Holland Publ. Co., p. 5-13.

KLIR, G.J., 1979. *Computer-aided system modeling.* In: Halfon, E. (ed.), *Theoretical Systems Ecology*, Academic Press, New York, p. 291-323.

KREUTZER, W., 1986. *System simulation: programming styles and languages.* Sydney a.o.: Addison-Wesley, 366pp.

MARR, G.R., 1989. *Better interaction with ACSL simulation programs.* In: Allen, R.W. (ed.), *Modeling and simulation on microcomputers*, 1989, The Society for Computer Simulation International, p. 69-73.

MAY, R.M. (ED.), 1981. *Theoretical ecology. Principles and applications.* Blackwell Scientific Publications, Osney Mead, Oxford, 2nd ed., 489pp.

NIEVERGELT, J. & WEYDERT, J., 1980. *Sites, modes and trails: Telling the user of an interactive system where he is, what he can do, and how to get to places.* In: R. A. GUEDJ et al. (eds.), *Methodology of Interaction*, North-Holland, Amsterdam, p.327-338.

NIEVERGELT, J. & VENTURA, A., 1984. *Die Gestaltung interaktiver Programme.* B. G. Teubner Stuttgart, 124pp.

ÖREN, T.I., 1982. *Computer-aided modeling systems.* In: Cellier, F.E. (ed.), *Progress in Modeling and Simulation*, Academic Press, London a.o., p.189-203.

ÖREN, T.I., 1984. *GEST - A Modeling and Simulation Language Based on System Theoretic Concepts.* In: Ören, T.I., Zeigler, B. P., Elzas, M.S.(eds.): *Simulation and Model-Based Methodologies: An Integrative View*, 651pp., Springer, Berlin a.o., p. 281-335.

ULRICH, M., 1987. *ModelWorks. An interactive Modula-2 simulation environment.* Post-graduate thesis, Project-Centre IDA, Swiss Federal Institute of Technology Zürich (ETHZ), Switzerland, 53pp.

VANCSO-POLACSEK, K., 1990. *Theory and practice of computer assisted simulation and modeling on professional workstations.* Diss. ETH No., 109pp.

VANCSO, K., FISCHLIN, A. & SCHAUFELBERGER, W., 1987. *Die Entwicklung interaktiver Modellierungs- und Simulationssoftware mit Modula-2.* In: Halin, J. (ed.), *Simulationstechnik*, Informatik-Fachberichte 150, Springer, Berlin, p. 239-249.

WILSON, E.O., 1975. *Sociobiology the new synthesis.* Belknap Press of Harvard University Press, Cambridge a.o., 697pp.

WIRTH, N., 1985. *Programming in Modula-2, Third, Corrected Edition.* Springer-Verlag, Berlin a.o., 202pp.

WIRTH, N., 1986. *Algorithms & data structures.* Prentice/Hall International, Inc., 288pp.

WIRTH, N., 1988. *The programming language Oberon.* Software - Practice and Experience, **18**: 171-90.

WIRTH, N., 1989a. *From Modula to Oberon.* Institut für Informatik ETHZ, Swiss Federal Institute of Technology Zürich, Switzerland, internal report 111: 3-10.

WIRTH, N., 1989b. *The programming language Oberon (revised report).* Institut für Informatik ETHZ, Swiss Federal Institute of Technology Zürich, Switzerland, internal report 111: 11-28.

WYMORE, A.W., 1984. *Theory of Systems.* In: Vick, C. R., Ramamoorthy, C. V.(eds.): *Handbook of Software Engineering*, Van Nostrand Reinhold Company, New York.

ZEIGLER, B.P., 1976. *Theory of modeling and simulation.* Wiley, New York a.o., 435pp.

ZEIGLER, B.P., 1979. *Multilevel multiformalism modeling: an ecosystem example.* In: Halfon, E. (ed.), *Theoretical Systems Ecology*, Academic Press, New York, p. 17-54.

ZEIGLER, B.P., 1984. *System theoretic foundations of modeling and simulation.* In: Ören, T.I., Zeigler, B. P., Elzas, M.S.(eds.), *Simulation and Model-Based Methodologies: An Integrative View*, 651pp., Springer, Berlin a.o.

Acknowledgements: I thank Dr. Olivier Roth for the reading of the manuscript, the sincere discussions, and many valuable suggestions for improvements. Many thanks are due to Markus Ulrich, especially for his original design of the simulation environment ModelWorks. I thank Dr. Olivier Roth, and Dimitrios Gyalistras for their substantial contributions to new ideas and the implementation of RAMSES sessions. Particular thanks also to the first users, namely Thomas Nemecek and also Harald Bugmann, who have been willing to serve as «guinea-pigs» for the testing and exploring of the numerous concepts and their validity in their daily research. Finally I owe heartily thanks to Prof. Dr. W. Schaufelberger, not only for his substantial support, but also for his unceasing encouragement, which made this research and the associated development only possible.

Methodologische Aspekte bei komplexen Simulationsaufgaben am Beispiel der Dynamik des Waldsterbens bearbeitet mit dem Simulationssystem SIMPLEX–II

Jochen Wittmann

Universität Erlangen–Nürnberg,
Institut für mathematische Maschinen und Datenverarbeitung IV,
Martensstraße 1, D-8520 Erlangen

Übersicht

Um die Dynamik komplexer Zusammenhänge in unserer Umwelt zu verstehen, bildet man sich Modelle. Bei wachsender Komplexität der Modelle wird man zu einer methodologisch sauberen Modellbeschreibung gezwungen. Der Artikel will nun die wesentlichen Anforderungen, die sich bei der Arbeit mit größeren Modellen stellen, an einem Beispiel aufzeigen. Diesen Anforderungen stellt er die entsprechenden Lösungsideen im Simulationssystem SIMPLEX–II gegenüber. Am Beispiel des Waldsterbens kann schließlich die Verwirklichung der Konzepte demonstriert werden.

1 Vorgehensweise bei der Modellerstellung — Anforderungen an die Modellbeschreibung

Bedingt durch eine zunehmend ganzheitliche Sicht unserer Umwelt stehen wir oftmals vor dem Problem, stark vernetzte und rückgekoppelte Wirkbeziehungen zu analysieren. Wir müssen feststellen, daß unsere Vorstellungskraft angesichts dieser Vorgänge stark beschränkt ist und bislang bevorzugt lineare Zusammenhänge erfaßt. Daher erscheint es sinnvoll, Expertenwissen aus unterschiedlichen Fachdisziplinen zusammenzutragen und in komplexen Modellen zu verknüpfen.

Dabei erfolgt die Entwicklung der Modelle jedoch nicht in einem einzigen Schritt, vielmehr sind die Modellbauer zu einer strukturierten Vorgehensweise gezwungen. Man kennt im wesentlichen zwei idealtypische Ansätze, größere Probleme anzugehen, nämlich den Top–Down– und den Bottom–Up–Ansatz. Diese Methoden sollen in Bezug zur Modellerstellung kurz erläutert werden, und zwar mit dem Ziel, aus der Beschreibung die Anforderungen an eine Experimentierumgebung abzuleiten.

Als Beispiel soll dabei die Analyse des Waldsterbens dienen. Dabei stützen wir uns auf ein Modell, das von Bossel u.a. vorgestellt wurde. Zweck dieser Studie war es, den Schadensverlauf an einem Waldstück nachzubilden. Einerseits um qualitative Aussagen über die Auswirkungen der Schadstoffeinwirkung zu erhalten und andererseits um Maßnahmen zur Schadensreduzierung bewerten zu können.

Top–Down–Ansatz:

Hierbei betrachtet man das zu simulierende System zunächst als Ganzes. Man untersucht die globalen Zusammenhänge und interessiert sich für diejenigen Zielgrößen, die zur Beantwortung der Fragestellung unmittelbar notwendig sind.

Ein derart grobes Modell stößt jedoch bald an die Grenzen seiner Aussagekraft. Eine differenziertere Sichtweise führt zur Aufspaltung des Gesamtsystems in kleinere Teilsysteme. Im Beispiel sind dies die Untersysteme Baum (mit Laub und Wurzeln), Wasserhaushalt im Boden, Bodenchemie sowie die im Boden stattfindenden Mineralisierungsvorgänge. Diese Untersysteme lassen sich voneinander unabhängig modellieren. Sie benötigen nur sehr wenige Information aus ihrer Umwelt; z.B. wird das Wachstum des Baumes durch den pH–Wert des Bodens oder das Ausmaß der Düngung beeinflußt. Der Wachstumsvorgang selbst kann jedoch in der Systemkomponente Baum unabhängig von den übrigen Teilsystemen nachgebildet werden.

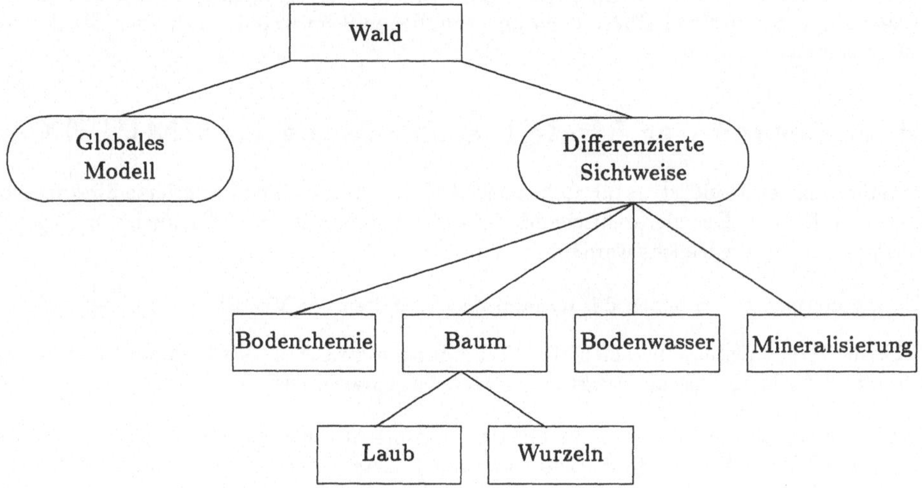

Abb. 1.1: Die Zerlegung des Systems in seine Teilsysteme

Bottom–Up–Ansatz:

Dieser Ansatz beginnt mit der Untersuchung von Teilsystemen. In parallel arbeitenden Arbeitsgruppen können die entsprechenden Teilmodelle entwickelt werden. Der Informationsfluß zwischen den Teilmodellen erfolgt über vorher abgestimmte Schnittstellen. An diesen Schnittstellen verbindet man schließlich die einzelnen Teile und erhält ein Modell des Gesamtsystems. Nun kann man die Experimente durchführen, die zur Beantwortung der Fragestellung notwendig sind.

Gemeinsamer Vorteil der beiden Ansätze ist die Zerlegung der Aufgabenstellung in Teilfragen. Diese können parallel und insbesonders aber von den jeweiligen Experten des jeweiligen Teilgebietes bearbeitet werden.

Für die Modellerstellung ergeben sich folgende Anforderungen:

- Mehrere Versionen zu einem Modell sollen möglich sein.

  Das ergibt sich aus der zunehmenden Differenzierung beim Top–Down–Ansatz. Sowohl die globale als auch die differenzierte Modellierung sollte dokumentiert werden. Die Versionen unterscheiden sich im allgemeinen in ihrer Struktur und sollten zum Experimentieren wahlweise verwendet werden können.

- Hierarchischer Modellaufbau

  Eine mehrfach geschachtelte Systemstruktur sollte in der Modellbeschreibung abgebildet werden können. Für die Kommunkation zwischen den Modulen muß ein Informationsfluß auf zweierlei Art definiert sein:

  a) innerhalb einer Hierarchieebene (z.B. pH–Wert von der Bodenchemie zum Baum)

  b) über Hierarchiestufen hinweg (z.B. sollte die Laubmenge im Gesamtmodell verfügbar sein)

- Die Teilmodelle müssen selbständig lauffähig sein.

  Um ein sinnvolles Arbeiten mit den Teilmodellen zu erlauben — besonders ist dabei an eine Validierung der einzelnen Komponenten gedacht — muß die Funktionalität der Experimentierumgebung bereits auf Komponentenebene vollständig verfügbar sein. Auf eine sinnvolle Behandlung der in dieser Phase noch unverbundenen Schnittstellen zwischen den Modulen ist zu achten.

2 Realisierung der Modellbeschreibung in SIMPLEX–II

Zur Formulierung der Modellstruktur steht in SIMPLEX–II eine objektorientierte Spezifikationssprache zur Verfügung. Der hierarchische Modellaufbau kann mit zwei verschiedenen Typen von Modellkomponenten verwirklicht werden:

- Basiskomponenten, in denen das dynamische Verhalten des Modells beschrieben ist.

- Höhere Komponenten, in denen die Verbindungen zwischen Basiskomponenten definiert sind, und die hierarchische Struktur des Modells repräsentiert ist.

Beispiele für Basiskomponenten sind in unserem Beispiel die Komponenten Baum, BoWasser, BoChemie und MineralN (für den Stickstoffhaushalt). Das Gesamtmodell heißt 'Wald' und beinhaltet die Verbindungen der Basiskomponenten untereinander.

Die Verbindungen bilden einen gerichteten Informationsfluß ab. Auf den Wert einer Modellgröße in einer bestimmten Komponente kann von allen anderen Komponenten, egal welcher Hierarchiestufe, lesend zugegriffen werden. Die Verbindung muß allerdings deklariert werden, sie steht nicht automatisch zur Verfügung. Es handelt sich um das 'Glaskastenprinzip'. Dieses Prinzip gewährleistet den unabhängigen Lauf von Teilmodellen. Durch das Verschalten zum Gesamtmodell kommen nämlich nur Leseoperationen hinzu, die die Dynamik der Teilkomponente jedoch nicht beeinflussen.

Zur Veranschaulichung dient eine Strukturskizze des Gesamtmodells (Abb. 2.1). Die Komponente, in der der Wert einer Modellgröße bestimmt wird, erhält an dem Verbindungspfeil ein ausgefülltes Kästchen. Die Komponente, die lesend zugreift, erhält ein leeres Kästchen.

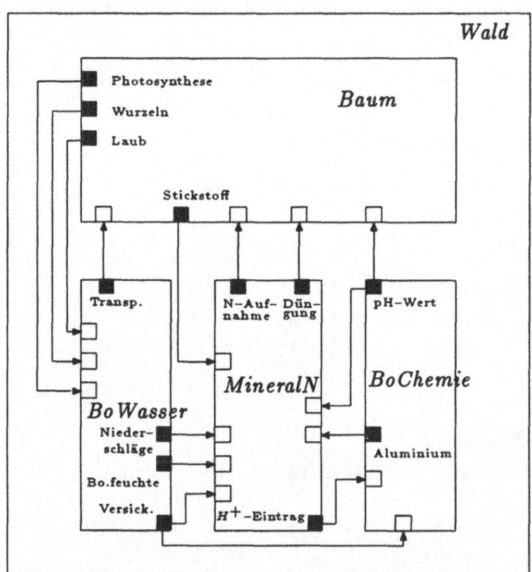

Abb. 2.1: Das Gesamtmodell 'Wald'

Diese Strukturskizze wird nun in den Code der Modellbeschreibungssprache umgesetzt:

```
HIGH LEVEL COMPONENT Wald
  SUBCOMPONENTS
    Baum,
    BoWasser,
    MineralN,
    BoChemie


COMPONENT CONNECTIONS
    BoWasser.tran --> Baum.tran ;        # Transpiration
    MineralN.aufN --> Baum.aufN ;        # Stickstoff-Aufnahme
    MineralN.md   --> Baum.md ;          # Düngungsfaktor
    BoChemie.ph   --> Baum.ph ;          # pH-Wert


    Baum.laubO    --> BoWasser.laub ;    # Laubmenge
    Baum.wurzO    --> BoWasser.wurz ;    # Wurzelmenge
    :

END OF Wald
```

Es ist lediglich eine Aufzählung der beteiligten Unterkomponenten nötig, die von einer Liste der gewünschten Verbindungen ergänzt wird. Die Bezeichner sind aus dem Originalmodell übernommen, genauere Erläuterungen der Bedeutung der Modellgrößen entnehme man /Boss 85/.

Hier geht es vielmehr um die durchgängige Darstellung des Weges von der Systemanalyse (Abb. 1.1) über die Modellstruktur (Abb. 2.1) bis hin zum Modellcode. Man beachte den einfachen und intuitiven Übergang von einem Schritt zum nächsten.

Die gesamte Modelldynamik ist wie erwähnt in den Basiskomponenten beschrieben. Als Beispiel soll das Teilmodell 'Baum' dienen. Gezeigt ist hier nur das Strukturschema, das die Möglichkeiten der Dynamikbeschreibung veranschaulicht.

```
BASIC COMPONENT Baum
  LOCAL DEFINITIONS

    < Tabellenfunktion über den Zusammenhang zwischen
      Außentemperatur und Wachstumsfaktor >

  DECLARATION OF ELEMENTS

    CONSTANTS            pi         (REAL) := 3.14159
    STATE VARIABLES      laub       (REAL) := 18.0
    DEPENDENT VARIABLES  wirkfaktor (REAL) := 1.0
    SENSOR VARIABLES     ph         (REAL) := 4.2

  DYNAMIC BEHAVIOUR
    DIFFERENTIAL EQUATIONS
      laub' := Aufbau-Abwurf-Schädigung;
        :
    END

    Schädigung := laub * wirkfaktor;

    < Ereignisse >
    < if - Konstrukte >

END OF Baum
```

Zunächst wird die Basiskomponente vereinbart. Danach folgen lokale Definitionen, die nur in dieser Komponente Gültigkeit besitzen. Anschließend werden die Modellgrößen deklariert. Es sind dies Konstanten, Zustandsgrößen (wie zum Beispiel die Laubmenge des Baumes), abhängige Größen (deren Wert berechnet sich aus Konstanten und Zustandsgrößen) sowie Sensorvariablen. Sensorvariablen sind die Größen, deren Werte von außen importiert werden. Es sind dies also die Modellgrößen aus anderen Komponenten, auf deren Wert lesend zugegriffen wird.

Bei der Deklaration in der lesenden Komponente sind diese Größen fest vorbesetzt. Diese Vorbesetzung bleibt solange gültig, bis die Verbindung mit einer zweiten Basiskomponente in einer höheren Komponente geschlossen wird. Auf diese Weise ist stets eine gültige Belegung der Variablen gewährleistet. Die Basiskomponente kann somit auch isoliert als eigenständiges Modell betrieben werden.

Anschließend erfolgt die Beschreibung der Modelldynamik. Beschreibungselemente sind Differentialgleichungen und diskrete Ereignisse für die Zustandsgrößen sowie Definitionsgleichungen für abhängige Variablen. Die Elemente können in beliebiger Reihenfolge aufgeführt werden. Eine Fallunterscheidung mithilfe von if-Konstrukten ist in allen drei Bereichen möglich.

Die Modellbeschreibungssprache in SIMPLEX-II stellt eine Spezifikationssprache dar. Die Formulierung erfolgt sehr problemnah mit frei wählbaren Bezeichnern. Dies erleichtert die Kommunikation zwischen den Mitarbeitern an dem Projekt, auch wenn diese aus verschiedenen Fachdisziplinen stammen. Außerdem verhilft die Sprache zu einer leicht verständlichen Dokumentation des Modells.

3 Anforderungen an die Modelldarstellungs– und Experimentierumgebung

Gemäß den vorausgegangenen Erläuterungen liegt nun eine Modellbeschreibung verteilt auf mehrere Komponenten vor. Die Modellerstellungsumgebung sollte es auch einem Benutzer, der mit dem Betriebssystem des Rechners nicht vertraut ist, erlauben, auf einfache Weise ein Simulationsprogramm zu generieren.

Das anschließende Experimentieren sollte über eine leicht bedienbare Oberfläche durchzuführen sein. Der Benutzer ist von der Aufgabe enthoben, sein Datenmaterial auf der Ebene des Betriebssystems zu verwalten, er arbeitet ausschließlich auf der für ihn interessanten logischen Ebene seiner Modelle.

Ausgehend von den Möglichkeiten bei der Modellbeschreibung bietet das Experimentiersystem eine sehr weitgehende Unterstützung für den Modellbauer. Die wesentlichen Punkte seien hier aufgeführt;

- Modellbeschreibung auf logische Fehler überprüfen

  Die Modellbeschreibung in SIMPLEX–II wird durch einen Compiler in C–Code umgesetzt. Dabei kann der Compiler manche logische Fehler im Modell erkennen und melden. Beispielsweise testet er, ob alle Modellgrößen im gesamten Zustandsraum genau einmal definiert sind und verhindert so eine zweideutige Modellbeschreibung.

  Bedingt durch den modularen Modellaufbau existieren Schnittstellen zwischen den einzelnen Komponenten. Bei der Modelldarstellung kann das System diese Verbindungen auf semantische Korrektheit überprüfen (z.B. Verbindungen nur zwischen Variablen gleichen Typs).

- rekursives Übersetzen

  Der hierarchische Modellaufbau erfordert, daß bei Aufruf einer höheren Komponente die gesamte Hierarchie nachverfolgt wird und aus allen beteiligten Modellteilen das globale Simulationsprogramm zusammengestellt wird. Diese Aufgabe erfüllt in SIMPLEX–II ein rekursiver Algorithmus zur Erstellung des lauffähigen Modells.

- Reaktion auf Quelltextänderungen

  Ändert der Benutzer den Text der Modellbeschreibung in einer Komponente der Hierarchie, so müssen Teile des Simulationsprogramms aktualisiert bzw. neu übersetzt werden. Diese Aufgabe übernimmt das System beim nächsten Aufruf des Modells. Insbesondere ist ein Schutz vor versehentlichem Löschen eingebaut. Das Löschen einer Komponente, die noch in der Hierarchie eines größeren Modells eingebaut ist, ist nicht möglich. Auf diese Weise unterstützt das System den Benutzer so weit als möglich gerade bei der Pflege von komplexen Modellen.

- Experimentdatenverwaltung

  Alle anfallenden Daten werden aus dem Experimentiersystem heraus verwaltet. Der Benutzer kann Modellzustände, Läufe und Experimente mit frei wählbaren Bezeichnern versehen und ablegen. Ein Aufsetzen auf bereits vorliegende Modellzustände ist möglich. Er kann beliebige graphische Darstellungsformen wählen (Funktionsdarstellung, Tortendiagramm, Säulendiagramm, ...) und die entsprechenden Bilder archivieren.

- Quellcodekonsistenz

  Das System führt Buch, welche Daten mit welchen Quelltexten der Modellbeschreibung erzeugt wurden. Existieren zu einem Modell nun noch archivierte Experimentdaten, läßt das System keine Änderungen am Quellcode zu. Auf diese Weise wird eine vollständige Dokumentation der Experimente erreicht. Stets ist ersichtlich, mit welcher Modellversion die Ergebnisse erzeugt wurden. Daten, die keinem Modell mehr zugeordnet werden können und damit auch nicht mehr reproduzierbar sind, fallen bei dieser Art der Verwaltung nicht an.

Alle diese Punkte verdeutlichen, bei welchen Arbeitsschritten der Anwender eine Unterstützung durch das Experimentiersystem erfahren kann und sollte. Entlastet von den lästigen Routineaufgaben bei der Modellerstellung und während des Experimentierens kann sich der Modellbauer wesentlich konzentrierter der eigentlichen Fragestellung der Studie widmen. So verhilft die Verwendung eines geeigneten Werkzeugs zu einer deutlich effizienteren Durchführung einer Simulationsaufgabe.

Literatur:

Bossel, H.; Metzler, W.; Schäfer, H. (Hrg.) :
   Dynamik des Waldsterbens
   Berlin 1985

Eschenbacher,P. :
   Entwurf und Implementierung einer formalen Sprache zur
   Beschreibung dynamischer Modelle
   Erlangen 1990, Dissertation am IMMD

Langer, K.-J. :
   Entwurf und Implementierung eines Simulationssystems mit
   intergrierter Modellbank- und Experimentdatenverwaltung
   Erlangen 1989, Dissertation am IMMD

Meyer, K.-P. :
   Modellierung der Schadstoffeinwirkung auf das Ökosystem Wald
   Erlangen 1989, Diplomarbeit am Inst. f. math. Maschinen u.
   Datenverarbeitung, Lehrstuhl IV

Analytische Untersuchung von Parametersensitivitäten eines
Nahrungskettenmodelles im stationären Zustand

Christoph Giersch
Institut für Biochemie der Pflanzen der Heinrich-Heine-Universität
Düsseldorf

D-4000 Düsseldorf, F.R.G.

Einleitung

Mathematische Modelle von Ökosystemen enthalten in der Regel Modellparameter wie etwa Umweltkapazitäten oder maximale Wachstums- oder Freßraten. Es ist klar, daß die Lösungen der Modellgleichungen auch von den Zahlenwerten dieser Parameter abhängen. Die Untersuchung der Parameterabhängigkeit der Lösung ist Gegenstand der Sensitivitätsanalyse, einer in der Ökologie seit langem etablierten Disziplin. Üblicherweise werden Sensitivitäten numerisch bestimmt, indem für vorgegebene Werte der einzelnen Parameter die durch die Änderung in einem Parameter verursachte Änderung der Lösung numerisch ermittelt wird. Die numerische Sensitivitätsanalyse kann für jeden Punkt im Parameterraum Zahlenwerte für die Sensitivitäten liefern (natürlich nur soweit die Stabilitäts- und sonstigen Eigenschaften des Modelles das erlauben).

Für Ökosysteme ist häufig die Vernetzung der Zustandsvariablen, d.h. die Topologie des Systemes, bekannt. Zur Struktur der einzelnen Terme der rechten Seiten der Modellgleichungen lassen sich oft gut begründbare Annahmen machen. Über Zahlenwerte für Parameter liegen aber in der Regel nur unzureichende Informationen vor. Es ergibt sich damit die Frage, ob auch ohne Kenntnis von oder Annahmen zu Zahlenwerten für Modellparameter Aussagen über Parametersensitivitäten möglich sind, ob also schon die topologische Struktur des Systems und allgemeine Eigenschaften der Modellgleichungen bestimmte Aussagen über Parametersensitivitäten zur Folge haben.

Hier soll dieser Frage am Beispiel eines einfachen Nahrungskettenmodells nachgegangen werden. Die topologische Struktur dieses Modelles ist sehr einfach, nämlich die einer unverzweigten Kette. Für die Modellparameter wird lediglich angenommen, daß sie alle größer als Null sind. Es wird untersucht, ob sich schon aus diesen Annahmen strukturelle Eigenschaften von Parametersensitivitäten ermitteln lassen, wobei mit "Struktur" solche Eigenschaften gemeint sind, die von den Zahlenwerten für die Parameter unabhängig sind.

Lotka-Volterra-Modell einer einfachen Nahrungskette

Die Dynamik der hier untersuchten Nahrungskette ist gegeben durch

$$dx_1/dt = x_1(a_{10} - a_{11}x_1 - a_{12}x_2) \qquad (1a)$$

$$dx_2/dt = x_2(a_{21}x_1 - a_{23}x_3 - a_{20}) \qquad (1b)$$

$$dx_3/dt = x_3(a_{32}x_2 - a_{30}) \qquad (1c)$$

x_j sind Populationsdichten, a_{kr} die Modellparameter. Es wird $a_{kr} > 0$ für alle Parameter angenommen. Population 1 zeigt logistisches Wachstum mit maximaler Wachstumsrate a_{10} und Umweltkapazität a_{10}/a_{11}. Wachstums- und Abbauprozesse für die Populationen 2 und 3 werden durch reine Lotka-Volterra-Terme beschrieben. Die Dynamik des Systems (1) wurde von Hallam [1] untersucht, der auch Bedingungen für die Persistenz der Populationen angab. Für das Folgende wird angenommen, daß sich die Nahrungskette im stationären Zustand befindet, daß also $dx_i/dt = 0$ für $i = 1,2,3$.

Die stationären Populationsdichten P_j erhält man, wenn man die triviale Lösung $P_1 = P_2 = P_3 = 0$ ausschließt und die rechten Seiten von (1) durch $P_j \neq 0$ dividiert:

$$P_1 = 1/a_{11}(a_{10} - a_{12}P_2) \tag{2a}$$

$$P_2 = a_{30}/a_{32} \tag{2b}$$

$$P_3 = 1/a_{23}(a_{21}P_1 - a_{20}) \tag{2c}$$

Da ökologisch unsinnige negative P_j-Werte ausgeschlossen werden sollen, erhält man aus (2a) mit P_2 aus (2b):

$$a_{10}a_{32} > a_{12}a_{30} \tag{3}$$

und aus (2c)

$$a_{21}(a_{10}a_{32} - a_{12}a_{30}) - a_{10}a_{20}a_{32} > 0 \tag{4}$$

Damit ergeben sich aus der Forderung, daß die stationären Populationsdichten alle positiv sein sollen, die Bedingungen (3) und (4) an die Modellparameter. Für das Folgende wird angenommen, daß diese beiden Bedingungen erfüllt sind.

Analytische Berechnung der Sensitivitäten stationärer Populationsdichten P_j

Unter der Sensitivität der stationären Populationsdichte P_j gegenüber dem Parameter a_{kr} im Punkte (P_1,P_2,P_3) versteht man den Wert der Ableitung $\partial P_j/\partial a_{kr}$ an der Stelle (P_1,P_2,P_3). Für manche Zwecke kann es auch günstig sein, mit normierten Sensitivitäten $S(P_j,a_{kr})$ zu arbeiten. Diese sind definiert als

$$S(P_j,a_{kr}) = \frac{a_{kr}}{P_j} \frac{\partial P_j}{\partial a_{kr}} \tag{5}$$

Für die einfache Nahrungskette (1) könnten analytische Ausdrücke für die Sensitivitäten direkt aus der analytischen Darstellung (2) der P_j bestimmt werden. Hier soll jedoch ein Weg begangen werden, der allgemeinere Strukturmerkmale der Parametersensitivitäten zu erkennen erlaubt.

Nach dem Satz über implizit definierte Funktionen [2] ist der Spaltenvektor $(\partial P_1/\partial a_{kr},..,\partial P_3/\partial a_{kr})^T$ der Sensitivitäten gegeben durch

$$(\partial P_1/\partial a_{kr}, \ldots, \partial P_3/\partial a_{kr})^T = - (\partial f_j/\partial x_i)^{-1} (\partial f_1/\partial a_{kr}, \ldots, \partial f_3/\partial a_{kr})^T,$$

$$j = 1, \ldots, 3; \quad i = 1, \ldots, 3; \quad a_{kr} \text{ beliebiger Parameter} \qquad (6)$$

$(\partial f_j/\partial x_i)^{-1}$ ist die Inverse der Jacobimatrix des Systems (1) am stationären Punkte (P_1, P_2, P_3); f_j bedeutet die rechte Seite der j-te Zeile im Modell (1). Es wird angenommen, daß $(\partial f_j/\partial x_i)^{-1}$ an der Stelle (P_1, P_2, P_3) existiert.

Jeder Parameter a_{kr} taucht im Modell nur einmal auf. Die Bezeichnung der a_{kr} ist so gewählt, daß $\partial f_j/\partial a_{kr} \neq 0$ für $j = k$ und $\partial f_j/\partial a_{kr} = 0$ für $j \neq k$. Weiter ersieht man aus (1), daß diese Eigenschaft unabhängig vom Wert des zweiten Index r in a_{kr} ist. Der Wert des Parameters r hat also keinen Einfluß auf die Struktur des Spaltenvektors $(\partial f_1/\partial a_{kr}, \ldots, \partial f_3/\partial a_{kr})^T$, d.h. die Positionen der von Null verschiedenen Einträge hängen nicht von r ab, sind also schon durch den Index k festgelegt. Durch Nebeneinanderfügen der Spaltenvektoren der Sensitivitäten bzw. der $(\ldots \partial f_j/\partial a_{kr} \ldots)^T$ erhalten wir die Sensitivitätsmatrix S bzw. die Koeffizientenmatrix B:

$$S = \left(\frac{\partial P_j}{\partial a_k} \right); \quad B = \left(\frac{\partial f_1}{\partial a_k} \right); \quad j = 1,2,3; \quad k = 1,2,3; \quad i = 1,2,3 \qquad (7)$$

Aus obiger Überlegung ergibt sich, daß alle Einträge in der Hauptdiagonalen von B von Null verschieden sind, während alle anderen Einträge verschwinden. Wegen (6) hat dann die Sensitivitätsmatrix S für die Nahrungskette (1) die gleiche Struktur wie die Inverse der Jacobimatrix von (1). Es läßt sich zeigen, daß diese Aussage nicht nur für zu (1) analoge Nahrungsketten beliebiger Länge (z.B. in [1]) gilt, sondern allgemein für alle Modelle, bei denen jeder Parameter nur einmal auftaucht und für die alle Elemente der Hauptdiagonalen von B von Null verschieden sind.

Die Jacobimatrix $J = (\partial f_j/\partial x_i)$ des Modelles (1) am stationären Punkte (P_1, P_2, P_3) ist

$$J = \begin{pmatrix} -a_{11}P_1 & -a_{12}P_2 & 0 \\ a_{21}P_1 & 0 & -a_{23}P_3 \\ 0 & a_{32}P_2 & 0 \end{pmatrix} = \begin{pmatrix} P_1 & 0 & 0 \\ 0 & P_2 & 0 \\ 0 & 0 & P_3 \end{pmatrix} \begin{pmatrix} -a_{11} & -a_{12} & 0 \\ a_{21} & 0 & -a_{23} \\ 0 & a_{32} & 0 \end{pmatrix},$$

$$(8)$$

deren Inverse

$$\left[\frac{\partial f_j}{\partial x_i} \right]^{-1} = \begin{pmatrix} -1/a_{11} & 0 & -a_{12}/(a_{11}a_{32}) \\ 0 & 0 & 1/a_{32} \\ -a_{21}/(a_{11}a_{23}) & -1/a_{23} & -a_{12}a_{21}/(a_{11}a_{23}a_{32}) \end{pmatrix} \begin{pmatrix} 1/P_1 & 0 & 0 \\ 0 & 1/P_2 & 0 \\ 0 & 0 & 1/P_3 \end{pmatrix},$$

$$(9)$$

wobei für die stationären Populationsdichten P_j die Werte aus (2) einzusetzen sind. Die Inverse der Jacobimatrix hat Winkelstruktur: die jeweils letzte Zeile und Spalte enthalten nur von Null verschiedene Einträge, während die vorletzte Zeile und Spalte jeweils verschwinden außer dort, wo sie die letzte Spalte bzw. Zeile treffen. Dann folgt wieder ein von Null verschiedener Eintrag. Es läßt sich zeigen [3], daß die Inverse der Jacobimatrix für die analoge Nahrungskette beliebiger Dimension m eine analoge Winkelstruktur besitzt. Das Element rechts oben in $(\partial f_j/\partial x_1)^{-1}$ ist Null für gerades m und von Null verschieden für ungerades m. Die Struktur der Inversen der Jacobimatrix von Modell (1) [und von analogen Modellen höherer Dimension] hängt davon ab, ob die Gesamtzahl der Populationen der Nahrungskette gerade oder ungerade ist.

Sensitivitäten stationärer Populationsdichten

Die Sensitivitäten $\partial P_j/\partial a_{kr}$ lassen sich nach (6) mittels der Inversen (9) leicht berechnen, wenn man jetzt konkrete Parameter einsetzt, d.h. auch dem zweiten Index r in a_{kr} einen konkreten Wert zuweist. Durch Multiplikaton mit a_{kr}/P_j erhält man daraus die normierten Sensitivitäten. Modell (1) hat drei Populationen P_j und insgesamt acht Parameter a_{kr}. Tabelle 1 führt die resultierenden 24 normierten Populationsdichten auf.

In Tabelle 1 wird durch explizite Rechnung illustriert, daß, wie oben gezeigt, die Struktur der Sensitivitätsmatrix nicht vom Wert für Index r des Parameters a_{kr} abhängt: für festes j und k sind die Sensitivitäten $S(P_j,a_{kr})$ entweder alle Null oder alle von Null verschieden, unab-

---

Tabelle 1: Normierte Sensitivitäten $S(P_j,a_{kr}) = (a_{kr}/P_j)(\partial P_j/\partial a_{kr})$ stationärer Populationsdichten P_j gegenüber Parametern a_{kr} für Modell (1). Weitere Erläuterungen im Text

<u>Sensitivität</u>				<u>gegenüber Parameter</u>				
<u>von</u>	a_{11}	a_{12}	a_{10}	a_{21}	a_{23}	a_{20}	a_{32}	a_{30}
P_1	-1	$-\alpha$	$1+\alpha$	0	0	0	α	$-\alpha$
P_2	0	0	0	0	0	0	-1	1
P_3	$-\beta$	$-\gamma$	$\beta+\gamma$	β	-1	$1-\beta$	γ	$-\gamma$

mit

$$\alpha = \frac{a_{12}\,a_{30}}{a_{10}\,a_{32} - a_{12}\,a_{30}} > 0 \qquad \beta = \frac{a_{21}\,a_{10}\,a_{32} - a_{21}\,a_{12}\,a_{30}}{a_{21}\,a_{10}\,a_{32} - a_{21}\,a_{12}\,a_{30} - a_{11}\,a_{20}\,a_{32}} > 1$$

$$\gamma = \frac{a_{21}\,a_{12}\,a_{30}}{a_{21}\,a_{10}\,a_{32} - a_{21}\,a_{12}\,a_{30} - a_{11}\,a_{20}\,a_{32}} > 0$$

---

hängig davon, welchen Wert r annimmt. Tabelle 1 zeigt auch gleich
zeitig, daß, wie erwartet, die Sensitivitätsmatrix die gleiche Struktur hat wie die Inverse (9) der Jacobimatrix. Die Struktur der Sensitivitätsmatrix von Model (1) ist also schon durch die beiden Annahmen
$a_{kr} > 0$ und $P_J > 0$ festgelegt.

Interessanterweise werden durch die Voraussetzungen $a_{kr} > 0$ und $P_J > 0$
sogar die Vorzeichen der Sensitivitäten in Tabelle 1 festgelegt: sie
sind unabhängig von den Zahlenwerten für die a_{kr}, denn wegen (3) gilt
$\alpha > 0$, aus (5) sieht man, daß der Nenner von β und $\gamma > 0$ ist, so daß β
> 1 und $\gamma > 0$. Damit sind aber die Vorzeichen aller 15 von Null verschiedenen normierten Sensitivitäten der Tabelle 1 unabhängig von
Zahlenwerten für die a_{kr} . Für Modell (1) ist somit nicht nur die
Struktur der Sensitivitätsmatrix schon durch eine einfache Plausibili-
tätsforderung (alle $P_J > 0$) bestimmt, sondern auch die Vorzeichen der
nicht verschwindenden Einträge. Ein systematischer Zugang zur Frage
der Vorzeichenstruktur ist allerdings durch dieses konkrete Beispiel
nicht eröffnet.

Sensitivitäten stationärer Biomasse-Flüsse

Die Modellgleichungen (1) lassen sich auch als Bilanzgleichungen für
Biomasse-Flüsse verstehen: Gleichungen (1a)-(1c) beschreiben jeweils
eine Quelle und eine Senke für die Biomassen der drei Populationen x_i.
Im stationären Zustand halten sich für jede einzelne Population die
Wirkungen von Quellen und Senken gerade die Waage. Für Modell (1) kann
man folgende stationäre Biomasse-Flüsse Φ_i definieren:

$$\Phi_1 = P_1 (a_{10} - a_{11} P_1) = a_{12} P_1 P_2 = \frac{a_{12} a_{30}}{a_{11} a_{32}} \left(a_{10} - \frac{a_{12} a_{30}}{a_{32}} \right) \tag{10a}$$

$$\Phi_2 = a_{21} P_1 P_2 = P_2 (a_{23} P_3 + a_{20}) = a_{21} \Phi_1 / a_{12} \tag{10b}$$

$$\Phi_3 = a_{32} P_2 P_3 = a_{30} P_3 = \frac{a_{30}}{a_{23}} \left(\frac{a_{32}}{a_{30}} \Phi_2 - a_{20} \right) \tag{10c}$$

Der erste Term auf der rechten Seite ist jeweils der Quellterm, der
zweite die Senke, und der dritte der Wert, den man erhält, wenn man in
einen dieser Terme die Werte für die P_J aus (2) einsetzt. Ganz analog
zu den Sensitivitäten der Populationsdichten lassen sich normierte
Sensitivitäten der Flüsse definieren:

$$S(\Phi_J, a_{kr}) = (a_{kr} / \Phi_J)(\partial \Phi_J / \partial a_{kr}) \tag{11}$$

Tabelle 2 gibt die 24 normierten Sensitivitäten der Flüsse Φ_J an.

Tabelle 2: Normierte Sensitivitäten $S(\Phi_J, a_{kr}) = (a_{kr}/\Phi_J)(\partial\Phi_J/\partial a_{kr})$ stationärer Biomasse-Flüsse Φ_J gegenüber Parametern a_{kr}; α, β und γ wie in Tabelle 1. Für die mit * gekennzeichneten nicht verschwindenden Sensitivitäten ist deren Vorzeichen schon durch die Annahmen $a_{kr} > 0$ und $P_J > 0$ festgelegt, siehe Text

Sensitivität	gegenüber Parameter							
von	a_{11}	a_{12}	a_{10}	a_{21}	a_{23}	a_{20}	a_{32}	a_{30}
Φ_1	$-1*$	$1-\alpha$	$1+\alpha*$	0	0	0	$\alpha-1$	$1-\alpha$
Φ_2	$-1*$	$-\alpha*$	$1+\alpha*$	$1*$	0	0	$\alpha-1$	$1-\alpha$
Φ_3	$-\beta*$	$-\gamma*$	$\beta+\gamma*$	$\beta*$	$-1*$	$1-\beta*$	$\gamma*$	$1-\gamma$

Diskussion und Schlußfolgerung

Für das Nahrungskettenmodell (1) läßt sich aus der Annahme, daß alle Modellparameter und alle stationären Populationsdichten positiv sind, die Struktur der Sensitivitätsmatrix der Populationsdichten [Matrix (9)] bestimmen. Doch nicht nur die Struktur der Matrix, auch die Vorzeichen der Parametersensitivitäten (Tabelle 1) ergeben sich schon aus diesen beiden Annahmen. So ist z.B. $S(P_3, a_{21})$ immer positiv, d.h. für jede Wahl der Parameter, die die obige Annahme befriedigt, führt eine Erhöhung von a_{21}, der maximalen Wachstumsrate von x_2, zu einer Erhöhung der stationären Populationsdichte von x_3. Zum Teil sind die Vorzeichen der Sensitivitäten intuitiv verständlich (so beispielsweise, daß eine Erhöhung von a_{10}, der maximalen Wachstumsrate von x_1, stets zu einer Erhöhung von P_1 führt).

Die Vorzeichenstruktur der Sensitivitäten der Flüsse (Tabelle 2) ist weniger klar. Nur bei 13 der von Null verschiedenen Sensitivitäten ist das Vorzeichen unabhängig vom den Zahlenwerten für die Parameter. Diese sind in Tabelle 2 mit einem Stern (*) gekennzeichnet. So erniedrigt sich z.B. stets der stationäre Biomasse-Fluß Φ_3 wenn Parameter a_{12} erhöht wird $\{S(\Phi_3, a_{12}) = -\gamma < 0\}$, während die Wirkung einer Änderung von a_{30} auf Φ_3 (Sensitivität $1-\gamma$, Tabelle 2) von der Wahl der Zahlenwerte für die a_{kr} abhängt.

Bemerkenswert im Hinblick auf Interpretation im ökologischen Zusammenhang ist die Winkelstruktur der Sensitivitätsmatrix der Populationsdichten [Tabelle 1 oder Matrix (9)]: während der Wert von z.B. P_3 durch alle 8 Modellparameter beeinflußt wird, hängt P_2 lediglich von zwei Parametern (a_{30} und a_{32}, Tabelle 1) ab. In der Nahrungskette benachbarte Populationen können nach Modell (1) drastische Unterschiede in ihrer Sensitivitätsstruktur aufweisen (s. a. [3]). Diese wird jedoch, genau wie die durch Matrix (9) gegebene und oben schon angesprochene Zweizähligkeit der Sensitivitätsmatrix, in der Natur nicht beobachtet. Während verschiedene nicht-ökologische Eigenschaften von Lotka-Volterra-Modellen (Dynamik, unendliche Umweltkapazitäten etc., z.B. [4]) gut untersucht sind, scheint der Sensitivitätsstruktur dieser Modelle bisher weniger Aufmerksamkeit geschenkt worden zu sein. Die hier durchgeführte analytische Berechnung von Parametersensitivitäten erlaubt es, Sensitivitätseigenschaften eines Modelles auf ihre Plausibilität hin zu überprüfen, ähnlich wie die qualitative Analyse der Dynamik eines Modelles erlaubt, desses Zeit- und Stabilitätsverhalten auf Plausibilität hin abzuklopfen. Die Analyse der Parametersensitivitäten stationärer Populationsdichten zeigt, daß das Nahrungskettenmodell (1) eine artifizielle, nicht mit Beobachtungen übereinstimmende Sensitivitätsstruktur aufweist.

Daß die Sensitivitätsanalyse der Nahrungskette die klaren Aussagen der Tabelle 1 liefert, liegt natürlich auch an der Einfachheit des Modelles (1). Aber auch für komplexere Modelle sind Aussagen zur Sensitivitätsstruktur möglich (Giersch, in Vorbereitung). Die zu erwartenden Aussagen zur Struktur der Sensitivitätsmatrix werden natürlich notwendigerweise in dem Maße schwächer, wie Koeffizientenmatrix B (7) und die Inverse (9) der Jacobimatrix dichter besetzt sind.

Der Deutschen Forschungsgemeinschaft danke ich für finanzielle Unterstützung

Literatur

1. Hallam, T.G. (1986) Community Dynamics in a Homogeneous Environment in Mathematical Ecology (T.G. Hallam & S.A. Levin, eds), pp.241-285, Springer, Berlin

2. Erwe, F. (1963) Differential- und Integralrechnung I, BI, Mannheim

3. Wennekers, T. & Giersch, C. (1990) Sensitivity Analysis of a Simple Model Food Chain, Ecological Modelling, eingereicht

4. Caughley, G. (1976) Plant-Herbivore Systems in Theoretical Ecology (R.M. May, ed.), pp. 94-113, Blackwell, Oxford

AN OBJECT-ORIENTATED SIMULATION APPROACH

TO ANALYZE THE BIONOMICS OF A PREDATORY SOIL MITE

Broder Breckling,   Andrea Ruf,   Karin Mathes

Arbeitsgruppe Ökosystemforschung und Bodenökologie
Universität Bremen, Fachbereich 2, Postfach 330 440, D-2800 Bremen 33

Zusammenfassung

Ein Objekt-orientiertes Simulationsmodell wurde konstruiert, um in Verbindung mit Laboruntersuchungen einen Beitrag zur Kenntnis der Biologie der Raubmilbe *Hypoaspis aculeifer* zu leisten. Bei der Objekt-orientierten Simulation werden die Individuen als einzelne, unabhängig agierende Datenobjekte dargestellt, die qualitativ und quantitativ unterschieden werden können. Es wurde der Einfluß von verschiedenen Parametern, die die Aktivität der Individuen charakterisieren, auf die resultierende Populationsentwicklung untersucht. Die Durchführung sehr vieler Wiederholungsläufe der Simulation zeigt ferner, daß ein Bereich von Nicht-Vorhersagbarkeit ökologischer Entwicklungen auch bei präziser Kenntnis der Eigenschaften der einzelnen beteiligten Komponenten verbleibt.

Summary

In order to investigate the bionomics of the gamasid mite *Hypoaspis aculeifer* studies in the laboratory were conducted. Parallelly an object-orientated simulation model was constructed to explain the observed phenomena. By using an object-orientated modelling concept individuals are represented as single, independently acting data-objects, which can be qualitatively and quantitatively differentiated. Thereby the influence of parameters describing the activity of the individuals onto the resulting dynamics at the population level were investigated. Moreover the simulation of a large number of repetitons shows that concerning the population development a necessary range of unpredictability remains although the characteristics of all participating components are known.

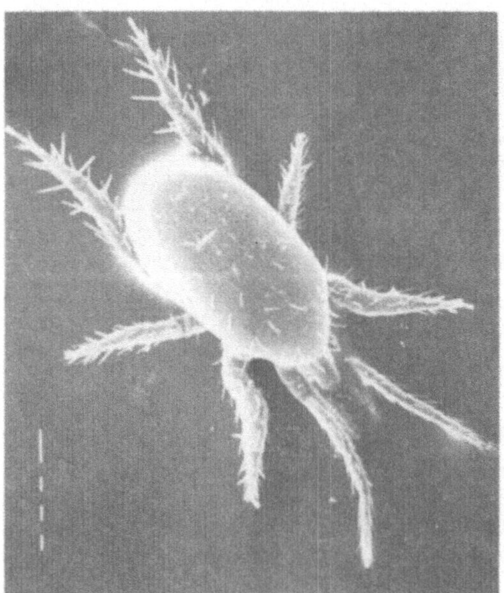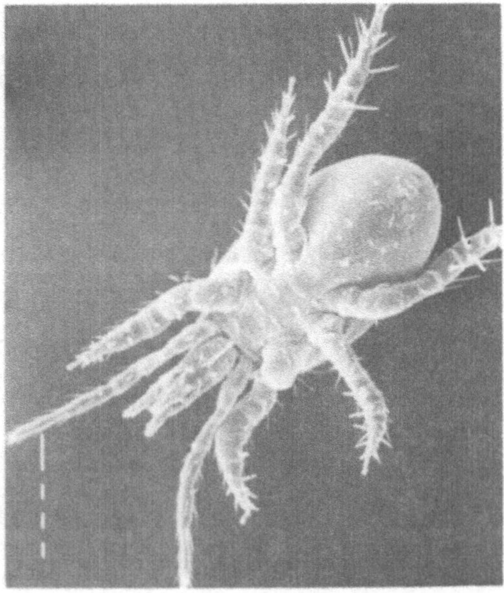

Fig. 1:   A female of the predatory soil mite *Hypoaspis aculeifer*; left: dorsal - / right: ventral view. (The upper bar in the left corner is 100 µm long. Scanning electro micrograph, A. Ruf)

Introduction

Human activities affect the soil-ecosystem in many ways. In the past there was the attempt to answer important questions by analyzing the soil as a 'black-box' and describing inputs and outputs. Looking into this 'black-box' you see a very differentiated and sensitive dynamic structure, which is not directly observable in many cases. At the microscopic or submicroscopic range the soil is a world of bizarre complexity. The significance of lots of processes for the persistence and productivity of terrestrial ecosystems is not known to full extent. The knowledge of soil-biology is still very limited in many fields. This applies especially to numerous species of the soil mesofauna. These organisms of about 0.01-1 mm size are living in the pore space of the soil. A large number of questions concerning their bionomics can not be studied under natural conditions for methodic reasons. Although most groups of the native soil mesofauna are investigated taxonomically, little is known concerning their bionomics and their quantitative contributions to the element cycling in the soil. The mesofauna individuals influence the spreading and population dynamics of soil-bacteria and -fungi due to their locomotory activities and food uptake. So they are directly and indirectly involved in the decomposition process of soil organic matter, which is a prerequisite to supply the plants with nutrients (VISSER, 1985, COLEMAN, 1986). This is the context in which the group of *gamasid* mites can be of importance. Our studies serve to extend the knowledge of the bionomics of the predatory soil mite *Hypoaspis aculeifer* (**Fig. 1**), a species which is abundant in pastures and arable lands.

We cultivated *Hypoaspis aculeifer* in the laboratory in order to help interpretating density fluctuations in the field. This knowledge is one element contributing to judge alterations of soil-ecosystems.

Cultures of *Hypoaspis aculeifer*

The mites which are of about 0.5 - 0.8 mm size as adult organisms were kept in transparent plastic vessels of cylindric shape with tightly fitting lids. The size of the vessels was 42 mm inside diameter and 17 mm depth. They were filled with a layer of about 0.5 cm plaster of paris / charcoal-mixture (9:1). During the experiments the surface of the mixture was moistened in regular intervals to give a high atmospheric humidity. Mites of the genus *Caloglyphus* served as food. At the beginning of the experiment one egg or one adult animal was put into each vessel. The development of the population resulting from this single individuum was studied for 3 months. RUF (1989) describes the dependence of the population density on the rate of egg laying, duration of development, etc.

Hypoaspis aculeifer is a parthenogenetic species. This means that the females are able to lay eggs without being inseminated. From these (haploid) eggs only males will develop. If the females have copulated once, they are able to produce both - fertilized (diploid) eggs from which females develop - and unfertilized (haploid) ones. Considering the population development two types of initial conditions can be distinguished. If females are taken out of a mixed population consisting of both sexes and are put into a culture vessel, they may lay diploid eggs right from the beginning of the population development. On the other hand, populations initiated from single females which were kept alone before the start of the study only males can arise during the first period of population development - as long as the sons have not become adult to inseminate their mother. This is the type of initial conditions our study is concerned with. **Fig. 2** shows some characteristic courses of these population developments.

At first, the females had only male descendants. As soon as these had become adult they inseminated their mother. Afterwards the number of females increased and we found that the number of males decreased though their life span was not over. From this a selective cannibalism of females on males was postulated. The plausibility of this hypothesis was examined by simulation before we were able to observe the cannibalistic behaviour in situ.

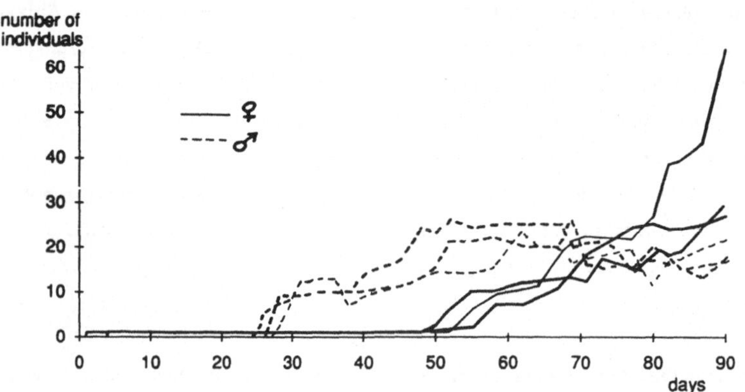

Fig. 2:

The observed development of 3 populations in different culture vessels, started with a single virgin female.

Object Orientated Simulation

It was intended to find out by simulation, whether the data obtained by observation of the real system and the assumptions, which had to be made about the further characteristics of the organisms were sufficient to reproduce the development of the laboratory populations. At the time, the model was built, we had no information about the mean frequency of cannibalistic attacks of females against males. Furtheron we had to estimate the sex ratio of the eggs of inseminated females on the basis of the observed development of only 18 eggs.

It is obvious that in the given situation the population size can not be varied continuously. The consequences of discrete events on the further development of the population (e.g. whether a certain egg is male or female, whether a male escapes predation by a female or not) would get lost by using a continuous mode of population representation in the simulation model. The representation of single individuals as data-objects, which can generate (activate) or terminate each other (BRECKLING 1990) allows a kind of model, which is very similar in structure to the experimental design. The basic principles behind this modelling approach are illustrated in **Fig. 3** and **Fig. 4**.

The simulation data presented here were generated by the program MULTIMIL *II*, which is written in SIMULA (DAHL et al. 1968). Compared to the previous version (RUF 1989) this program is extended by additional output-options. The computing was done on the SIEMENS 7881 Mainframe of the University of Bremen (Regionales Rechenzentrum).

SIMULA-Program

Declarations:

```
Process-Class Individuum


    While alive  do

        ...

        (Activities)

        If ... Activate New Individuum

        ...

    End of while
```

Mainprogram :

———— Activate New *Individuum*

... Output-Statements

Fig. 3: Object-orientated description of individuals in SIMULA. A CLASS containing the instructions which specify the model organism is declared. During the run of the simulation a copy of the class is activated from the mainprogram. One of the possible activities can be the subsequent activation of other copies of the class representing other individuals.

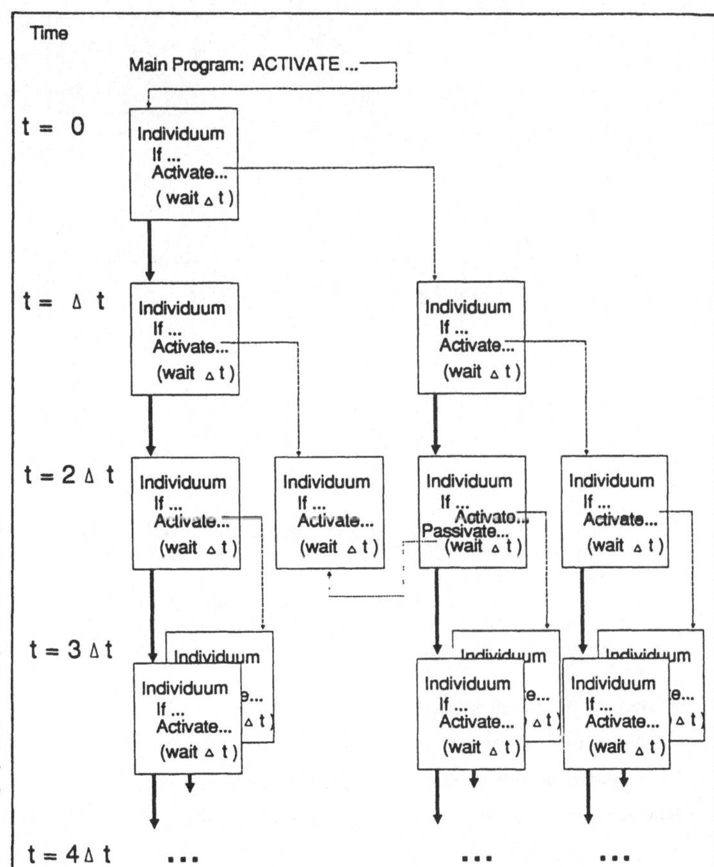

Fig. 4: Generation and deletion of individuals during the progression of simulation. Single individuals operate in interaction with others specified by their own (inner) state or according to their state diagnosis concerning the environment. The model representation is not continuous but calculates successive time intervals.

According to the experiment the simulation was started with a representation of a single adult virgin (<u>w-adult</u>). To model egg-deposition <u>w-adult</u> is able to activate processes of the type <u>m-ei</u>(egg) in adequate temporal distances. These processes activate <u>m-unreif</u> processes representing immature organismic states, followed by <u>m-adult</u>. When the first males have become active, a call of the procedure 'Befruchtung'(which simulates insemination) by the active process <u>w-adult</u> causes to set the variable 'befruchtet'(inseminated) to true. After this, processes of type <u>m-ei</u> and <u>w-ei</u>(egg of both sexes) can become started in a randomly selected mode. An active female executes the procedure 'Kannibal' which can terminate a <u>m-adult</u> process or <u>m-/w-unreif</u> processes with much lower probability. The interaction mode between the individuals is shown in **Fig. 5**.

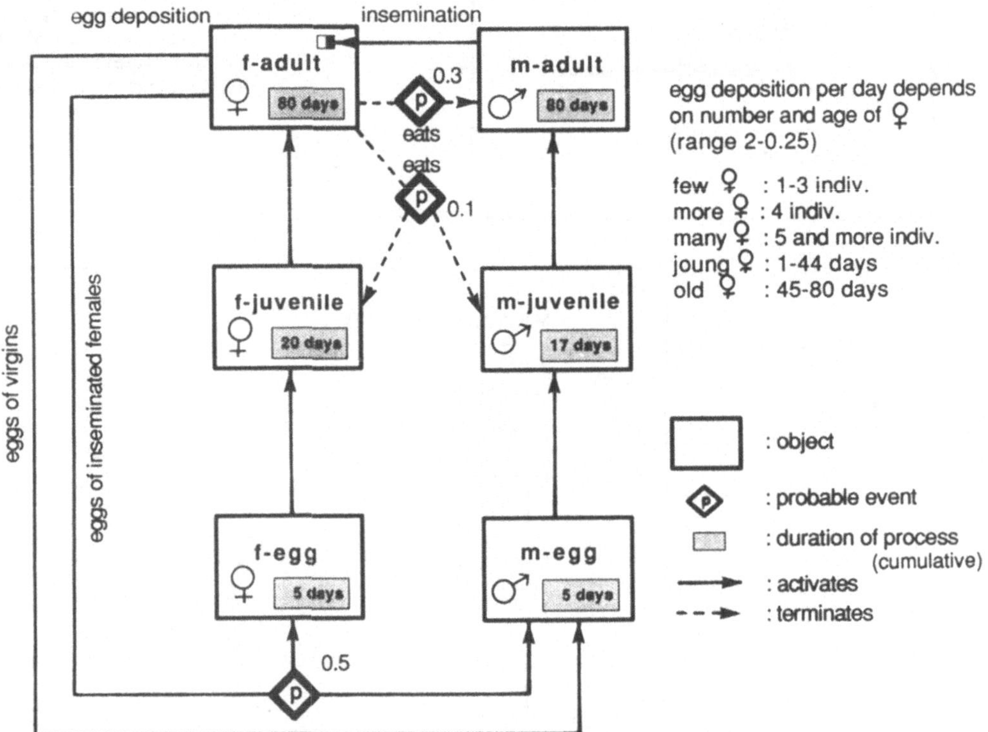

Fig. 5:   The model representation of our knowledge about population dynamics of the mite *H. aculeifer*. 3 life stages are represented, egg, juvenile, and adult with different characteristics. Males are able to inseminate females. The number of deposited eggs per female per day is density dependent. Random processes are: the cannibalism of the females against the males and juveniles and the sex ratio in the eggs of inseminated females. Most of the values of the parameters were estimated by the laboratory cultures, the others were found by parameter fitting.

Because of the involved random-generator a single run of the simulation model does not describe a *mean* population development, but it represents a unique situation just like a single experimental set. SIMULA allows the execution of automatically repeated runs with different sets of random numbers analogous to executing parallel experiments in the laboratory. Thereby it is possible to produce sets of simulation data, which can be evaluated statistically. It is obvious that reruns of the simulation-model can be carried out more often then real experiments. We shall give some examples of how the interaction among individuals of this soil animal population can be analyzed.

RUF (1989) investigated, how the population development is affected by changes of parameters characterizing the activities of the individuals. **Fig. 6** shows, how a combination of changes of the primary sex ratio of eggs and different rates of successful cannibalistic attacks affect the dynamics at the population level. The data were generated by a single execution of the program with multiple automatic restarts with different parameter-sets.

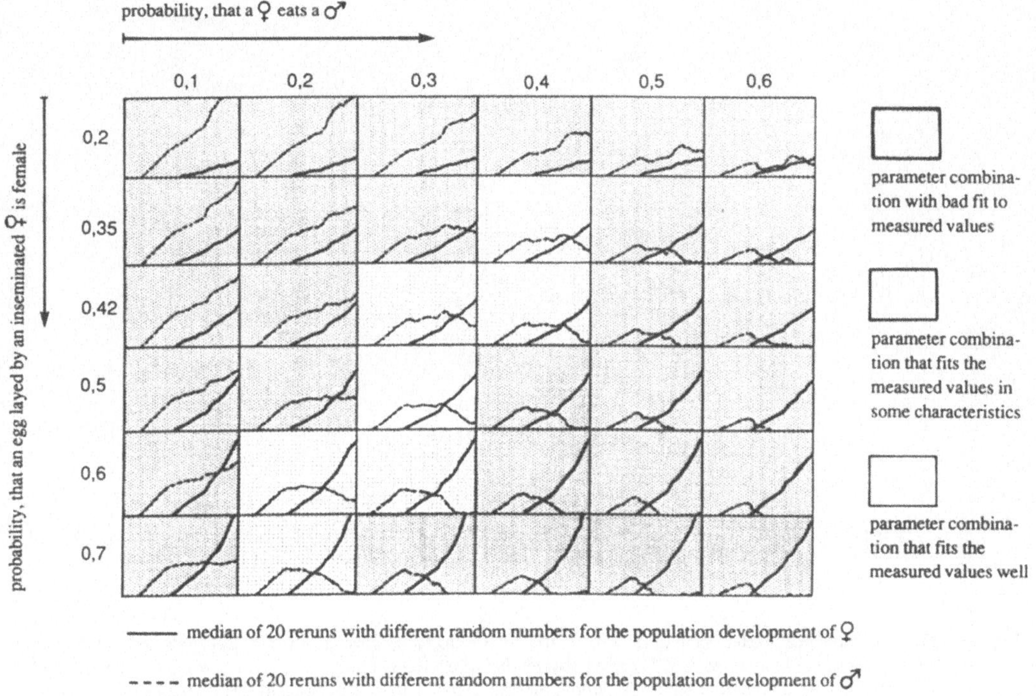

Fig. 6:   Combinations of two estimated parameters to get the best fit with the observed data. Only a narrow range of parameter combinations gives results which fit the measured population development.

The version *II* of MULTIMIL allows a further investigation of the variability of different runs. **Fig. 7** shows the medians of 10 and of 1000 reruns with different sets of random numbers that serve to specify the random dependent events (see **Fig. 5**). Although single runs may differ remarkably from each other, even a relatively small number of reruns gives a rather good impression of the range of states, most of the parallels will reach. This can give hints, how many identical experiments should be carried out in order to get an adequate impression of the mean population development.

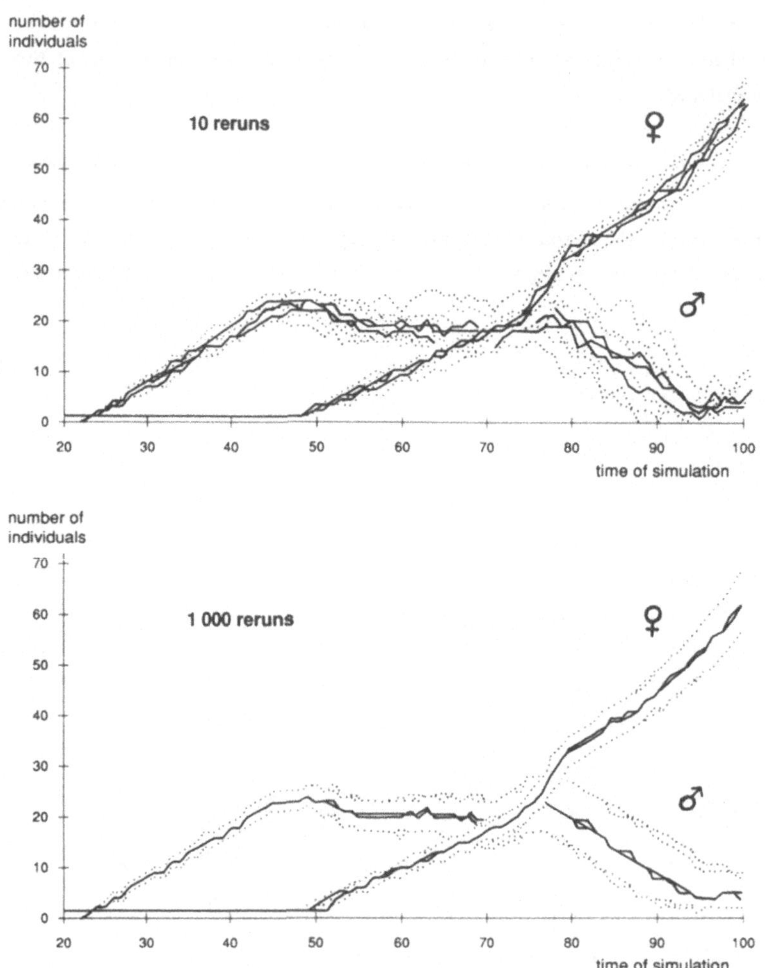

Fig. 7: The quartiles calculated out of 10 (top) and 1000 (bottom) reruns with 3 different sets of random numbers each. The area with 50% of the values (between the dotted lines) is quite well approximated by 10 reruns. At 1000 reruns this area is almost independent of the set of random numbers used.

In order to illustrate the mode of rerun variability the following type of output was realized: For any simulation time, for which a rerun coincides with one of the previous runs, the number of active adult individuals and the number of coincidations is protocollated (multiple coincidations are added). This yields a three-dimensional array, which can be visualized as a kind of 'mountain area'. The distribution of coincidations can differ significantly from the shape of a normal distribution at least at certain periods of the simulation time (**Fig. 8**). Even by increasing the number of reruns the coincidations do not approximate a normal distribution of the type, which can be obtained by many simple random processes. Even with 10 000 reruns we still obtain a 'landscape-like' surface with asymmetries and a certain kind of obvious irregularity.

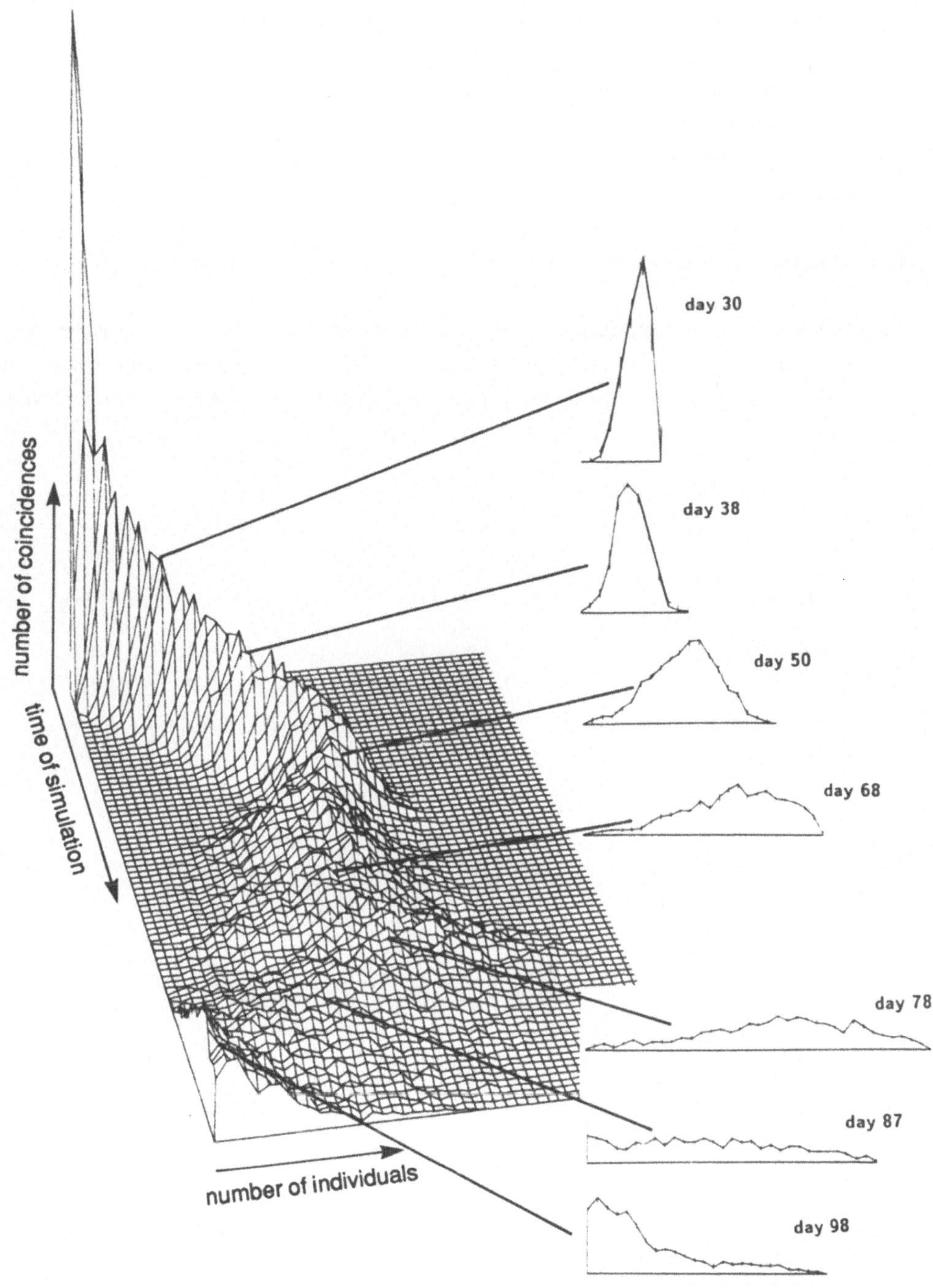

Fig. 8: A landscape of coincidations formed by the cumulation of 1000 reruns with slices at different simulation times. The coincidations are plotted over the simulation time and the number of male individuals.

In order to quantify this 'roughness', we defined the following index of folding:

We calculate the sum of all local maxima of the coincidations at a certain time of simulation and subtract the sum of the local minima at the same point of time. We double the resulting value and then have a term for the vertical component of the curve length. From this we obtain a folding measure by subtracting two times the modal value.

This folding index can be calculated for the coincidations of successive simulation times (see **Fig. 9**).

The folding does not completely disappear even for high numbers of simulations. This is another hint, that the frequency-distribution we obtain while investigating a large number of problems in soil-biology can not be approximated by a normal (gaussian) distribution but may show a large variety of context specific characteristics.

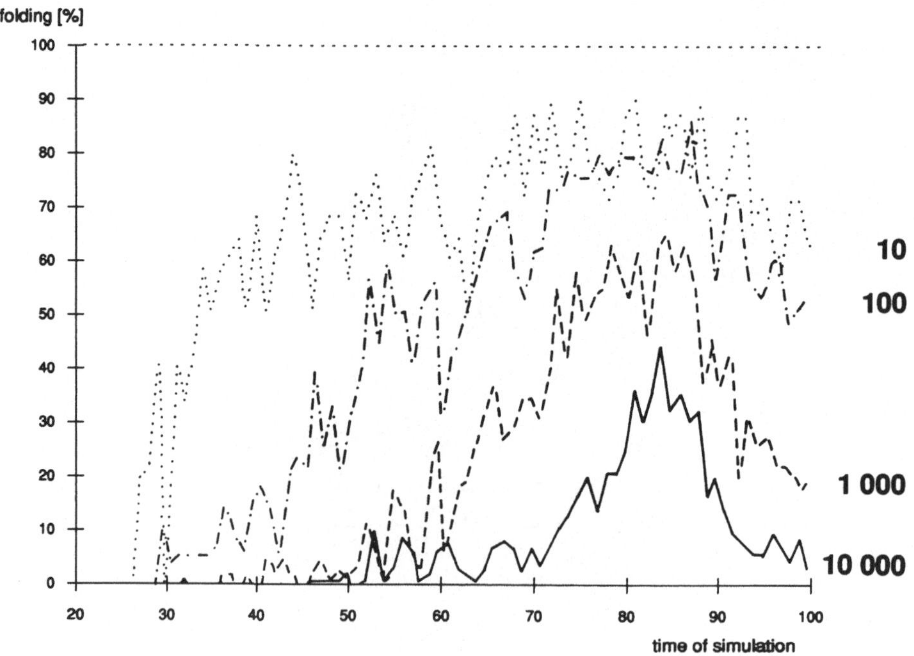

Fig. 9: The folding index with 10, 100, 1 000, and 10 000 reruns in percent of the total length of the cross section curve. Still with 10 000 reruns the folding reaches almost 50 % of the total curve length in a particular period of the simulation time.

Discussion

Variability is an essential characteristic of natural processes. By building models this is often reduced to some deterministic "main issue". Concerning real ecological events an important aspect is lost. Our simulations have demonstrated one necessary reason for the variability in population dynamics. Organisms appear and act as single, discrete, integer-valued unities. Some events can have lasting effects because the changed state has consequences for the subsequent situations. The result of this is an *unevitable range of unpredictability of ecological developments*. This can be the case even if characteristics of the involved components are fully known. The individual-orientated simulation shows this very clearly. As a consequence it is necessary to study the domain of uncertainties we have as a constituting part of more or less any observable development of complex ecological connexes (BRECKLING 1990). Even in the case of the genetically rather uniform laboratory populations of *Hypopaspis* which are kept under controlled conditions this effect is obvious. The individual-orientated simulation turned out to be an excellent theoretical tool to represent the irregularity of patterns of ecological processes.

References

BRECKLING, B. 1990
    Singularität und Reproduzierbarkeit in der Modellierung ökologischer Systeme
    <Singularity and Reproducability in Modelling Ecological Systems> (in German)
    Dissertation Universität Bremen, FB 2

COLEMAN, D.C. 1986
    The Role of microfloral and faunal interactions in affecting soil processes
    in: M.J. Mitchell, J.P. Nakas (Eds.) Microfloral and faunal interactions in natural and agroecosystems
    (p. 317 - 348)
    Martinus Nijhoff, Dr. W. Jungk Publishers (Dordrecht)

DAHL, O.J.; MYRHAUG, B.; NYGAARD, K. 1968
    SIMULA.
    Common Base Language
    Oslo (Norwegian Computing Centre)

RUF, A. 1989
    Die Populationsentwicklung der Raubmilbe *Hypopaspis aculeifer* (Canestrini 1883)
    Die Bedeutung von Arrhenotokie und Kannibalismus in Laborzuchten und deren Simulation im Computermodell.
    <Population Development of the Predatory Mite *Hypoaspis aculeifer* (Canestrini 1883)
    The Importance of Arrhenotoky and Cannibalism in Laboratory Cultures and their Simulation in a Computer Model; Diploma Thesis> (in German)
    Diplomarbeit im FB2 (Biologie/Chemie) der Universität Bremen

VISSER, S. 1985
    Role of the soil invertebrates in determining the decomposition of soil microbial communities.
    in: A.H. Fitter (Ed.) Ecological interactions in soil: p. 297 - 317
    Blackwell (Oxford)

A MODEL TO SIMULATE THE POPULATION DYNAMICS
OF THE CODLING MOTH *(CYDIA POMONELLA)*:
REPRODUCTION

Heike Lischke, SFB 123 Stochastische Mathematische Modelle
Im Neuenheimer Feld 294, D-6900 Heidelberg, Federal Republic of Germany

ZUSAMMENFASSUNG. Es wird ein Modell zur Fortpflanzung des Apfelwicklers *(Cydia pomo-nella)* vorgestellt. Das Modell ist als ein System gewöhnlicher Differentialgleichungen formuliert, wobei jede Differentialgleichung einen Zustand der Tiere während der Fortpflanzung repräsentiert. Der der Kopulation vorausgehende Flug der Faltermännchen und die Eiablage sind klimaabhängig modelliert. Ein Modell für den Flug der Männchen konnte an Pheromonfallendaten angepasst werden; es ergibt sich eine starke Temperatur- aber keine signifikante Feuchteabhängigkeit. Erste Simulationen mit dem gesamten Fortpflanzungsmodell unter Benutzung von Parameterwerten, die durch "tuning" des Modells erhalten wurden, zeigen eine befriedigende Übereinstimmung mit Daten abgelegter Eier.

SUMMARY. We present a model for the reproduction of the codling moth *(Cydia pomonella)*. It is formulated as a system of ODE's, each ODE representing a certain status of the animal during reproduction. The flight of the male moths before copulation and the oviposition are modelled as climate dependent. The male moth flight model was fitted to pheromone trap data; it shows a strong temperature but no significant humidity dependence. First simulations with the complete reproduction model, using parameter values obtained by "tuning", show a satisfying correspondence to data of oviposited eggs.

1. Introduction

In West Germany the codling moth is one of the major pests in apple production. It is planned to control the pest by means of the integrated pest management, using the granulosis virus CpGV (DICKLER, HUBER, 1984). To forecast the hatch of the first larval instar of the pest, which is the virus sensitive one, a prognosis model is of great interest. A model describing the population dynamics of the animal will be used as a base for such a prognosis system. One submodel of the population dynamics model is the model of reproduction. It includes a detailed modelling of ecophysiological and behavioral effects in the life of the insect. After a short overview of the structure of the entire model this submodel will be presented.

2. Structure of the complete model

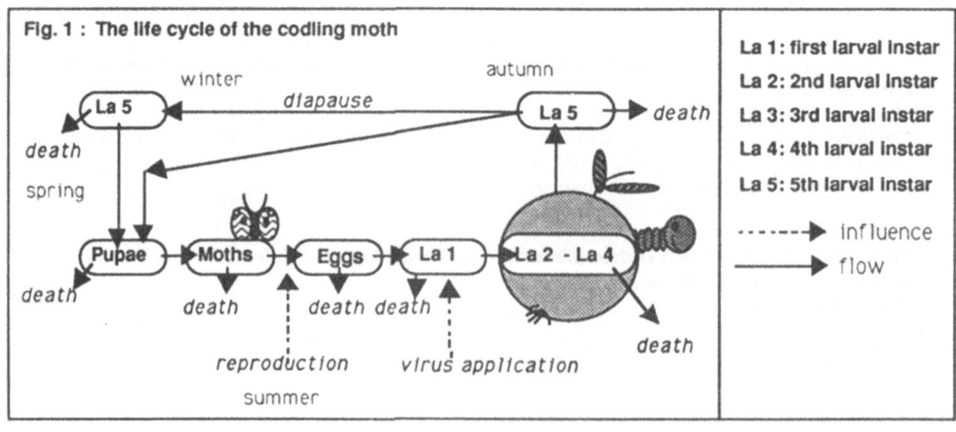

As shown in fig. 1, the life cycle of the codling moth (AUDEMARD, 1976) is determined by 4 main elements :
- the development through the developmental stages
- the reproduction
- the mortality
- the diapause (i. e. hibernation)

This scheme suggests a modular structure of the population dynamics model: The frame consists of a model describing the development (LISCHKE, BLAGO,1990). In this frame submodels for reproduction, mortality and diapause will be implemented. In the following the submodel for the reproduction will be presented.

3. Reproduction model

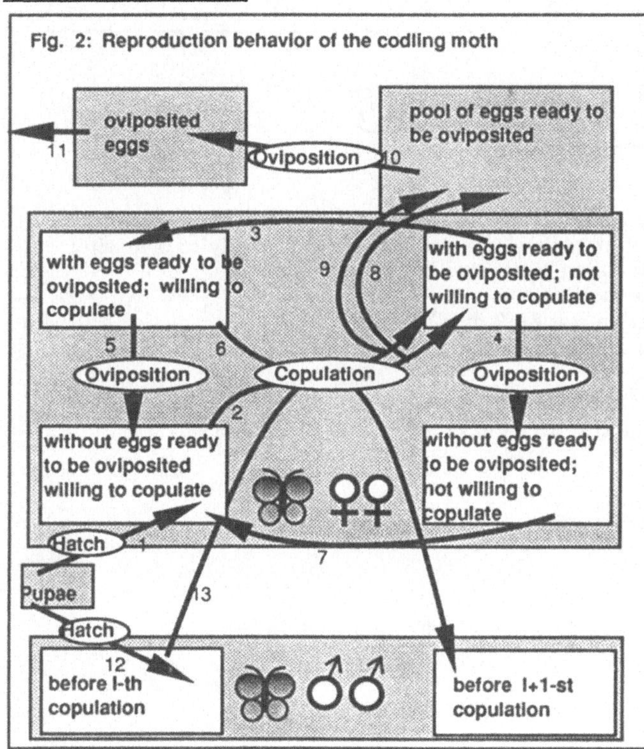

This model (fig. 2) aims to describe the complex reproduction behavior of the codling moth as precisely as possible. Reproduction is divided into the processes of copulation and oviposition. The moths copulate in a time period around sunset, when the males fly to the pheromone omitting females. The male moth flight appears to be strongly climate dependent (BATISTE et al., 1972), (HAGLEY, 1977); this effect is taken into consideration by the male moth flight submodel presented later. Male and female moths are able to copulate several times, the males loosing fertility with each copulation (HOWELL, HUTT, 1978). In the model they pass from one copulation class to the next with each copulation. We associate with each copulation class a certain fertility. The female moths stop omitting pheromone for three days after each copulation (FLURI, 1974), (HOWELL, HUTT, 1978).

In the model they are either "willing to copulate" or not. During each copulation the females receive a spermatophore, a certain amount of sperms, which they use to fertilize their oocytes during oviposition. In the model they either got "eggs ready to be oviposited", i. e. sperms to fertilize their oocytes, or they have "no eggs ready to be oviposited". The latter occurs, if they didn't yet copulate or have used all their sperms before. Oviposition takes place during nighttime. The females fly from one leaf to the next and lay their eggs one by one. The "pool of eggs ready to be oviposited" increases with each copulation. During each oviposition the eggs pass from this pool to the oviposited eggs.

In each of these compartments the individuals die with a certain mortality.

The links to the development model are given by the oviposited eggs and the pupae, from which the moths emerge.

This compartment model was formulated as the following system of ODE's and process functions. (The numbers below the equations refer to the arrows in fig. 2)

Female moths:

$$(3.1) \quad \frac{dF_{n,w}(t)}{dt} = -\mu_f(t) \cdot F_{n,w}(t) + \underbrace{\frac{1}{\rho} \cdot F_{n,u}(t)}_{7} + \underbrace{p_{ovi}(t) \cdot F_{e,w}(t)}_{5} - \underbrace{cop_f(t) \cdot F_{n,w}(t)}_{2}$$

$$+ \underbrace{sxr \cdot hatch_p(t)}_{1}$$

$$(3.2) \quad \frac{dF_{e,w}(t)}{dt} = -\mu_f(t) \cdot F_{e,w}(t) + \underbrace{\frac{1}{\rho} \cdot F_{e,u}(t)}_{3} - \underbrace{p_{ovi}(t) \cdot F_{e,w}(t)}_{5} - \underbrace{cop_f(t) \cdot F_{e,w}(t)}_{6}$$

$$(3.3) \quad \frac{dF_{e,u}(t)}{dt} = -\mu_f(t) \cdot F_{e,u}(t) - \underbrace{\frac{1}{\rho} \cdot F_{e,u}(t)}_{3} - \underbrace{p_{ovi}(t) \cdot F_{e,u}(t)}_{4} + \underbrace{cop_f(t) \cdot (F_{e,w}(t) + F_{n,w}(t))}_{6,2}$$

$$(3.4) \quad \frac{dF_{n,u}(t)}{dt} = -\mu_f(t) \cdot F_{n,u}(t) - \underbrace{\frac{1}{\rho} \cdot F_{n,u}(t)}_{7} + \underbrace{p_{ovi}(t) \cdot F_{e,u}(t)}_{4}$$

$F_{i,j}(t)$	: population density of female moths in compartment i,j; i=e: with eggs ready to oviposite; i=n: without eggs ready to oviposite; j=w: willing to copulate ; j=u: unwilling to copulate
$\mu_f(t)$	: mortality of female moths
ρ	: duration of copulation rest ($= 3$ days)
sxr	: sexratio (female moths / all moths)
$hatch_p(t)$	: hatch from pupae
$cop_f(t)$	: copulation rate of female moths
$p_{ovi}(t)$	: portion of female moths, which have oviposited all their eggs

Copulation rate of female moths:

$$(3.1.1) \quad cop_f(t) = \gamma \cdot \frac{min(M_{fl}(t), F_{e,w}(t) + F_{n,w}(t))}{F_{e,w}(t) + F_{n,w}(t)}$$

γ	: contact rate
$M_{fl}(t)$	: flying male moths (see male moth flight submodel, (4.2))

Portion of female moths, which have oviposited all their eggs:

$$(3.1.2) \quad p_{ovi}(t) = \frac{E_{ovi}(t)}{PE(t)}$$

$$(3.1.2.1) \quad E_{ovi}(t) = ovi(t) \cdot (F_{e,w}(t) + F_{e,u}(t))$$

$$(3.1.2.1.1) \quad ovi(t) = \omega \cdot \psi_c(t)$$

$E_{ovi}(t)$	: eggs oviposited at time t
$ovi(t)$	: oviposition rate
ω	: oviposition constant
$\psi_c(t)$	: daytime and climate dependence of oviposition (analog to (4.2.1))

For simplicity in equation (3.1.2) we make the following assumption: With each egg oviposited by one female, a certain portion of this female passes to the females "without eggs ready to be oviposited". Thus, the portion of all female moths "with eggs ready to be oviposited" lost by each oviposition equals the portion of the eggs in the "pool of eggs ready to be oviposited " which is laid at time t.

Eggs:

$$(3.5) \qquad \frac{dPE(t)}{dt} = -\mu_f(t) \cdot PE(t) + \underbrace{cop_f(t) \cdot (F_{e,w}(t) + F_{n,w}(t)) \cdot fert(t)}_{8,9} - \underbrace{E_{ovi}(t)}_{10}$$

$$(3.6) \qquad \frac{dE(t)}{dt} = -\mu_e(t) \cdot E(t) + \underbrace{E_{ovi}(t)}_{10} - \underbrace{hatch_e(t)}_{11}$$

$PE(t)$: population density of pool of eggs ready to be oviposited
$E(t)$: population density of oviposited eggs
$\mu_e(t)$: mortality of eggs
$hatch_e(t)$: hatch from eggs
$fert(t)$: average fertility

Average fertility:

$$(3.5.1) \quad fert(t) = fert_f \cdot \frac{\sum_{l=1}^{n_c-1} M_l(t) \cdot fert_l}{\sum_{l=1}^{n_c-1} M_l(t)}$$

$fert_f$: fertility of female moths
$fert_l$: fertility of male moths before l^{th} copulation
$hatch_e(t)$: hatch from eggs

Male moths:

$$(3.7) \qquad \frac{dM_1(t)}{dt} = -\mu_m(t) \cdot M_1(t) - \underbrace{cop_m(t) \cdot M_1(t)}_{13} + \underbrace{(1 - sxr) \cdot hatch_p(t)}_{12}$$

$$(3.8) \qquad \frac{dM_k(t)}{dt} = -\mu_m(t) \cdot M_k(t) + \underbrace{cop_m(t) \cdot (M_{k-1}(t) - M_k(t))}_{13}, \quad k = 2,...,n_c - 1$$

$$(3.8) \qquad \frac{dM_{n_c}(t)}{dt} = -\mu_m(t) \cdot M_{n_c}(t) + \underbrace{cop_m(t) \cdot M_{n_c-1}(t)}_{13}$$

$M_l(t)$: population density of male moths before l^{th} copulation, l=1,...,n_c
$\mu_m(t)$: mortality of male moths
$cop_m(t)$: copulation rate of male moths
n_c : number of copulation classes of male moths (=5)

Copulation rate of male moths:

$$(3.7.1) \quad cop_m(t) = \gamma \cdot \psi_c(t) \cdot \frac{min(M_{fl}(t), F_{e,w}(t) + F_{n,w}(t))}{M_{fl}(t)}$$

4. Male moth flight submodel

Equation (4.1) describes the population dynamics of all male moths...

$$(4.1) \qquad \frac{dM}{dt} = -\mu_m(t) \cdot M(t) + (1 - sxr) \cdot hatch_p(t)$$

$M(t)$: population density of all male moths

...whereas in equation (4.2) the flying male moths are modelled:

$$(4.2) \quad M_{fl}(t) = M(t) \cdot \psi_c(t)$$

$M_{fl}(t)$: population density of flying male moths
$\psi_c(t)$: daytime and climate dependence of male moth flight

Daytime and climate dependence of male moth flight:

$$(4.2.1) \quad \psi_c(t) = \Omega(DT(t)) \cdot \Omega(T(t)) \cdot \Omega(H(t))$$

$DT(t)$: daytime
$H(t)$: relative humidity
$T(t)$: temperature
$\Omega(.)$: optimum function

The optimum function Ω (4.2.1.1) takes values between 0 and 1. It assumes 1 in the optimum range of the argument and 0 in that range, where biological activity stagnates.

$$(4.2.1.1) \quad \Omega(x) = \frac{arctan((x - x_{incr}) * q_{incr}) + \frac{\pi}{2}}{\pi} \cdot \frac{arctan((x_{decr} - x) * q_{decr}) + \frac{\pi}{2}}{\pi}$$

x_{incr} : point of inflection of the increasing step
x_{decr} : point of inflection of the decreasing step
q_{incr} : slope of increasing step
q_{decr} : slope of decreasing step

5. Calibration and validation

The entire reproduction model was programmed in FORTRAN77. All phenological data were taken from (BLAGO, 1987) and (LISCHKE, BLAGO, 1990), all climatic data from the Biologische Bundesanstalt für Land- und Forstwirtschaft, Institut für Pflanzenschutz im Obstbau, Dossenheim.

The male moth flight submodel could be fitted seperately to data of male moths caught in pheromone traps during the first codling moth generation in 1985. The numerical method used was a least-square algorithm with the conjugate gradient minimizing procedure ZXCGR (IMSL). Hourly temperature and humidity data and data of moth hatch were used as input. Fig. 3 shows the fit of the male moth flight model, which gives a good qualitative correspondence between pheromone trap data and simulation. In figs. 4 and 5 the functions for temperature and humidity dependence resulting from the parameter estimation are plotted. In the range of the observed weather data a strong influence of temperature but no significant influence of humidity on flight activity can be recognized. As a test for the quality of the fit the climatic data of 1988 were fed into the model. Fig. 6 shows the simulation as well as observed pheromone trap data. The entire reproduction model was linked to the development model to describe the development of the eggs. Because the female moths oviposite their eggs by flying from one leaf to the other the parameter values from the male moth flight were taken for the climate dependence of oviposition. Other parameter values of reproduction and egg development were obtained by "tuning"the model with input data from the year 1985. "Tuning" means changing the parameter values in a "biologically reasonable" way and comparing between simulation and data during a large number of simulation runs. A numerical parameter identification was impossible because of the small amount of data. Fig. 7 shows the simulation of the eggs for 1985, together with data of the same year.

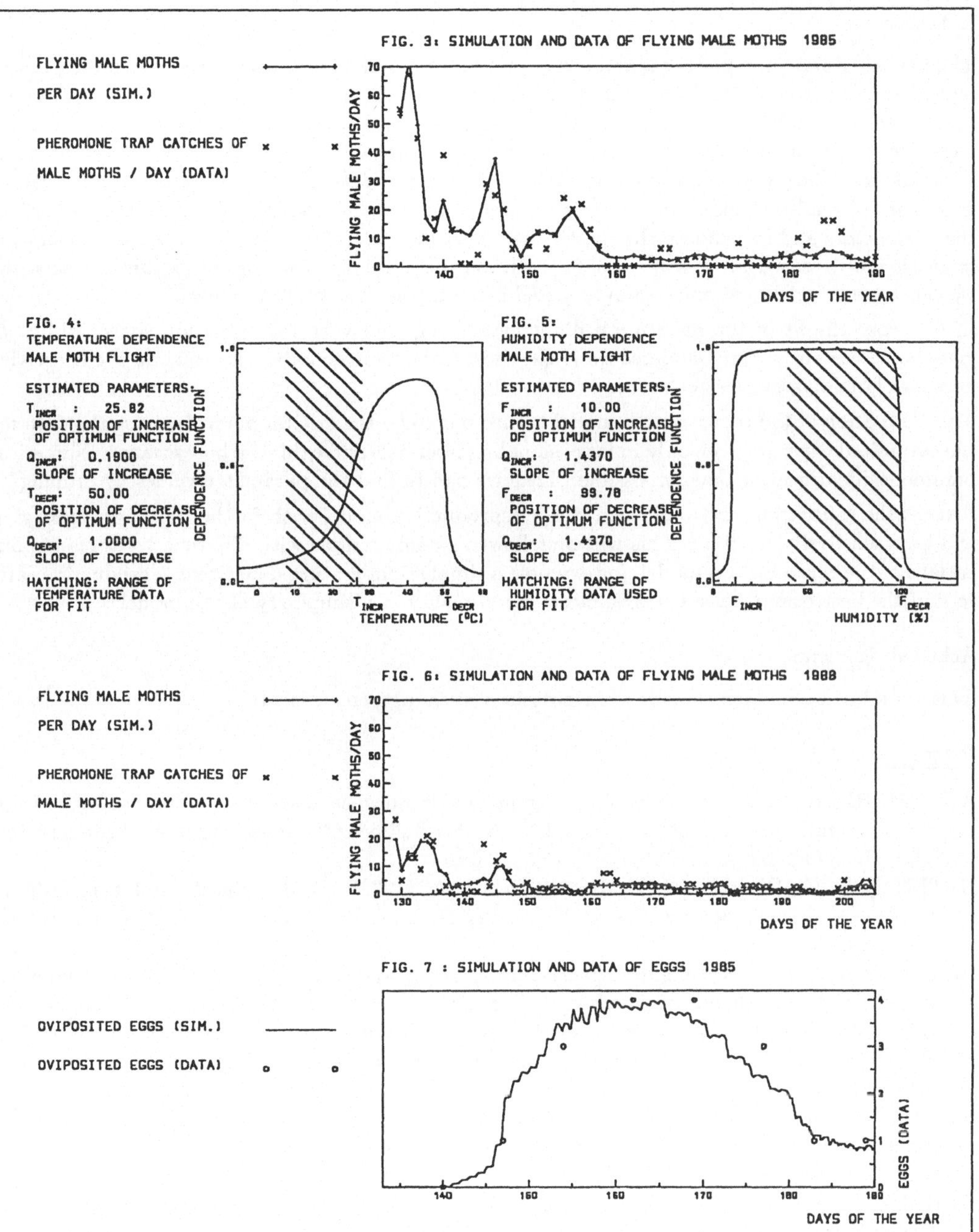
FLYING MALE MOTHS
PER DAY (SIM.)
PHEROMONE TRAP CATCHES OF x
MALE MOTHS / DAY (DATA)

FIG. 3: SIMULATION AND DATA OF FLYING MALE MOTHS 1985
FLYING MALE MOTHS/DAY
DAYS OF THE YEAR
140 150 160 170 180 190

FIG. 4:
TEMPERATURE DEPENDENCE
MALE MOTH FLIGHT
ESTIMATED PARAMETERS:
T INCR : 25.82
POSITION OF INCREASE
OF OPTIMUM FUNCTION
Q INCR: 0.1900
SLOPE OF INCREASE
T DECR: 50.00
POSITION OF DECREASE
OF OPTIMUM FUNCTION
Q DECR: 1.0000
SLOPE OF DECREASE
HATCHING: RANGE OF
TEMPERATURE DATA
FOR FIT
DEPENDENCE FUNCTION
T INCR T DECR
TEMPERATURE [°C]

FIG. 5:
HUMIDITY DEPENDENCE
MALE MOTH FLIGHT
ESTIMATED PARAMETERS:
F INCR : 10.00
POSITION OF INCREASE
OF OPTIMUM FUNCTION
Q INCR: 1.4370
SLOPE OF INCREASE
F DECR : 99.78
POSITION OF DECREASE
OF OPTIMUM FUNCTION
Q DECR: 1.4370
SLOPE OF DECREASE
HATCHING: RANGE OF
HUMIDITY DATA USED
FOR FIT
DEPENDENCE FUNCTION
F INCR F DECR
HUMIDITY [%]

FLYING MALE MOTHS
PER DAY (SIM.)
PHEROMONE TRAP CATCHES OF x
MALE MOTHS / DAY (DATA)

FIG. 6: SIMULATION AND DATA OF FLYING MALE MOTHS 1988
FLYING MALE MOTHS/DAY
DAYS OF THE YEAR
130 140 150 160 170 180 190 200

OVIPOSITED EGGS (SIM.)
OVIPOSITED EGGS (DATA) o o

FIG. 7 : SIMULATION AND DATA OF EGGS 1985
EGGS (DATA)
DAYS OF THE YEAR
140 160 160 170 180 180

6. Discussion

The presented model includes many behavioral and ecophysiological aspects of the codling moths reproduction, some of which like the decreasing fertility of the male moths, have been observed up to now only in laboratory studies. Especially for the practical use as a prognosis system our model may appear much too complex. We chose intentionally a strategy of model development different from the usual KISS (keep it small and simple) strategy. The first step of our strategy is to build a realistic model, including all possible influences we know. Then we try to identify the parameters and to validate the model. The next step is to find out in which components the model is sensitive, where we can simplify it without neglecting something important. By this way we can hope to get in the end a simple model including all important influences.

In our case, the fit of the male moth flight model, together with its validation shown in fig. 6, gives some certainty that temperature is the most important climatic influence on flight activity and that humidity can be neglected.

The calibration of the reproduction model turned out to be the critical point, because the data for the oviposited eggs are given only in large sampling intervalls and exhibit a big variance. Numerical parameter identification was impossible. Thus we had to find the parameter values by "tuning".

Besides the calibration and validation of the reproduction model with further data, our next step will be a sensitivity analysis. This will be followed by simulations with different types of simplifications of the reproduction model, for example a simple temperature dependent reproduction rate, to find the best compromise between closeness to reality and simplicity of the model.

Acknowledgements

This work has been supported by the Deutsche Forschungsgemeinschaft.

References

AUDEMARD, H. (1976): Étude démoécologique du Carpocapse (Laspeyresia Pomonella) en vergers de pommiers de la basse vallée du Rhône. Possibilitées d'organisation d'une lutte intégrée. Thse Doc. Ing. Univ. F. Rabelais Fac. Sci. Tours

BATISTE, W. C., OLSON, W. H., BERLOWITZ, A. (1972): Codling Moth: Influence of Temperature and Daylight Intensity on Periodicity of Daily Flight in the Field. J. Econ. Entom. 66(4), 883-892

BLAGO ,N. (1987): Anpassung des Amerikanischen Prognosemodells "BUGOFF 2" für Cydia Pomonella L. an die Mitteleuropäischen Bedingungen durch Mikroklimatische und Phänologische Untersuchungen. Diplomarbeit Univ. Neapel, Fak. f. Agrarwiss., Porticj, Inst. f. Entomol.

DICKLER, E., HUBER, J. (1984): Das Apfelwicklergranulosevirus: Stand der Forschung und Möglichkeiten seiner Einführung in die Obstbaupraxis. Gesunde Pflanzen 36, 285-289

FLURI, P. (1974): Untersuchung über das Paarungsverhalten des Apfelwicklers (Laspeyresia Pomonella) und ber den Einflußvon künstlichem Sexuallockstoff auf die Kopulationshäufigkeit. Mitt. Schweiz. Ent. Ges. 47(3,4),252-259

HAGLEY, E. A. C. (1977): Effect of Rainfall and Temperature on Codling Moth Oviposition. Environ. Entomol. 5(14), 967-969

HOWELL, J. F., HUTT, W. . (1978): Codling Moth: Mating Behavior in the Laboratory. Ann. Entomol. Soc. Am. 71,891-895

IMSL LIBRARY (1984): User's Manual. - Vol. 1-4. - Edition 9.2.- International Mathematical and Statistical Libraries, Inc.

LISCHKE, H., BLAGO, N. (1990): A Model to Simulate the Population Dynamics of the Codling Moth (Cydia Pomonella): Development and Male Moth Flight. Acta Horticulturae (in press)

Consequences of intraspecific predation:
A stage structured population model approach

by

Wilfried Gabriel[1] **and Frank van den Bosch**[2]

[1] Max Planck Institute for Limnology
Department of Physiological Ecology
P.O.Box 165, D - 2320 Plön, Fed. Rep. of Germany
and
[2] Institute of Theoretical Biology
P.O.Box 9516, NL - 2300 RA Leiden, The Netherlands

ZUSAMMENFASSUNG: DGLs mit zeitverzögerten Argumenten beschreiben ein Räuber-Beute-System, in dem nur die älteren Individuen räuberisch sind, aber neben der Beutepopulation auch Jungtiere der eigenen Art jagen. Mit zunehmendem innerartlichem Räuberdruck verringern sich zunächst die Amplituden der Räuber-Beute-Oszillationen, bevor mit weiterer Zunahme gedämpfte Schwingungen auftreten. Bei einer kannibalistischen Aktivität, die Labormessungen bei räuberischen Copepoden entspricht, werden die Oszillationen so stark gedämpft, daß sich schnell ein stabiles Gleichgewicht einstellt. Dadurch verringert sich das Aussterberisiko erheblich.

SUMMARY: By differential equations with time delayed arguments, a predator-prey system is described where only older animals are predacious but hunt, besides a prey population, also juveniles of their own species. Increasing intraspecific predation first reduces the amplitudes of predator-prey oscillations; a further increase then causes damping. With a cannibalistic activity as measured for copepods, the system turns very fast into a stable equilibrium. This reduces the risk of extinction drastically.

INTRODUCTION

Predator-prey-relationships are important biotic interactions in ecosystems. They are often modelled by various types of Lottka-Volterra-equations. With enormous effort mathematicians gave many valuable insights into principal mechanisms. The application to field populations, however, and also to laboratory experiments often failed. This augmented the prejudice against mathematical modelling among many biologists.

To improve the applicability of Lottka-Volterra-systems, it is indispensable to include more detailed information on the individuals into the mathematical description. Not only age structures are of great importance to population dynamics but also other traits and variabilities of the individuals within a population, e.g. behavior, physiological acclimation, genotypic variation, phenotypic plasticity, reaction norms. One of the attempts to include more of the individuals' characteristics into mathematical population models is done by the theory of structured populations (Nisbet et al. 1985, Metz and Diekmann 1986). In the following we apply this approach to the problem of intraspecific predation. We present a model for a species that becomes predacious during life time, feeding on smaller animals of other species as well as on its own juveniles.

Consequences of intraspecific predation on population dynamics will be discussed by calibration of the model to experimental data on a freshwater cyclopoid copepod.

INTRASPECIFIC PREDATION

Intraspecific predation is a widespread phenomenon especially when food becomes scarce. Well known examples are fish populations wherein adults become carnivorous and feed on their own young. It is interesting to study the population when adults have no other food sources than their own species. One may also

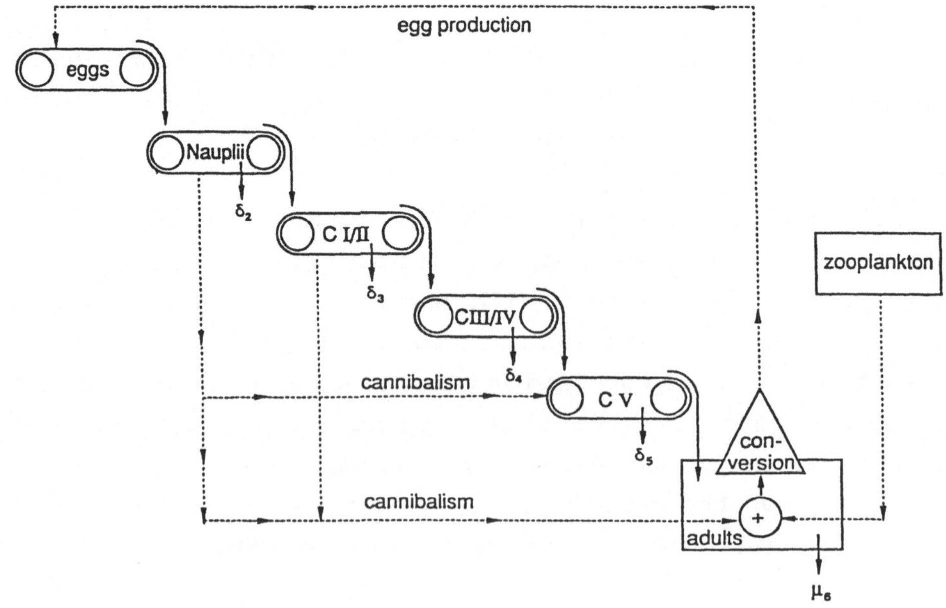

Fig. 1:   Stage structured population model visualized by conveyor belts.

ask which amount of cannibalism is tolerable for a population if adults (and older juveniles) live partially on pre-adults which have access to other food sources. This problem has been analyzed for cyclopoid copepods by a detailed simulation model based on difference equations (Gabriel 1985a, Gabriel and Lampert 1985). It has been concluded that copepod populations can hardly survive cannibalism without sufficient alternative prey for the adults. For periods of food shortage in the alternative prey, cannibalism increases drastically the survival probability of a population (Gabriel 1985b). These models, however, did not allow to draw more general conclusions. Applying the methods of structured populations (van den Bosch et al. 1988), a general mathematical description for population survival under cannibalism has been obtained. With this analysis it was possible to estimtate the boundary conditions under which populations can survive in spite of heavy intraspecific predation.

A MODEL FOR PHYSIOLOGICALLY DISTINCT AGE CLASSES
AND ITS APPLICATION TO CYCLOPOID COPEPODS

We use the method of stage structured populations and expand the previous approach (van den Bosch et. al. 1988) to a more specific model. The aim is to describe a species in which the individuals change their feeding behavior during lifetime. They start as herbivores and become carnivorous or omnivorous in later stages. The various stages differ in physiology, energy content and food selectivity. The model is applicable not only to the cyclopoid copepod species we had in mind during construction. In copepods, the various life stages can be identified because they are distinguishable in morphology. This makes it easier to measure the model parameters in experiments.

We differentiate five stages: the first three stages are herbivorous, stage four is partially predacious and stage 5 is obligatory carnivorous. The prey population (other zooplankter than the predator) is assumed to follow the usual logistic population growth in absence of the predator. Predation is modelled by a Holling type II functional response which determines an additional death rate (see equation 1). The dynamics of a stage

Table: Equations and parameters

$$(1) \quad dn_1/dt = r\, n_1(t)\, [1 - n_1(t)/K] - n_1(t) \sum_{j=5}^{6} F_{j1}(t) n_j(t)$$

$$(2\text{-}5) \quad dn_i/dt = R_i(t) - R_{i+1}(t) - \delta_i(t)\, n_i(t) \qquad \text{for } i=2,\ldots,5$$

$$(6) \quad dn_6/dt = R_6(t) - \mu_6(t)\, n_6(t)$$

$$(7\text{-}10) \quad dP_i/dt = P_i(t)\, [\delta_i(t-\tau_i) - \delta_i(t)] \qquad \text{for } i=2,\ldots,5$$

$$(11) \quad R_2(t) = \epsilon\, n_6(t-\tau_E) \sum_{i=1}^{5} E_i\, F_{6i}(t-\tau_E)\, n_i(t-\tau_E)$$

$$(12\text{-}15) \quad R_i(t) = R_{i-1}(t-\tau_{i-1})\, P_{i-1}(t) \qquad \text{for } i=3,\ldots,6$$

$$(16\text{-}19) \quad \delta_i(t) = \sum_{j=5}^{6} F_{ji}(t) n_j(t) + \mu_i \qquad \text{for } i=2,\ldots,5$$

$$(20\text{-}29) \quad F_{ji}(t) = C_{ji} \Big/ \Big[1 + \sum_{i<j} C_{ji}\, H_{ji}\, n_i(t)\Big] \qquad \text{for } i=1,\ldots,5 \text{ and } j=5,6$$

n_1	alternative prey (zooplankton)
n_2	predator stage 1 (nauplii)
n_3	predator stage 2 (copepodites CI and CII)
n_4	predator stage 3 (copepodites CIII and CIV)
n_5	predator stage 4 (copepodites CV)
n_6	adult predator (copepodes)
P_i	probability to survive through stage i
R_i	rate of recruitment into stage i
δ_i	per capita intrinsic death rate of stage i
F_{ji}	Holling type II functional response of stage j on i
r	intrinsic growth rate of alternative prey
K	carrying capacity of alternative prey (n_1)
μ_6	natural intrinsic mortality rate of adult predator
ϵ	conversion efficiency of uptaken energy into eggs
τ_E	development time of eggs
E_i	energy content of an individual prey n_i
τ_i	development time of stage i
C_{ji}	attack rate of stage j on stage i
H_{ji}	handling time of stage j on stage i

of the predator population is described by the difference between the rate of recruitment into a stage and its following stage and by an additional loss term due to natural and predator induced mortality in the stage (equations 2-6).

It is important to mention that this description of the stages by differential equations is not done in the usual way of compartment models in which the information of cohorts inside the compartments is lost. An appropriate picture of model is a system of conveyor belts (see Figure 1), each stage represented by its own belt: individuals falling onto the belt from a previous stage are traced during aging with modifications due to death until they reach the end of the belt and fall onto the next belt. The time spent on a belt - or equivalently the stage duration - enters, as time delay, into the equations. Thereby, necessary information is kept in time delayed variables. The input onto the belts are equivalent to the rates of recruitment into the stages. These rates are obtained by multiplying the recruitment rate of the previous stage with the probabilities to survive through the stage (equations 12-15). The survival probability of a stage is described by a differential equation using the death rate in the stage at actual time besides its value delayed by the stage duration (equations 7-10). Stage specific death rates are calculated by the summation over the natural death rates and the mortality due to cannibalism (equations 16-19). Cannibalism is described by a Holling type II functional response which depends on prey densities (alternative prey and cannibalized stages of the own species). To calculate the recruitment into the first stage, one has to sum over the food uptakes and to convert it into eggs. The production of newborns occurs with a time delay of the egg development period (equation 11).

With reference to previous work (Nisbet et al. 1985, van den Bosch et al. 1988) the justification of the whole equation system (see Table) is straight forward. The quite complex system of delayed differential equations can be implemented and integrated on personal computers by a special software (Maas et al. 1982). One has to specify the starting conditions of the variables as well as its time delayed values very carefully in a consistent manner. For our studies of the dynamical behavior of the system, it was appropriate to inoculate the system by a well defined pulse (we used a Gamma distribution) of nauplii and then let evolve the system. The model parameters have been partially taken from literature data and partially from experiments (van den Bosch and Santer, in prep.) which have been designed especially for this model.

RESULTS AND DISCUSSION

Intraspecific predation does not only stabilize the population dynamics for short periods of food shortage (Gabriel 1985b) but also in the long run. For cyclopoid copepodes, this striking effect has been demonstrated by van den Bosch and Gabriel (in press) applying the stage structured model approach presented in this paper.

Figure 2 shows a typical cycle of the predator prey interaction in case of zero intraspecific predation. The fact that the abundance of the CIII/CIV-stage ($=n_4$) is larger than the CI/CII-stage ($=n_3$) just reflects that the duration time for n_4 is longer than for n_3. In Figure 3, only the time courses of alternative prey ($=n_1$), nauplii ($=n_2$), and adult predators ($=n_6$) are shown. The various panels display the population dynamics for different degrees C of intraspecific predation. C=1 means intraspecific predation according to the attack rates measured in the laborato-

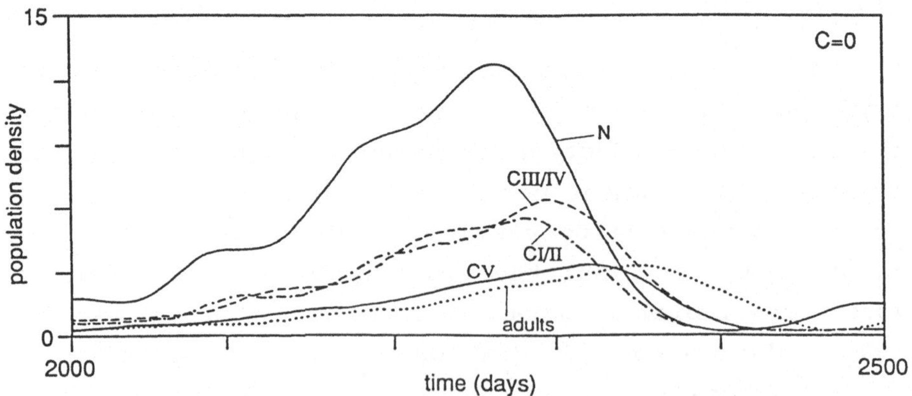

Fig. 2:   Time course of various stages of the predator during one predator prey cycle without cannibalism (cannibalistic activity C=0). The stages are indicated in the figure (N = nauplii, CI...CV juvenile stages).

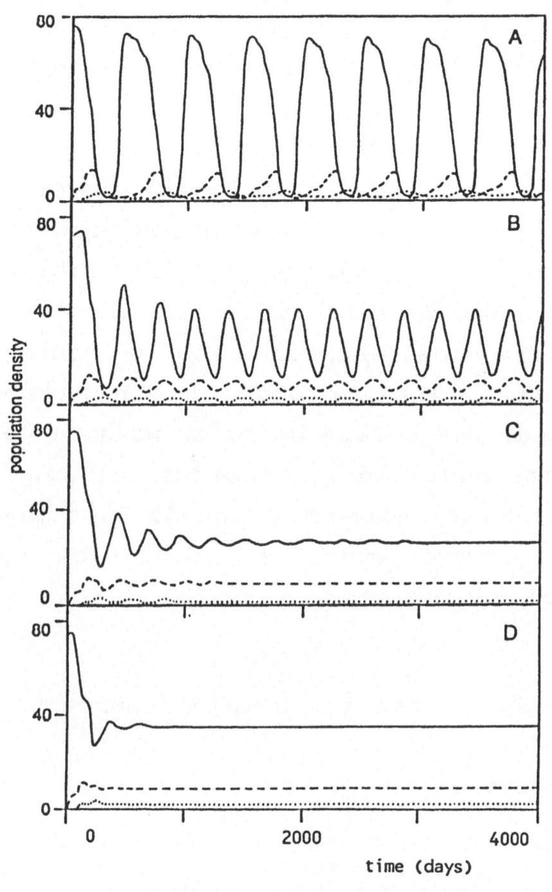

A) Without cannibalism. (C=0)

B) Attack rates of the predator on conspecifics are
40% of the values measured
in the labatory. (C=0.4)

C) Attack rates are 60% of
the measured values.
(C=0.6)

D) Attack rates are equal to
the measured values. (C=1)

Fig. 3:   Time course of predator and prey population for various
degrees of cannibalistic activity C of the predator.
Continuous line: prey (zooplankton). Dashed line:
predator nauplii. Dotted line: predator adults.

ry experiments, C=0 (upper panel) means no cannibalistic activity
so that only the alternative prey is consumed. With increasing
intraspecific predation, amplitudes of the predator-prey
oscillations decrease but frequency increases (panel with C=0.4),
then, for C > 0.5 we observe damped oscillations resulting in a
stable equilibrium of the system. The transition time from the
initial state to the stable equilibrium is very short for C=1.
With this cannibalistic activivty, the predacious adults still
live mainly on the alternative prey population: less the 10% of
the uptaken energy is gained by cannibalism.

Large oscillations of the predator prey system endanger the
species because of the high extinction risk by random fluctua-

tions. Therefore, the observed stabilizing effect of intraspecific predation may be of great importance for species survival.

The model presented is valid only for the unrealistic case of stable environment. For example, the developmental times vary with temperature, but temperature is far from being constant during the growing season of the predator. In addition, growth rate and carrying capacity of the alternative prey population show high variability during season. To apply the model to field populations, one has to modify the model equations e.g. by making several parameters time dependent. The approach of structured populations can still be applied, but before doing so we have to perform experiments to determine additional parameter values.

Such complicated experiments are already planned. This may indicate that the approach of stage structured population is promising also for experimental biologists.

Acknowledgement : We thank Baraba Santer for helpful comments.

Literature cited

Gabriel, W.(1985a): Simulation komplexer Populationsdynamik. - In: D.P.F.Möller (ed.): Simulationstechnik. Informatik Fachberichte Bd.109. pp 318-324, Springer, Berlin Heidelberg.

Gabriel, W.(1985b): Overcoming food limitation by cannibalism: A model study on cyclopoids. Arch.Hydrobiol.Beih.21: 3164-3168.

Gabriel, W. and Lampert,W.(1985): Can cannibalism be advantageous in cyclopoids? A mathematical model. Verh.Internat.Verein.-Limnol. 22: 3164-3168.

Maas, P., W.S.C. Gurney, and R.M. Nisbet (1982): SOLVER - An adaptable program template for initial value problem solving. Applied Physics Industrial Consultants. Glasgow: Univ. of Strathclyde. (see also later and recent updates)

Metz, J.A.J. and O. Diekmann (eds.) (1986): The dynamics of physiologically structured populations. Lecture Notes in Biomathematics 68. Springer-Verlag, Heidelberg.

Nisbet, R.M., S.P.Blythe, W.S.C.Gurney, and J.A.J. Metz (1985): Stage structured models of populations with discrete growth and development processes. IMA Journal of Mathematics Applied in Medicine and Biology 2: 57-68.

Van den Bosch, F., A.M. de Roos, and W. Gabriel (1988): Cannibalism as a life boat mechanism. J.Math.Biol. 26: 619-633.

Van den Bosch, F. and W. Gabriel (1990): The impact of cannibalism on the population dynamics of cyclopoid copepods. Verh.Internat.Verein.Limnol. 24: (in press).

Van den Bosch, F. and B. Santer: Cannibalism in Cyclops abyssorum. (submitted)

THE COEXISTENCE OF THREE SPECIES OF *DAPHNIA* IN THE KLOSTERSEE:
III. THE SIMULATION MODEL CODA

A. Seitz

AG Populationsbiologie, Institut für Zoologie der Universität Mainz
Saarstraße 21, D-6500 Mainz, FRG

Abstract

The population dynamics of three coexisting *Daphnia* species are described by means of a discrete event simulation model. The model is formulated in the algorithmic language SIMULA which allows very convenient object-oriented programming. Different compartments of a lake are simulated at different levels of integration, each represented by a PROCESS CLASS in SIMULA: The properties of the lake are deposited in a set of meteorological data which were measured in the field. Fish are simulated as a population with properties which are derived from field observations and *Daphnia* are represented as individuals. The model shows that hypotheses about the stabilizing mechanisms (seasonal change of environmental conditions, vertical migration of *Daphnia*, size selective and abundance dependent feeding of planktivorous fish) of the coexistence of the *Daphnia* species are sufficient to explain the stable coexistence observed.

Zusammenfassung

Die Koexistenz dreier Daphnienarten wird mit Hilfe eines objektorientierten Simulationsprogrammes beschrieben. Als Beschreibungssprache dient SIMULA, das eine sehr einfache Realisierung von parallelen Prozessen und diskreten Ereignissen ermöglicht. Die verschiedenen Kompartimente eines Sees werden auf verschiedenen Integrationsniveaus simuliert, die jeweils durch eine PROCESS CLASS in SIMULA repräsentiert werden: Der See mit seinen physikalischen und chemischen Eigenschaften wird als ein Datensatz tatsächlich im Freiland gemessener Größen realisiert. Fische werden durch eine Population dargestellt, deren Eigenschaften aus Freilandbeobachtungen bestimmt wurden. Daphnien werden als Individuen simuliert. Das Modell zeigt, daß Hypothesen über die Stabilisierungsmechanismen der Koexistenz der Daphnien (jahreszeitlicher Wechsel der Umweltbedingungen, Vertikalwanderung der Daphnien, größen- und abundanzabhängiges Fressen planktivorer Fische) ausreichen, um eine stabile Koexistenz zu ermöglichen.

1. Introduction

Competition between individuals of the same or different species is considered to be one of the most important forces of evolution (HUTCHINSON 1959). It leads to character displacement of species with overlapping resource utilization. The direct proof of this hypothesis is difficult in practice in spite of the clear logic of the theoretical arguments (JACOBS 1985). Nevertheless the "competition exclusive principle" (HARDIN 1960) is one of the fundamentals of ecological and evolutionary research.

In plankton ecology, the "size-efficiency-hypothesis" leads to an inconsistancy between theory and observations in the field: species with strong resource overlapping coexist in an apparently uniform environment. This is explained by a balance between the better food utilization of larger zooplankton species and their higher mortality induced by size selective fish. Whereas the effect of fish is commonly accepted (ZARET 1980, GREEN 1983, NORTHCOTE 1988), the differential competitive ability of different sized zooplankton is still under discussion (Dodson 1974, Lynch 1977, Tillmann & Lampert 1984).

Several years ago, a series of investigations was performed on the coexistence of different *Daphnia* species (*Daphnia hyalina* Leydig, Hellich *Daphnia galeata gracilis* and *Daphnia cucullata* Sars) in the Seeoner Klostersee north of Lake Chiemsee in Upper Bavaria, FRG (Jacobs 1977a,b; Seitz 1977, 1980a,b). This coexistence is characterized by a seasonal change in species composition with a relatively higher abundance of the larger species *D. hyalina* and *D. galeata* during spring and early summer which is followed by a dominance of *D. cucullata* during late summer and autumn (Fig. 1).

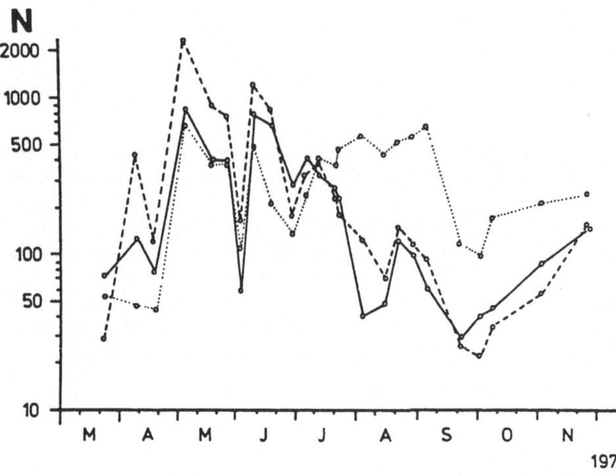

Figure 1: Population dynamics of *Daphnia hyalina* (solid line), *D. galeata* (broken line) and *D. cucullata* (dotted line) in the Lake Seeoner Klostersee during 1974.

All observations showed that a set of different conditions and interactions facilitates the coexistence of three *Daphnia* species. The key factors are shown in fig. 2.

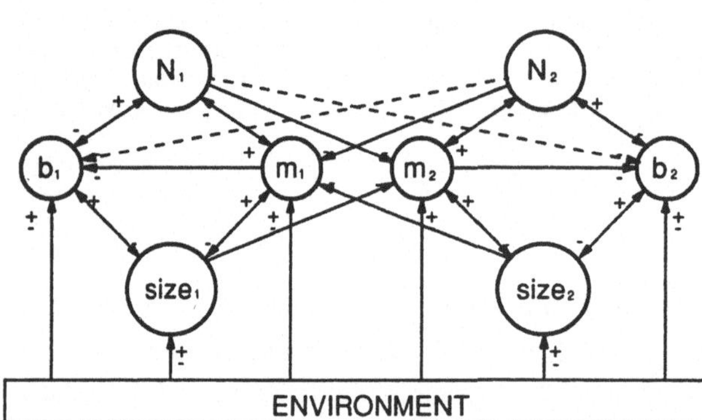

Figure 2: Schematic diagram of the interactions between two of three species and the environment. N= abundance, b= instantaneous birth rate, m= instantaneous rate of mortality, size= body length of animals. The signs indicate positive or negative interactions. The third species interacts in a similar way with the two others. The solid lines show interactions which were found by a regression analysis of field data (Seitz 1980a,b). The relations which are represented by the dotted lines were additionally proved by laboratory experiments (Seitz 1984).

These observations and conclusions coincide with the present knowledge about the population dynamics of zooplankton and the species compositon of zooplankton communities (HUTCHINSON 1961, GLIWICZ et al. 1981; DEMOTT 1983). The quantitative investigation of these inter-relationships by means of a simulation model aims at two goals.

In a first step it will be tested whether the field data are quantitatively and qualitatively sufficient to explain the permanent stability of the coexistence. In the second step, key factors will be extracted. Facing anthropogenetic environmental changes, these key factors can be used to estimate the robustness of the system against man-made disturbances.

2. The Model

Our knowledge about the system allows modelling its different compartments at different levels of integration. The structure of the simulation program is shown in fig. 3.

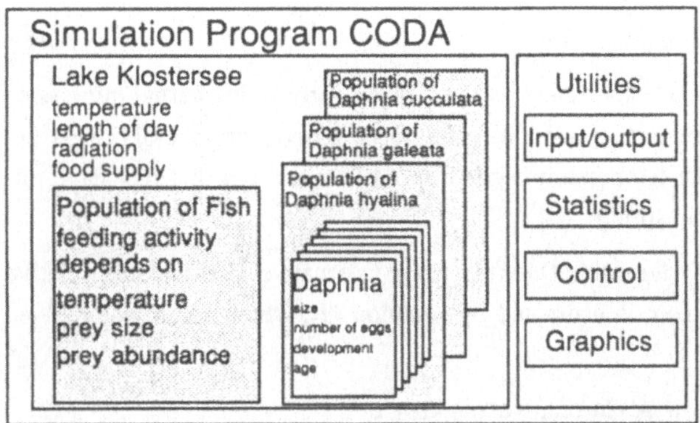

Figure 3: The structure of the simulation model. The program consists of the same components as a research program: The objects, which are the **lake**, the **fish population** and three **populations of** *Daphnia*. These components are observed and data about their development are collected by a number of utilities. Each component is realized by a PROCESS CLASS of SIMULA. The *Daphnia* populations are linked lists of descriptions of individual *Daphnia*. Again, each *Daphnia* is represented by a PROCESS CLASS.

The lake is a simple list of its properties (e.g. temperature distribution, day length, food supply) which is updated by the program to reflect the situation at different seasons. These data were collected during a two-year period 1973 and 1974. Food supply is not modelled in detail. Because the biomass of *Daphnia* is less than 20% of total pelagic zooplankton, the lake is treated as an open system with respect to food supply. There is no loop back from zooplankton to phytoplankton. A fixed function is used to describe the seasonal change of the relation between body length of *Daphnia* and the number of eggs they produce. This function has been derived from field data by regression analysis for each species. An example of this function is presented in fig. 4.

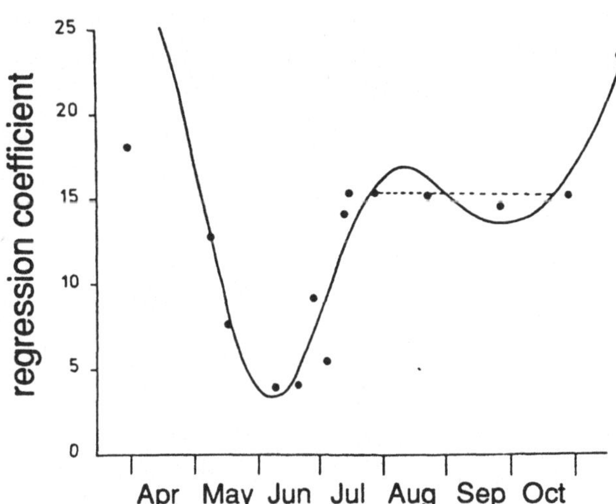

Figure 4: Seasonal changes of the regression coefficient of clutch size on the body length of a female. The filled circles are coefficients which were calculated by field data analysis of *Daphnia hyalina*. The smooth line is an estimation by Fourier regression on this data. The broken line shows the data which were used for the simulation during summer.

The **fish population** is treated as a single unit. Its properties were estimated by field data analysis. It was assumed that fish were the only source of *Daphnia* mortality. The PROCESS CLASS FISH daily inspects the populations of the *Daphnia* and decides for each individual whether it is eaten (that means removed from the system). The probability is calculated in several steps:

(a) Calculate the overall mortality rate of zooplankton from a function which describes the effects of water temperature and abundance of plankton. This function is given in equation (1)

$$m = (\text{Temperature} - 9)^T \cdot \text{Abundance}^A \cdot K \tag{1}$$

where T= 1.4086, A= 0.5764, K= 0.00009. These values were derived by partial regression analysis from field data. Temperature is measured in centigrade, abundance in individuals per 100 l. The reduction of the actual lake temperature by 9°C takes into account that fish have this threshold temperature to start feeding in spring.

(b) Calculate a selection coefficient (W_i) for each *Daphnia* from its relative size and the relative abundance of its species. This coefficient is estimated by equation (2), which again was derived from field data analysis.

$$\ln(W_i) = S \cdot \ln(\text{relative size}) + A \cdot \ln(\text{relative abundance}) \tag{2}$$

where S= 1.767 and A= 0.388.

(c) Calclulate the individual death rate (m_i) by (3):

$$m_i = W_i \cdot m \tag{3}$$

(d) The individual probability of dying during one day (p_m) is calculated from this death rate by (4):

$$p_m = 1 - \exp(-m) \tag{4}$$

Daphnia were simulated as individuals. The most important property of an individual is its size. Its fertility (number of eggs which it can produce) and its mortality depend on this. Additionally, size influences the position of an individual within the water column and therefore the duration of development. This is caused by a strong vertical temperature gradient in the lake during the summer and a size dependent vertical distribution and migration of the *Daphnia*. All these components are modelled within the PROCESS CLASS DAPHNIA.

There are several functions which describe individual growth and fertility. Only two will be described in detail: The duration of development from one instar to the other is estimated by a polynomial equation (5) which was fitted to data from HALL (1964).

$$D = \exp(K0 + K1 \cdot T + K2 \cdot T^2 + K3 \cdot T^3 + K4 \cdot T^4 + K5 \cdot T^5) \tag{5}$$

In this equation T is the natural logarithm of the temperature of the animal in °C, D= duration of development from one instar to the other in days, K0= -1.732, K1= 14.297, K2= -16.28, K3= 8.998, K4= -2.464, K5= 0.2589. For juvenile instars this duration of development is reduced by 25%.

The duration of development for each individual was estimated by integration of the incremental development during daily vertical migration. For this migration behaviour the length of day and the body length of the *Daphnia* are the most important parameters. Length of day determines the time span during which *Daphnia* stay at warm upper water layers (during night) or cold lower water layers (during day).

The body length influences the depth and the amplitude of migration. The depth function has been estimated for each species from field data. A simple linear relationship can be applied (6):

$$\text{depth} = a + b \cdot \text{size} \qquad\qquad (6)$$

where depth= position of an individual within the water column in meters, size= body length in mm. The values for a and b are given in tab. 1.

Table 1: Parameters for the estimation of the vertical distribution of *Daphnia* depending on species, size and day time					
	Daphnia hyalina		*Daphnia galeata*		*Daphnia cucullata*	
	a	b	a	b	a	b
during day	3.319	4.320	0.786	5.594	0.937	6.489
during night	-1.484	4.337	-1.430	3.735	1.449	0.534

Size is estimated by a growth function which takes into account instar, temperature and season. Clutch size of each individual *Daphnia* is estimated from body length and the seasonal regression coefficient which is provided by the PROCESS CLASS LAKE.

3.   Simulation Experiments and Results

In a first step all parameters were used without modification. These basic simulations showed a good correspondence with the field data (see fig. 5), but not all properties of the populations in the field can be found in the model, e.g. the fast breakdown of the population densities of the larger species during June is not found to the same extent. Birth and death rates correspond with the field observations in amplitude and seasonal trend. Short-term fluctuations of all growth parameters resemble the field data. This corroborates the interpretation that they are not caused by sampling errors in the field but result from continuous changes of the environment which prevent the system from reaching a steady state. This too is reflected in the seasonal trend of the age structure of populations which can be easily studied with the model.

The results of the simulations show that our knowledge of the system is sufficient to understand the qualitative and most of the quantitative causes of the forces which influence the population dynamics of the *Daphnia* species. This result is remarkable, since the observations of the behaviour of the model were made at one or two levels above the processes which were modelled. Therefore it can be argued that not only the phenomenon of the coexistence has been described, but also that the underlying processes have been understood.

This results enable different experiments to be performed with the model in order to seek key factors which are responsible for the stable coexistence. In these simulation experiments environmental condition as well as properties of *Daphnia* and fish were changed.

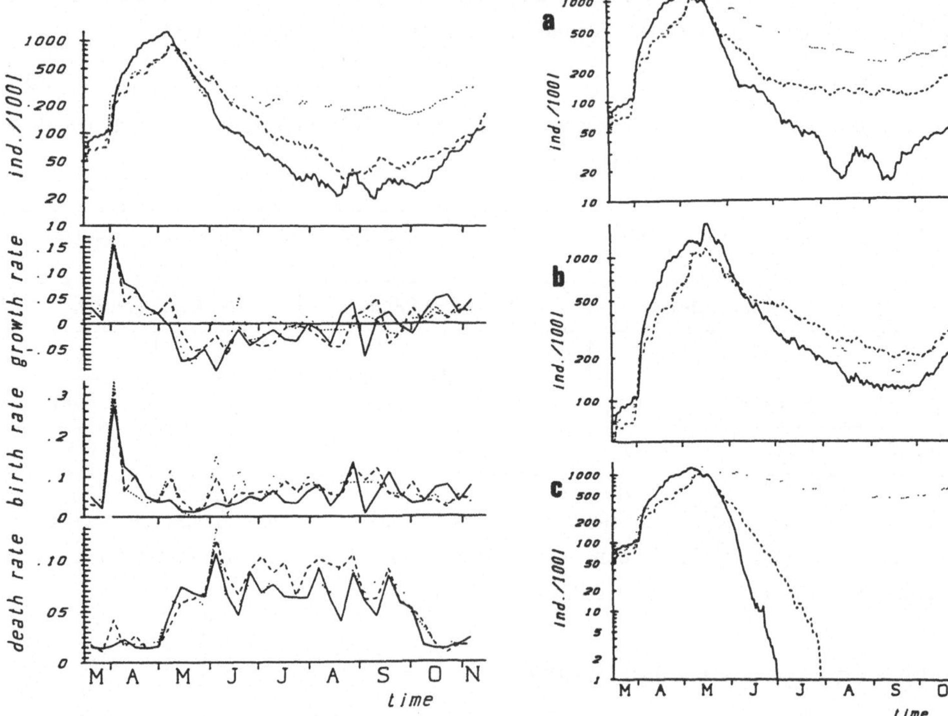

Figure 5: Result of the simulation of the population dynamics of the three *Daphnia* species and the growth, birth and death rates of the populations during one simulated year. Solid lines= *Daphnia hyalina*, broken lines= *D. galeata*, dotted lines= *D. cucullata*.

Figure 6: Population dynamics of the *Daphnia* in the model under the conditions of altered selectivity of fish. (a) fish do not distinguish between species but select only depending on size and abundance. (b) fish feed only abundance dependent. (c) fish select only size dependent.

No spectacular effects were found as long as the selective behaviour of fish was not altered and as long as the seasonal trend of environmental conditions changed in the usual sequence. Consequences of the manipulated selectivity are shown in fig. 6. As expected, the abundance dependent feeding stabilizes the coexistence and the size selectivity favours the smaller species.

Under homothermic conditions, where differential vertical distribution and migration have no consequences, the larger *D. hyalina* becomes dominant throughout the year because of its higher fertility. This effect is not found if fish are allowed to feed selectively.

The influence of environmental change was tested under two conditions: (a) constant food supply and seasonal variation of fish-induced mortality and (b) fixed values for fish and environment. These simulations again proved the strong effect of fish. In case (a), the population dynamics were only quantitatively different from the basic model. Case (b) demonstrated the effect of the continuously changing environment. Only under conditions of "constant August" and "constant October" was a balance between birth and death rates reached. Other conditions resulted in an unrealistic continuous growth which is limited by a (not simulated) reduction of food. The seasonal change of species composition is reflected in these data: *D. cucullata* becomes dominant during August and October, *D. hyalina* and *D. galeata* are dominant during spring and early summer.

4. Conclusions

The technique of discrete event simulation proved to be a very convenient tool to model complex interactions between animals and their biotic and abiotic environment. The model allowed a quantitative test of the conditions under which a stable coexistence of the three *Daphnia* species was reached. The driving forces were selective feeding of fish and seasonal changes of environmental conditions.

Since the model requires only data about animals which can be measured in the lab and field data which can be easily collected, it can be used as a tool to predict effects of anthropogenetic influences on the populations and communities.

5. References

DEMOTT W.R. (1983): Seasonal succession in a natural *Daphnia* assemblage. Ecol. monogr. **53**: 321-340

DODSON, S.I. (1974): Zooplankton competition and predation: an experimental test of the size-efficiency hypothesis. Ecology **55**: 605-613

GLIWICZ, Z.M., GHILAROV, A. & PIJANOWSKA J. (1981): Food and predaation as major factors limiting two natural populations of Daphnia cucullata Sars. Hydrobiologia **80**: 205-218

GREEN C.H. (1983): Selective predation in freshwater zooplankton communities. Int. Rev. ges. Hydrobiol. **68**: 297-315

HALL D.J.(1964): An experimental approach to the dynamics of natural populations of *Daphnia galeata mendotae*. Ecology **45**: 94-112

HARDIN G. (1960): The competitive exclusion principle. Science **131**: 1292-1297

HUTCHINSON G.E. (1959): Homage to Santa Rosalia or why are there so many kinds of animals? Amer. Nat. **93**: 145-159

- (1961): The paradox of the plankton. Amer. Nat. **95**: 137-145

JACOBS J. (1977a): Coexistence of similar zooplankton species by differential adaptation to reproduction and escape, in an environment with fluctuating food and enemy densities. I. A model. Oecologia (Berl.) **29**: 233-247

- (1977b): Coexistence of similar zooplankton species by differential adaptation to reproduction and escape, in an environment with fluctuating food and enemy densities. II. Field data analysis. Oecologia (Berl.) **30**: 313-329

- (1985): Konkurrenz und Einnischung - Hat Konkurrenz um Ressourcen eine evolutionsbiologische Bedeutung für die Artenmannigfaltigkeit der Tiere? Z. zool. Syst. Evolut.-forsch. **23**: 243-258

LYNCH M. (1977): Zooplankton competition and plankton community structure. Limnol. Oceanogr. **22**: 775-777

NORTHCOTE T.G. (1988): Fish in the structure and function of freshwater ecosystems: A "top-down" view. Can. J. Fish. Aqua. Sci. **45**: 361-379

SEITZ A. (1980a): The coexistence of three species of *Daphnia* in the Klostersee: I. Field studies on the dynamics of reproduction. Oecologia (Berl.) **45**: 117-130

- (1980b): The coexistence of three species of *Daphnia* in the Klostersee: II. The stabilizing effect of selective mortality and conclusions for the stability of the system. Oecologia (Berl.) **47**: 333-339

- (1984): Are there allelopathic interactions in zooplankton? Laboratory experiments with *Daphnia*. Oecologia (Berl.) **62**: 94-96

TILLMANN U. & LAMPERT W. (1984): Competitive ability of differently sized *Daphnia* species: An experimental test. J. Freshwater Ecol. **2**: 311-323

ZARET T.M. (1980): Predation and freshwater communities. Yale University Press, New Haven, London. 187 p.

RISK ASSESSMENT OF TOXICANTS TO PELAGIC FOOD-WEBS:
A SIMULATION STUDY

H.J. Poethke, D. Oertel & A. Seitz
AG Populationsbiologie, Institut für Zoologie der Universität Mainz
Saarstraße 21, 6500 Mainz, FRG.

Abstract

For a generalized mathematical model of the pelagic compartment of a lake ecosystem we show that a small variance in input parameters may result in a large variance in predicted biomasses. Since input parameters for this kind of models always show a great variance, we conclude that deterministic prediction of biomass-trajectories is rather useless and demonstrate how stochastic analysis of the model may lead to a helpful risk analysis.

Zusammenfassung

An einem einfachen Modell für das Pelagial eines Sees zeigen wir, daß bereits geringe Änderungen in den Eingangsparametern zu beträchtlichen Änderungen bei den prognostizierten Biomassen führen können. Da die Eingangsparameter für solche Modelle jedoch stets eine große Streuung aufweisen, schließen wir, daß deterministische Voraussagen für derartige Systeme zwecklos sind und demonstrieren, daß eine stochastische Analyse einer großen Zahl von Monte-Carlo Simulationen die geeignete Form einer Risiko-Analyse für solche Systeme sein kann.

1. Introduction

Ecotoxicological assessment of the impact of chemical substances on the aquatic environment is still one of the unsolved problems of environmental management and protection (Schlosser 1988). Though there are test procedures which allow the detection of acute effects of such substances on single organisms or even sublethal effects and influences on the level of populations (e.g. FITSCH & KAISER 1987), there is still a lack of knowledge on how particular substances may influence whole biota. Since it will usually not be possible to examine such effects in situ (i.e. by experiments with toxic substances in lakes, rivers or large reservoirs), most research is restricted to model-ecosystems (micro- and mesocosm experiments) and mathematical models of ecosystems that enable extrapolation from the effects on single species to the effects on the level of ecosystems.

Simulation models have been widely applied to aquatic systems for predicting the effect of cultural eutrophication (e.g. DiTORO et al. 1975, JORGENSEN 1976, 1983) of power plant operation (SWARTZMAN et al., 1978) or the effect of toxic substances introduced into the environment (e.g. PARK et al. 1980, JORGENSEN 1983). In the following we will focus our interest on the problem resulting from a relatively high variance in data supporting these models.

2. The model

In our work we used a model that is principally based on the model structure developed for the pelagic compartment of CLEAN, a generalized lake-ecosystem model (PARK et al. 1975, SCAVIA et al. 1975). Nutrient recycling is modelled in accordance with ROSE (1985) dividing nitrogen and phosphorus into various organic and inorganic component forms. First order kinetics are used. Phosphorus decay is modelled as a three step process (detrital phosphor - dissolved organic phosphor - dissolved inorganic phosphor) and nitrogen decay is modelled as a five step process (detrital nitrogen - diss. organic nitrogen - ammonia - nitrite - nitrate). For the sake of simplicity and to reduce computing time the model-ecosystem was made up of five species only: two phytoplancton species, two zooplankton species and a planktivorous fish (see Fig.1) This reduction resulted in a model of 16 state variables: eight for the concentrations of the different component forms of nitrogen and phosphor, six for the biomasses, nitrogen concentrations and phosphor concentrations of the two algal species and two for the biomasses of the two zooplankton species. Since we performed simulation experiments over a time span of one year only, we did not explicitly model fish growth but assumed the fish stock to be constant (resulting in a constant predation pressure on the zooplankton).

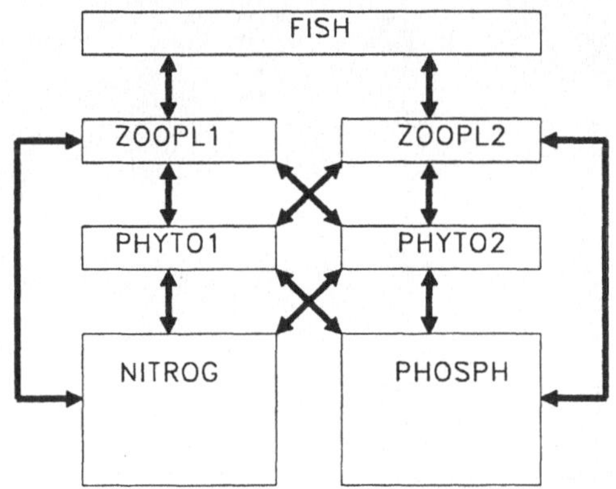

Figure 1.
Structure of the
simulation model

3. Simulation results

If simulation models are used as predictive tools severe problems may arise from a high variance of of input data. For pelagic ecosystem models most data can be found in the literature, but the variance in these data is enormous. Depending on the literature used, values for the maximum rate of photosynthesis of Scenedesmus for example vary between 0.1 and 3.4 /day (for further examples see ROSE 1985). This variance may result e.g. from between-lake fluctuations in environmental variables not included in the model, like acidity or concentration of trace elements, or from genetic differences between populations. But, even if data are collected for one site and one specific year, spatial heterogeneity and limited accuracy of measurements may result in a substantial variance of model parameters.

To give an impression of the influence of variable input parameters on model outputs we did the following: The simulation model was started with a "standard" set of parameters. To reach an equilibrium state simulation was carried out for ten years. Biomass data at the end of the tenth year were taken as "standard initial condition" for the subsequent simulations. Starting from this initial condition, 4000 simulation runs were carried out covering one year each. Input parameters to these simulations were generated from the "standard" set by multiplying each of the 76 input parameters with a normal distributed

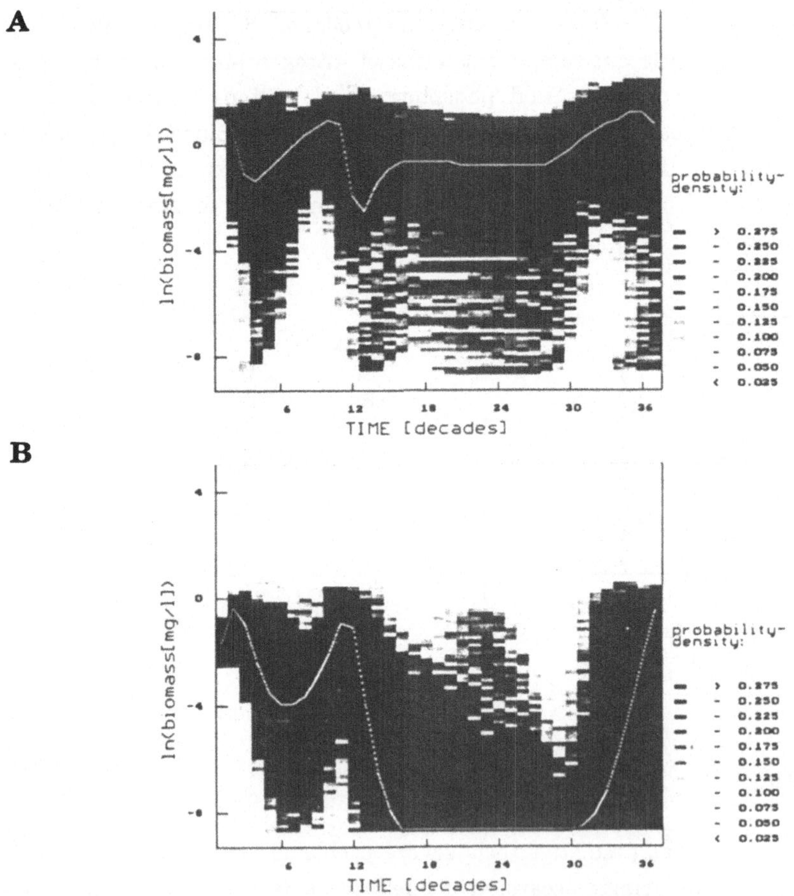

Figure 2.
Density distribution of total phytoplankton (A) and total zooplankton (B) resulting from 4000 simulation runs with variable input parameters (for details see text).

random number (mean=1.0; standard deviation=0.10). Every ten days biomasses of all species were classified using a logarithmic scale comprising 100 classes ranging from exp(-9.0) to exp(+5) [mg dry weight per litre]. Thus the simulation runs resulted in density distributions over these 100 classes for each of the 36 decades of a year (see Fig.2).

From the results of these simulation runs we can see that even for this rather simple model structure and the assumed small variance of input parameters the variance in the predicted courses of biomasses over the year is enormous. Since the variation of input parameters between sites and between years will usually be larger then the 10% assumed in our simulation runs, we conclude that deterministic predictions of the abundance of a particular species at a particular time of the year are rather meaningless. This conclusion is supported by a great number of field observations which show that between-year differences of biomass trajectories of algal biomass of the same lake (for an example see ECKARTZ-NOLDEN 1988) or between pond differences between "identical" experimental ponds (for an example see PELZER 1988) may be enormous. In an ecotoxicological analysis the influence of fuzzy input parameters will thus likely disguise small longterm effects of chemicals.

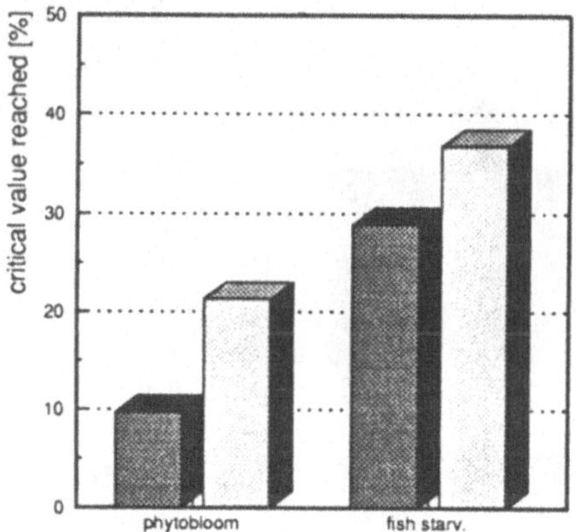

Figure 3.
Frequency of simulation runs that produce a phytoplankton bloom (peak phytoplankton density exceeds a critical value) and frequency of simulation runs that result in fish starvation (mean zooplankton density is below a critical value) for the randomized standard parameter set (dark shaded bars) or for a randomized parameter set with respiration rate of both zooplankton species increased by 10% (light shaded bars)

The problems arising from a great variance of input parameters may be tackled by a stochastic analysis of a large number of simulation runs started with input parameters which are drawn from a probability distribution (which is given by the range of parameters found in the literature or derived from laboratory experiments). To reduce the enormous amount of output data produced by thousands of different simulation runs, the analysis should focus on well defined undesirable ecosystem states (catastrophes). Such catastrophes will depend on the actual use of the lake under study. For a large water reservoir it may be a phytoplankton bloom (phytoplankton biomass rises above a critical level). For another lake it may be a reduction of zooplankton biomass that leads to severe starvation of a particularly valuable fish species (for some examples of water quality parameters and management variables see SEIP & IBREKK 1988).

Fig.3 gives an example of such a risk analysis [O'NEILL et al., 1983; GINTZBURG & AKCAKAYA, 1990]. Simulations were carried out with the standard parameter set and for a polluted pelagic system. The effect of the pollutant was assumed to be an increase in respiration losses (10%) for both zooplankton species. From 4000 simulation runs performed with the randomized (multiplied with a normal distributed random number, mean=1; standard deviation=0.1) standard parameter set only ten percent produced peak phytoplankton values that exceeded a critical value of 7.5 mg dry weight per litre. However, when zooplankton respiration losses were increased by 10%, more than 20% of the simulation runs produced phytoplankton peaks above this threshold. The probability of fish starvation was also increased by the introduction of the pollutant.

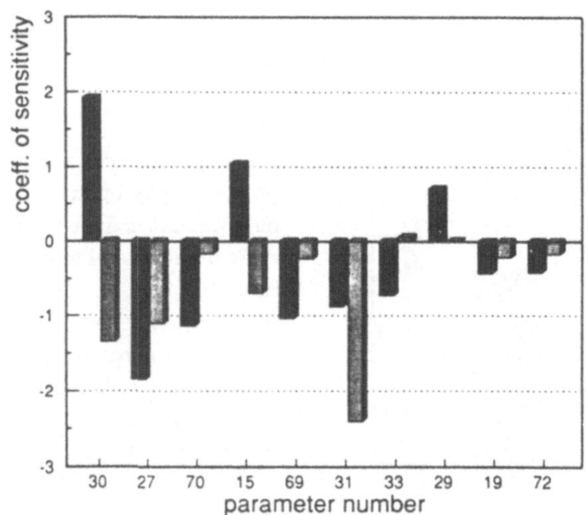

Figure.4
Sensitivity coefficients for the ten most influential input parameters of the model. Dark bars refer to the change in peak phytoplankton biomass, light bars to the change in mean zooplankton biomass (for details see text).

Ecosystem risk analysis may also be helpful in pinpointing species-effects which might be particularly harmful for the ecosystem under study. Fig.4 shows the result of a sensitivity analysis for the ecosystem model. Sensitivity coefficients are calculated dividing the change of the output value (dark bars - output is peak phytoplankton biomass; light bars - output is mean zooplankton biomass) by the change of the input parameter (see JORGENSEN 1986). Coefficients are given for the ten parameters having the strongest effect on the change in peak phytoplankton biomass. They are organized according to their absolute value.

Numbers refer to : 30 - maximum respiration rate of zoopl.spec.1; 27 - maximum filtration rate of zoopl.spec.1; 70 - preference of zoopl.spec.1 for phytopl.spec.1; 15 - maximum rate of photosynthesis of phytopl.spec.2; 69 - assimilation coefficient of phytopl.spec.1 for zoopl.spec.1; 31 - optimum temperature for zoopl.spec.1; 33 - maximum filtration rate of zoopl.spec.1; 29 - coefficient of saturation for filtering efficiency of zoopl.spec 1; 19 - optimum nitrogen conc. for growth of phytopl.spec.2; 72 - preference of zoopl.spec.1 for phytopl.spec.2.

It can be seen that the choice of output value (peak phytoplankton density vs. mean zooplankton density) has a great influence on the sensitivity coefficient of the different parameters. Respiration rate of zooplankton species 1 has a strong positive influence on peak phytoplankton density but has the tendency to decrease mean zooplankton density. Mean zooplankton density is most strongly influenced by the optimum temperature of zooplankton species 2. This parameter only slightly influences peak phytoplankton density. The structure of the ecosystem under study will influence the sensitivity coefficients too. To demonstrate this effect we strongly reduced phosphor-input to the ecosystem and compared the sensitivity coefficients of this "oligotrophic" lake with the standard "eutrophic" lake (Fig.5; eutrophic lake - dark bars; oligotrophic lake - light shaded bars). Not only is model sensitivity strongly reduced by reduced nutrient concentration but also the relative importance of different parameters is changed.

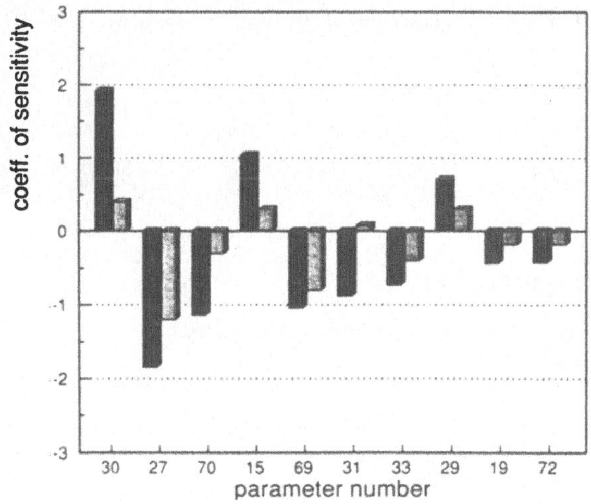

Figure 5.
Sensitivity coefficients of the ten most influential input parameters. Dark bars refer to the eutrophic lake, light bars refer to the oligotrophic lake (for details see text).

In the simple linear sensitivity analysis as presented in Figs. 4 and 5 it is implicitly assumed that the ecosystem reacts linearly to a change in input parameters. Since our question is not "which parameter has the strongest influence on certain output values near our standard parameter set?" but "the manipulation of which parameter may result in a phytoplankton bloom or in severe fish starvation?" this approach may not be adequate. For the ecosystem under study we thus studied the correlation between the incidence of these "catastrophes" and parameter modification in a statistic analysis of 4000 random simulations. The results confirm the tendency of the simple sensitivity analysis; the number of parameters, however, which are really important for the model output seem to be much smaller in this kind of analysis. But we need more simulation experiments to confirm this result.

The model presented above is far from describing the whole complexity of a pelagic ecosystem. But even for this simple model the variance in input parameters resulting from limited accuracy of measurements or between year and between lake variance of organisms creates substantial variance in the predicted course of species biomasses in the course of the year. Because of this vagueness deterministic modelling is a rather weak tool for the generation of general statements (covering more than one year and more than one particular lake) in ecotoxicology. Inaccuracy of parameters and vagueness of some of the algorithms used make it necessary to be aware of our ignorance concerning the "real" interactions and to incorporate it in the form of a stochastic analysis covering the whole probability space of parameters and algorithms. With increased availability of fast computers stochastic analysis by Monte Carlo simulation is possible even for rather complex systems, and this kind of analysis of ecological risk will be a helpful tool in ecosystem management.

References

Eckartz-Nolden, G. (1988) : Untersuchungen zur jahreszeitlichen Veränderung der Phytoplanktonpopulationen des Laacher Sees unter Berücksichtigung der Beziehungen zum Zooplankton. Diss. Univ. Bonn, 93 p.

Fitsch, V., Kaiser, H. (1987) : Population dynamics of Daphnia
magna - Simulations using the individuals approach. in Möller, D.P.F. (Ed) System analysis of Biological Processes. Vieweg, Braunschweig, Wiesbaden.

Gintzburg, L.R., Akçakaya, H.R. (1990) : Ecological risk analysis for single and multiple populations. In Seitz, A. and Loeschcke V. (Eds.) Species Conservation: A population biology approach. Birkhäuser Verlag, Basel

Jorgensen, S.E. (1976) : A eutrophication model for a lake. Ecol.Model. 2: 147-165.

Jorgensen, S.E. (1983) : Modelling the distribution and effect of toxic substances in aquatic ecosystems. In Jorgensen, S.E. (Ed.): Application of ecological modelling in environmental management, Part A. Elsevier Scientific Publications, Amsterdam.

Jorgensen, S.E. (1986) : Fundamentals of ecological modelling. Elsevier, Amsterdam

O'Neill, R.V., Bartell, S.M., Gardner, R.H. (1983) : Patterns of toxicological effects in ecosystems: a modelling study. Environmental Toxicology and Chemistry 2, pp. 451-461

Park, R.A., O'Neill, R.V. et al (1975) : A generalized model for simulating lake ecosystems. Simulation : 33-50.

Park, R.A., Connolly, C.I. et al. (1980) : Modelling transport and behaviour of pesticides and other toxic organic materials in aquatic environments. Center for ecological Modelling Report No. 7, Rensselaer Polytechnic Institute.

Pelzer, M. (1988) : Zeigen parallel angesetzte Ökosysteme eine gleiche Dynamik? Diplomarbeit. RWTH Aachen

Rose, K.A. (1985) : Evaluation of Nutrient-Phytoplankton-Zooplankton models and simulation of the ecological effects of toxicants using laboratory microcosm ecosystems. PhD. Thesis, University of Washington.

Scavia, D., Bloomfield, J.A. et al (1975): Documentation of CLEANX: a generalized model for simulating the open-water ecosystems of lakes. Simulation: 51-56.

Schlosser, H.J. (1988): Auswertung ökotoxikologischer Forschungen zur Belastung von Ökosystemen durch Chemikalien. Projektleitung Biologie, Ökologie, Energie (PBE) der Kernforschungsanlage Jülich.

Seip, K.L., Ibrekk, H. (1988): Regression equations for lake management - how far do they go? Verh.internat.Verein.Limnol.23, pp. 778-785.

Swartzman, G.L., Deriso, R.B., Cowan, C. (1978) : Comparison of simulation models in assessing the effects of power-plant-induced mortality on fish populations. Nuclear regulatory Commission Techn. Report CR-0474.

DiToro, D.M., O'Connor, D.J., Thomann, R.V., Mancini, J.L. (1975) : Phytoplankton-zooplankton-nutrient interaction model for western Lake Erie. in: B.C. Patten (Ed.) Systems analysis and simulation in ecology, Vol. III., Academic Press, New-York, pp. 423-474.

UMWELTQUALITÄT

Ein pharmakokinetischer Ansatz zur Untersuchung der Aufnahme von ^{137}Cs
durch Kinder nach dem Reaktorunfall von Tschernobyl

U. Wellner

Institut für klinische und experimentelle Nuklearmedizin der
Medizinischen Einrichtungen der Universität zu Köln
Direktor: Prof.Dr.med. H. Schicha

Zusammenfassung

Der Verlauf der Radioaktivität von ^{137}Cs in Kindern nach dem Reaktorunfall von Tschernobyl, der mit Hilfe eines Ganzkörperzählers gemessen wurde, konnte auf der Grundlage eines pharmakokinetischen An-
satzes theoretisch nachgebildet werden. Die Kinder des Betriebskindergartens der Universitätskliniken Köln nahmen während der Vegetationsperiode 1986/87 86,9, 1987/88 114,4 und 1988/89 24,4 Bq ^{137}Cs pro
kg Körpergewicht auf.

Einleitung

Zweck der Studie war die Untersuchung der Folgen des Reaktorunfalls
von Tschernobyl für Kinder im Kölner Raum mit Hilfe von Aktivitätsmessungen über Ganzkörperzähler.

Methodik und Ergebnisse

Bei der Untersuchung (*) der Auswirkungen des Reaktorunfalls von
Tschernobyl auf die Bevölkerung mit Hilfe von Aktivitätsmessungen über
Ganzkörperzähler wurde eine Längsschnittstudie an Kindern des örtli-
chen Betriebskindergartens durchgeführt.

Die Ergebnisse der Messungen sind in Tabelle 1 dargestellt.

---

* Gefördert vom Bundesminister für Umwelt, Naturschutz und Reaktorsicherheit

Tabelle 1: Ergebnisse der Ganzkörpermessungen			
Monat.Jahr Messung	Zahl der Kinder	Mittelwert $Bq \cdot kg^{-1}$	Streuung $1\ \sigma$
7.86	18	10,45	2,51
9.86	15	10,99	2,95
11.86	9	7,77	0,96
12.86	11	6,82	0,77
6.87	15	14,52	3,97
12.87	7	5,20	0,64
4.88	7	2,34	0,57
8.88	4	3,72	0,77
10.88	2	1,06	0,60
2.89	4	1,44	0,70

Eine erste maximale Anreicherung wurde 1986 ca. 3 Monate nach dem Reaktorunfall von Tschernobyl beobachtet. Die fortgesetzten Untersuchungen bis jetzt zeigten, daß jeweils im Frühjahr ein Anstieg der Radioaktivität in den Kindern zu beobachten war.

Die effektive Halbwertszeit $(T_{\frac{1}{2}eff})$, mit der das ^{137}Cs aus dem Körper der Kinder verschwindet, beträgt 27 Tage (3,4).

Die Betrachtung des kindlichen Organismus als Einkompartimentsystem, das pro Zeiteinheit und kg Körpergewicht eine Aktivität $a(t)$ aufnimmt und mit der effektiven Eliminationskonstante $\lambda_{eff} = ln2/T_{\frac{1}{2}eff} = \beta$ verliert, war naheliegend (1,2). Für den obigen Wert von $T_{\frac{1}{2}eff}$ ist $\beta = 0,770$ Monat^{-1}.

Für die ^{137}Cs-Zufuhr wurde der Ansatz

$$a(t) = a_0 \cdot e^{-\alpha \cdot t} \qquad [Bq \cdot kg^{-1} \cdot Monat^{-1}]$$

gewählt.

Für die Aktivitätsänderung pro kg Körpergewicht gilt:

$$\frac{dx}{dt} = -\beta \cdot x + a_0 \cdot e^{-\alpha \cdot t} \qquad [Bq \cdot kg^{-1} \cdot Monat^{-1}]$$

Diese Bilanzgleichung hat die Lösung:

$$x = \frac{a_0}{\alpha - \beta} \cdot (e^{-\beta \cdot t} - e^{-\alpha \cdot t})$$

Die maximale Anreicherung von ^{137}Cs im Organismus wird zum Zeit-punkt t_{max} beobachtet. Dieser Zeitpunkt ist nur von den Koeffizi-enten α und β abhängig:

$$t_{max} = \frac{ln\alpha - ln\beta}{\alpha - \beta}$$

Zu diesem Zeitpunkt erreicht die maximale Anreicherung den Wert

$$x_{max} = \frac{a_0}{\alpha} \cdot e^{-\beta \cdot t_{max}}$$

Der Zeitpunkt t_{max} und die maximale Anreicherung x_{max} können aus der Darstellung der Meßergebnisse abgelesen werden. Da β bekannt ist, und für die beobachtete Gruppe $t_{max} = 2,6$ Monate ist, können α [Monat^{-1}] und a_0 [Bq$\cdot$kg$^{-1}\cdot$Monat^{-1}] berechnet werden.

Die pro Vegetationsperiode, jede Vegetationsperiode beginnt im Frühjahr, dem Menschen zugeführte Radioaktivität ist

$$A = a_0 \cdot \int_0^\infty e^{-\alpha.t} \, dt = a_0/\alpha$$

Für die Vegetationsperioden 1986/87, 1987/88 und 1988/89 wurden folgende Werte ermittelt und bei den theoretischen Ansätzen für alle Vegetationsperioden verwendet:

$\alpha = 0,150$ Monat^{-1} ($T_{\frac{1}{2}z}$ = 4,6 Monate) und

$\beta = 0,770$ Monat^{-1} ($T_{\frac{1}{2}eff}$ = 0,9 Monate = 27 Tage).

$T_{\frac{1}{2}z}$ ist die Halbwertszeit, mit der die Aktivitätszufuhr abnimmt. Sie wird aus α berechnet.

Die maximale Anreicherung von ^{137}Cs in den Kindern wurde durch Subtraktion der Erwartungswerte aus dem theoretischen Ansatz von den Meßwerten zu diesem Zeitpunkt ermittelt.

Der Zeitpunkt $t = 0$ ist für den Kölner Raum der 1. Mai 1986. Dies ist der Zeitpunkt der primären Kontamination der Biosphäre durch den Reaktorunfall von Tschernobyl. Die zweite Vegetationsperiode begann im Februar 1987 (II) und die dritte im April 1988 (III).

Der Aktivitätsverlauf in den Kindern läßt sich durch die folgenden Gleichungen beschreiben:

Vegetationsperiode I:

$$A(t) = \frac{13,0}{\alpha - \beta} \cdot (e^{-\alpha \cdot t} - e^{-\beta \cdot t})$$

Vegetationsperiode II:

$$A(t') = \frac{13,0}{\alpha - \beta} \cdot (e^{-\alpha \cdot (9,3 + t')} - e^{-\beta \cdot (9,3 + t')}) +$$

$$+ \frac{17,2}{\alpha - \beta} \cdot (e^{-\alpha \cdot t'} - e^{-\beta \cdot t'})$$

Vegetationsperiode III:

$$A(t'') = \frac{13,0}{\alpha - \beta} \cdot (e^{-\alpha \cdot (22 + t'')} - e^{-\beta \cdot (22 + t'')}) +$$

$$+ \frac{17,2}{\alpha - \beta} \cdot (e^{-\alpha \cdot (12,7 + t'')} - e^{-\beta \cdot (12,7 + t'')}) +$$

$$+ \frac{3,7}{\alpha - \beta} \cdot (e^{-\alpha \cdot t''} - e^{-\beta \cdot t''})$$

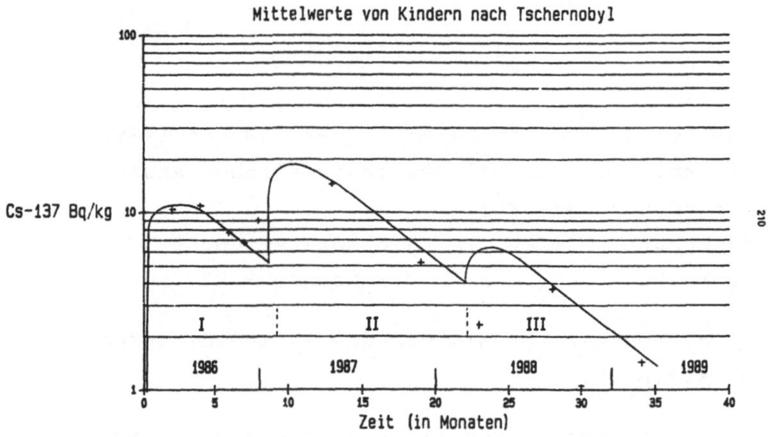

Abb. 1: Gemessene ^{137}Cs-Aktivität pro kg Körpergewicht in Kindern verglichen mit der theoretisch abgeleiteten Aktivität gem. I, II und III

Diskussion

Die Auswertung der Längsschnittstudie an Kindern des Betriebskindergartens der Medizinischen Einrichtungen der Universität zu Köln hat ergeben, daß in jedem Jahr nach dem Reaktorunfall von Tschernobyl ein relatives Maximum der ^{137}Cs-Aktivität pro kg Körpergewicht aufgetreten ist.

Dieser Aktivitätsverlauf kann durch eine Hypothese über die Aktivitätszufuhr beschrieben werden:

$$a(t) = a_0 \cdot e^{-\alpha \cdot t} \qquad [Bq \cdot kg^{-1} \cdot Monat^{-1}]$$

Die Zeitkonstante $1/\alpha$, mit der die Aktivitätszufuhr abnimmt, hat in jeder Vegetationsperiode den gleichen Wert. Auch die biologische Halbwertszeit, mit der ^{137}Cs von den Kindern ausgeschieden wird, wurde für die Dauer der Studie als unveränderlich angesehen.

Dadurch ist die effektive Eliminationskonstante β ebenfalls eine konstante Größe für dieses untersuchte Kollektiv.

Die Differentialgleichung, die die Bilanz für ^{137}Cs pro kg Körpergewicht beschreibt, hat eine Lösung, die die Aktivität zu jedem Zeitpunkt nach Beginn der Aktivitätszufuhr beschreibt. Die maximale Aktivität pro kg Körpergewicht tritt bei den vorgegebenen Werten für α und β 2,6 Monate nach Beginn der Aktivitätszufuhr auf.

Die Analyse des Aktivitätsverlaufs diente der Bestimmung der Aktivitätszufuhr a und des Zeitpunktes t, zu dem die Aktivitätszufuhr in jedem Jahr mit a_0 begann.

Mit diesem pharmakokinetischen Ansatz konnte die ^{137}Cs-Aktivität pro kg Körpergewicht für die Vegetationsperioden I, II und III nach Tschernobyl beschrieben werden (Abb.1). Allein durch die Anpassung der maximalen Aktivitätszufuhr a_0 und des Zeitpunktes, zu dem diese Akivitätszufuhr in jedem Jahr begann, gelang die Anpassung des Modellverlaufs an die Meßwerte. Die kinetischen Konstanten des Systemes, α und β, standen bei dieser Anpassung nicht zur Disposition. Diese Tatsachen können als Stützung der obigen Hypothesen gewertet werden.

Tabelle 2: Aussagen zur ^{137}Cs-Aufnahme durch Kinder in Köln			
	Vegetationsperiode		
	1986/87	1987/88	1988/89
Nr.	I	II	III
Beginn	1.05.86	10.02.87	1.03.88
a_0 Bq·kg^{-1}·Mon.$^{-1}$	13,0	17,2	3,7
a_0/α Bq·kg^{-1}	86,9	114,4	24,4
Zeit t Monate	0	9,3	22,0
Zeit t' Monate	–	0	12,7
Zeit t" Monate	–	–	0

In Tabelle 2 sind die gefundenenen Werte für a_0 und die Zeitpunkte angegeben, zu denen in jedem Jahr nach dieser Hypothese die Aktivitätszufuhr mit einem neuen Schub begonnen hat. Die Zahlenwerte a_0/α stellen die Gesamtzufuhr von ^{137}Cs [Bq·kg^{-1}] für jede Vegetationsperiode dar.

Diese Untersuchung hat gezeigt, daß im Jahr 1987 mehr ^{137}Cs inkorporiert wurde als im Jahr des Reaktorunfalls. Eine Erklärung für dieses Phänomen ist, daß die Diskussionen über den Reaktorunfall von Tschernobyl die Auswahl der Nahrungsmittel offenbar nicht mehr im gleichen Ausmaß beeinflußt haben.

Durch eine exponentielle Extrapolation der gefundenen ^{137}Cs-Inkorporation in den Vegetationsperioden II und III nach I (zum Zeitpunkt

t = 0) wurde versucht, abzuschätzen, mit welcher Inkorporation in der
Vegetationsperiode I gerechnet werden mußte, wenn die Ernährungsgewohnheiten der Bevölkerung sich nach dem 1. Mai 1986 nicht verändert
hätten (Abb.2). Danach wäre während der ersten Vegetationsperiode nach
Tschernobyl eine Aufnahme von 355 Bq ^{137}Cs pro kg Körpergewicht zu erwarten gewesen. Tatsächlich wurde im Rahmen dieser Studie die Aufnahme
von 87 Bq ^{137}Cs pro kg Körpergewicht gefunden.

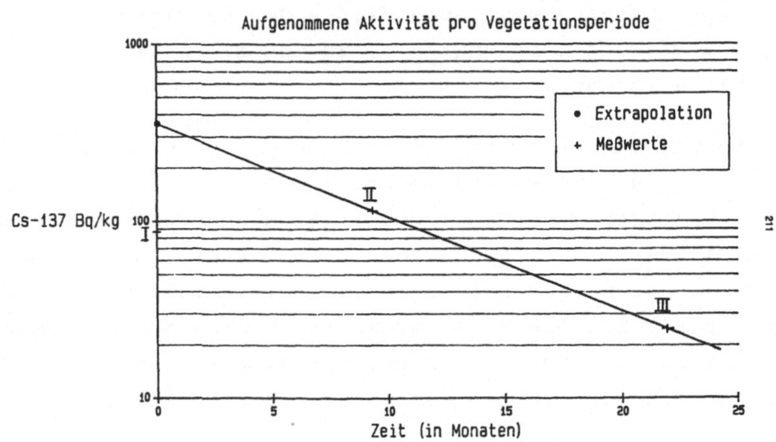

Abb. 2: Exponentielle Extrapolation zur Ermittlung der denkba-
ren ^{137}Cs-Aktivitätszufuhr im ersten Jahr nach dem Reaktorunfall von Tschernobyl

Die Differenz zwischen erwarteter und gemessener Aktivitätszufuhr im
ersten Jahr (I) nach dem Reaktorunfall von Tschernobyl kann als Einfluß der Empfehlungen (5) und der vielfältigen Diskussionen in der Öf-
fentlichkeit auf die Ernährungsgewohnheiten der Bevölkerung angesehen
werden. Vermutlich wäre die Aufnahme radioaktiver Stoffe mit der Nahrung durch die oberflächliche Kontamination der Nutzpflanzen, die im
Rahmen dieser Studie nicht untersucht werden konnte, noch viel größer
als der oben ermittelte Wert gewesen.

Für fünfjährige Kinder wird für ^{137}Cs die effektive Äquivalentdosis
pro zugeführte Aktivität durch Ingestion mit $8,6 \cdot 10^{-9}$ Sv/Bq (3), für
zehnjährige Kinder mit $9,3 \cdot 10^{-9}$ Sv/Bq (4) angegeben. Die Kinder, an
denen diese Untersuchungen durchgeführt wurden, waren im Mittel
7 Jahre alt und wogen im Mittel 25 kg. Durch lineare Interpolation erhält man für siebenjährige Kinder den Dosisfaktor
$8,9 \cdot 10^{-9}$ Sv/Bq.

Die mit diesen Werten berechnete effektive Äquivalentdosis durch Inkorporation von ^{137}Cs ist für diese Kinder in der folgenden Tabelle 3
zusammengefaßt (10 µSv = 1 mrem). Dabei wurden die Daten des
Bundesgesundheitsamtes, Institut für Strahlenhygiene (ISH) (3,4), zur

Berechnung der effektiven Äquivalentdosis verwendet.

Tabelle 3

Jahr	H
	µSv
1986	19,3
1987	25,5
1988	5,4
µSv	50,2
mrem	5,0

Aus der Zahl der ^{137}Cs-Zerfälle und einer angenommenen Energie-
absorption von 0,35 MeV pro Zerfall wurde die Energiedosis für die
Kinder bestimmt. Mit dem Qualitätsfaktor Q = 1 Sv/Gy für β- und Gam-
mastrahlung folgt für die Kinder in Köln die Äquivalentdosis 48 µSv
(4,8 mrem) als Folge des Unfalls von Tschernobyl, was mit dem oben angegebenen Wert, 50 µSv, gut übereinstimmt.

Da die natürliche Äquivalentdosis in den drei Jahren nach dem Unfall
von Tschernobyl im Kölner Raum ca. 3 mSv (300 mrem) betragen hat, die
durch den Reaktorunfall verursachte Dosisbelastung jedoch nur 0,050
mSv (5,0 mrem), ist eine meßbare Schädigung der Gesundheit der Kinder
und der übrigen Bevölkerung nicht zu erwarten.

Key words: Human body counter, ^{137}Cs, kinetic model, Tschernobyl

Stichwörter: Ganzkörperzähler, ^{137}Cs, pharmakokinetisches Modell,
            Tschernobyl

Literatur

(1)   WELLNER, U.: Tracer in Lebewesen: Quantitative Tracerkinetik
      für physiologische und phamakologische Substanzen -
      Stuttgart, Hippokrates-verlag 1982, ISBN 3-7773-0502-2

(2)   DOST, F.H.: Grundlagen der Pharmakokinetik
      Stuttgart, Georg Thieme Verlag 1968

(3)   ISH-Heft 79: Dosisfaktoren für Inhalation oder Ingestion von
      Radionuklidverbindungen (Altersklasse 5 Jahre)
      Bundesgesundheitsamt, München 1985, ISBN 3 - 924403 - 69 - 4

(4)   ISH-Heft 80: Dosisfaktoren für Inhalation oder Ingestion von
      Radionuklidverbindungen (Altersklasse 10 Jahre)
      Bundesgesundheitsamt, München 1985, ISBN 3 - 924403 - 70 - 8

(5)   Bundesminister für Umwelt, Naturschutz und Reaktorsicherheit:
      Auswirkungen des Reaktorunfalls in Tschernobyl in der
      Bundesrepublik Deutschland: Empfehlungen d. Strahlenschutz-
      kommission zur Abschätzung, Begrenzung und Bewertung
      (Red. D. Gumprecht; A. Kind)
      Stuttgart, New York: Fischer 1986
      ISBN 3-437-11084-5

(6)   Verordnung über den Schutz vor Schäden durch ionisierende
      Strahlen (Strahlenschutzverordnung - StrlSchV)
      BGBl. Teil I, Nr.125 (1976), S. 2905

PARAMETERSCHÄTZUNG IN NICHTLINEAREN MODELLEN
ANWENDUNG AUF EIN WASSERGÜTEMODELL

Karl-Friedrich Albrecht, Peter Rudolph
Zentralinstitut für Kybernetik und Informationsprozesse
Kurstraße 33
Berlin
1086

1. Einleitung

Bei dynamischen Systemen, die durch gewöhnliche Differentialgleichungen (DGS) gegeben sind, ist das statistische Modell durch

$$y(t_{k+1}) = x(t_k) + \int_{t_k}^{t_{k+1}} f(x(s),p)\,ds + \varepsilon_k$$

gegeben. Hierbei bedeuten y die Datenzeitreihe, ε_k unabhängige
zufällige , meist normalverteilte Fehler und x die Lösung des
parameterabhängigen DGS's. Hängt die Kovarianzmatrix von ε_k noch vom
Zustand des Modells an einem früheren Zeitpunkt ab, so ist y als
Realisierung eines stochastischen Differentialgleichungssystems
zu betrachten (w Wiener Prozeß):

$$y(t_{k+1}) = x(t_k) + \int_{t_k}^{t_{k+1}} f(x(s),p)\,ds + \int_{t_k}^{t_{k+1}} g(x(s))\,dw(s)$$

Die unterschiedliche Betrachtungsweise führt zu unterschiedlich
motivierten Abstandsfunktionen, die durch die Art der Daten und der
Fehlereinwirkung begründet sind. Notwendige Voraussetzung, in
dynamischen Modellen Parameter erfolgreich bestimmen zu können, ist
die Festlegung der Parametergrenzen und die Auswahl geeigneter
Integrationsverfahren. Diese umrissenen Probleme, die wesentlich sind
für die Qualität einer Modellanpassung werden für ein
Eutrophierungsmodell (Braun [1]) mit über 50 freien Parametern
betrachtet.

2. Das Modell

Untersucht wird ein von Braun [1] entworfenes Modell, das die
Dynamik der Eutrophierung in Flachlandgewässern beschreibt. Es wird

das Wachstum planktischer Algengruppen, die Grün-,Blau- und Kieselalgen in Abhängigkeit von physiologisch unverzichtbaren Nährstoffen, wie Orthophosphat, Ammonium- und Nitratstickstoff sowie Kieselsäure beschrieben. Die Algengruppen zeichnen sich jeweils durch charakteristische Umwelteinflüsse aus (Temperatur,Lichtangebot etc.). Mit diesen Algenarten sind ca 80% der Biomasse des Phytoplanktons erfaßt. Ausgegangen wird von der lokalen Massenbilanz, die als partielle Differentialgleichung gegeben ist. Wird ein als homogen durchmischt angenommenes Flußsegment betrachtet, entsteht ein System gewöhnlicher Differentialgleichung (DGS) ,das die inneren Stoffumsätze und die äußere Massenbilanz beschreibt:

$$x(t) = f(t,x(t),p) + \frac{Q}{V}(x_{input}(t) - x(t))$$

(x_{input} ist hier der Vektor der Einleitungen vom vorangehenden Flußsegment).

Der Parametervektor besteht aus 65 Elementen, von denen 53 zu adaptieren sind. Als typiches Wachstum für das Phytoplankton ist eine substratabhängige Monod-Kinetik angesetzt, in die die Abhängigkeit von den Triebkräften, Strahlung und Temperatur, mit berücksichtigt ist. Die nichtlineare Bilanzgleichung für eine Algenart x_i lautet dann   ASS - (RESP + MORT + EXKR) - FRAß - SEDI mit den funktionalen Bezeihungen

ASS Assimilation

$$\mu_{i,max}\Pi\ f_{ij}(N_j)FT_i(T)FL_i(L)$$

N_j Nährstoff$_j$

$f_{ij}(N_j)$ $\dfrac{N_j}{N_j + a_{4+j,i}}$

FT Temperaturabhängigkeit der Wachstumsrate

$$\exp(-(T-a_{2,1})^2/a_{7,i})$$

FL Strahlungsabhangigkeit der Wachtumsrate

$$\text{Fotoperiod } (\arctan(L/a_{1,i})-\arctan(L\ g(x)/a_{1,i}))$$

LOSS summarische Verlustterme

$$x_i\ a_{9,i}\exp(-\ (T-a_{6,i})^2/a_{8,i})$$

SEDI Sedimentation

$$a_{11,i}\ \exp(-Q\ x_i/P_{20})$$

Q Durchfluß

Als Modell für das Schätzproblem dient der autochthone Stoffumsatz, der durch das Anfangswertproblem $\dot{x}=f(t,x,p)$, $x(t_o)=x_{input}(t_o)$ beschrieben ist. Die Differenzdatenreihe zwischen Input zum und Output vom See liefert die dazugehörige Datenreihe.

3. Parameterschätzung

Die 53 freien Parameter variieren in ökologisch begründeten Grenzen. Die Startwerte und die mögliche Bandbreite wurden der Literatur entnommen bzw. sind Ergebnis des Sachverstandes (Tafel 1 Spalte 3-5). Die Abbildung 1 zeigt die resultierenden Modellverläufe mit diesen Parametern und die entsprechenden Datenwerte.

Die Zustandsvariablen sind 14-tägig über 7 Jahre gemessen worden. Es ist leicht zu sehen, daß das Modell mit den Startparametern die Daten schlecht beschreibt. Eine bessere Parameterkombination ist durch zeitaufwendiges Probieren gefunden worden ([1]). Das Modell adaptiert die Daten schon besser. Es bleiben aber noch unakzeptable Differenzen bei Grün- und Blaualgen, bei Kieselsäure und O-Phosphat. Zuletzt haben wir ein interaktives Softwaresystem zur Parameterschätzung verwendet, um die Parameter zu finden, die die Dynamik der Meßdaten besser beschreibt. Das System basiert auf MINUIT ([2]). Die zusätzlichen Möglichkeiten

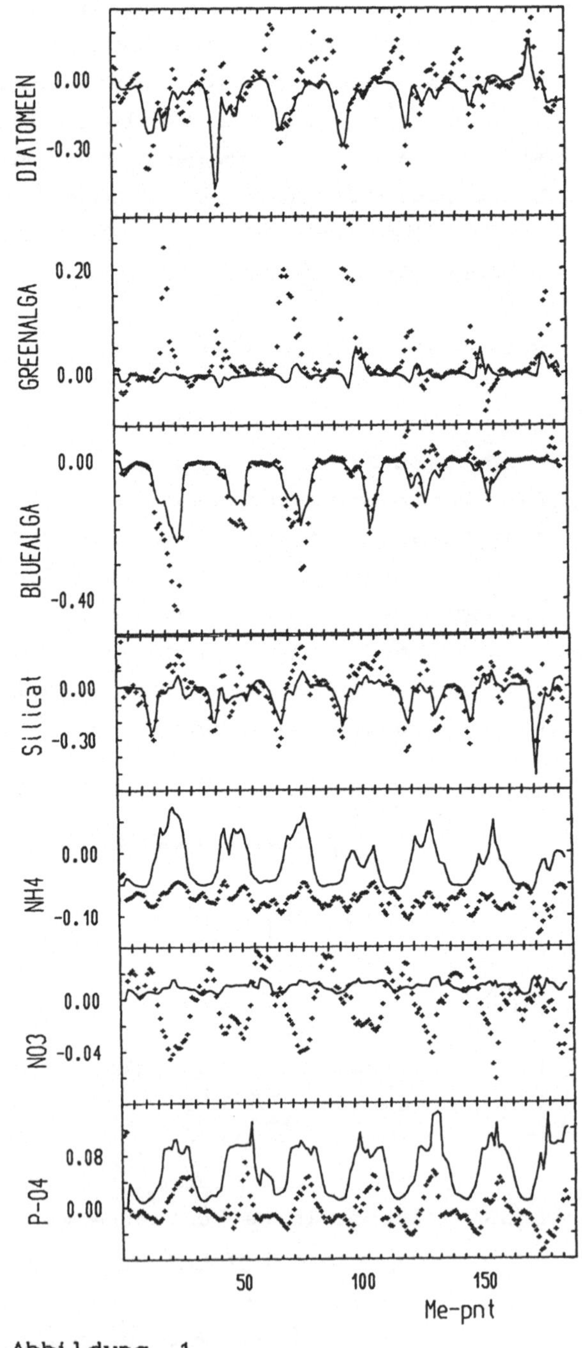

Abbildung 1
Das Modell ERNA mit den
Startparametern

insbesondere die des graphischen Vergleichs, waren wesentlich bei der Suche nach besseren Parameterkombinationen im 53-dimensionalen Parameterraum. Bei solchen Systemen ist es hoffnungslos, ein absolutes Minimum mit akzeptablem Aufwand zu erreichen. Deshalb besteht die Aufgabe eher darin, schnell bessere Parameterwerte zu finden. Für dieses Ziel haben wir 2 unterschiedliche Methoden verwendet und ihre Effizienz verglichen: Zuerst wurde die Anzahl der freien Parameter auf die 5 Parameter mit maximalen Gradienten reduziert. Nach 50 MIGRAD-Iterationen wurden die Gradienten neu berechnet und wiederum 5 neue Parameter für die Optimierung festgelegt usw.. Der Nachteil der Methode ist, daß für jede Berechnung des Gradienten 106 Modellaufrufe nötig sind. Die zweite Methode war effektiver. Sie besteht in der Parmetersuche mit SIMPLEX ([2]). Es läßt sich methodisch abschätzen, daß in diesem Fall die Anzahl der notwendgen Modellberechnungen ungefar das Vierfache der Parameteranzahl beträgt. Deshalb war es am Ende möglich, mit dieser Methode bei nur 200 Modellaufrufen gute Ergebnisse zu erzielen bezüglich der endgültig gewählten Abstandsfunktion. Die generelle Form der Abstandsfunktion war

$$F = \sum_j \frac{1}{N_j R_j} \sum_j w_{jk} (x_j^{Daten}(t_k) - x_j^{Modell}(t_k))^2$$

N_j ist hier zur Normalisierung benutzt während R_j als Gewicht die Genauigkeit der Daten bzw. der Modelldarstellung bewertet. w_{jk} schließlich drückt die Streuung der Daten aus. Die Gewichte können so gleichzeitig dazu benutzt werden, ausgewählte Daten für die Optimierung auszuschließen (w_{jk}=0). Verschiedene Kombinationen von R und w sind benutzt worden, um eine möglichst gute Modelladaption zu erreichen. Während sowohl w_{jk}=1/x_{jk}^{Modell} als auch w_{jk}=1/x_{jk}^{Daten} zu schlechten Ergebnissen führten, wurde mit dem Gauß-Kriterium schon eine bessere Anpassung erreicht. Am wirkungsvollsten erwies sich

$$w_{jk} = (x_j^{Daten}(t_k) - \bar{x}_j)^2 / x_j^{Ampl^2}$$

als Gewichtsfunktion. Hier bedeutet $\bar{x}_j$ den Mittelwert und x_j^{Ampl} die mittlere Amplitude der j-ten Komponente. Das bedeutet, daß das Gewicht eines Datenpunktes umso größer ist, je größer sein Abstand vom Mittelwert ist. Außer für Detritus und NO_3-Stickstoff sind die entsprechenden Gewichte R gleich 1 gesetzt worden. Da Detritus nicht gemessen wurde, haben wir einen kleine Wert für R angesetzt ($1/\sqrt{2}$). Für NO_3 wurde ein größerer Wert genommen ($\sqrt{1,5}$), mit dem eine

bessere generelle Anpassung der Modellkurven an die Daten erreicht
werden konnte. Fast die gleiche Akzeptanz wurde mit der auf 5 Jahre
reduzierten Datenmenge erreicht. Die erreichten Parameterwerte sind
in Tabelle 1 Spalte 8 zu sehen. Alle Parameter sind innerhalb der
vorgegebenen, inhaltlich begründeten Grenzen. Die Dynamik der
Modellkurven für diese Parameterkombination ist in Abbildung 2b im
Vergleich mit den Daten dargestellt. Es ist ersichtlich, daß die
Hauptcharakteristika der Daten gut durch das Modell widergespiegelt
werden. Der Vergleich der beiden Abbildungen 2 zeigt, daß die
Anwendung der interaktiven Optimierungssoftware (INMINGRA) eine
deutliche Verbesserung in der Übereistimmung der Modell- mit der
Datendynamik liefert. Die Übereinstimmung des vorhergesagten
Modellverlaufes mit den Daten der letzten 2 Jahre ist nicht
signifikant schlechter als die Anpassung der vorherigen 5 Jahre, so
daß das Modell ERNA für die Vorhersage gut geeignet ist.

4. Literatur

[1] Braun, P.: Das Eutrophierungsmodell ERNA, Dissertation, TU
     Dresden, 1984
[2] James, F,Roos, M.: MINUIT,Computer Physics Communication 10 (75)

Nr.	name	value	error	lower limit	upper limit	'hand made'	INMINGRA
1	p1	.75	15	.5	1.	.7	0.59261
2.	p2	.076	0028	002	15	.005	0.00341
3.	p3	.00525	00045	.0005	01	.001	0.00960
4.	p4	15	.08	.05	25	.15	0.24375
5.	p5	2.	.2	.2	5	.5	1.83037
6.	p6	.055	.028	.01	.1	.04	0.09207
7.	p7	1.05	.3	.1	2	.5	0.21840
8.	p8	14	.00001	.13998	.14002	.14	0.34202
9.	p9	.075	.00001	.07498	.07502	.075	0.25559
10.	a121	.775	13	.05	1 5	.2	1.49609
11.	a11	225.	120.	50.	400.	200.	36.61870
12.	a21	12.5	3.	9.	16.	13.	9.14710
13.	a31	2.1	.25	1.8	2.4	2.1	2.14768
14.	a41	.05	.025	.02	.08	.05	0.06861
15.	a51	275	.13	.05	.5	.2	0.57883
16.	a61	19.5	1.5	17	22	20.	21.31549
17.	a71	10.	2.5	7.	13	10.	8.65491
18.	a81	15.	2.5	12	18.	15.	12.40032
19.	a91	.055	.025	01	1	.04	0.12584
20.	a122	65	3	1	1 2	44	0.31115
21.	a12	175	60.	50.	300	120.	97 00476
22.	a22	22 5	1 8	20	25	23	15 10497
23.	a32	1 85	18	1 6	2 1	1.8	2.09687
24	a42	055	025	02	08	.05	0 04106
25	a52	275	13	05	5	2	0.06186
26.	a62	20.	1.8	18	22.	20.	21 56784
27.	a72	10.	2.8	7	13	10.	12.81580
28.	a82	15	2 8	12	18	15.	12.63725
29.	a92	095	.035	.04	15	.08	0.04016
30.	a123	.3	05	.1	.5	.16	0.81208
31.	a13	225.	80.	50.	400.	300.	397.82823
32	a23	22.5	.8	21	24.	23	21.80896
33.	a33	1.85	18	1 6	2.1	1.8	1.70024
34.	a34	.055	.028	.02	.08	.05	0.09964
35	a53	.275	.14	.05	5	.2	0.79520
36.	a63	20.5	1.8	18.	23.	20.	22.16703
37.	a73	10.	2.8	7.	13.	10.	12.62052
38.	a83	15.	2.8	12.	18	15	20.48608
39.	a93	.095	.038	.04	.15	08	0.12431
40.	p10	.275	.13	.05	5	.2	0.25269
41.	a101	1 75	.45	5	3.	1.	2.68961
42.	p12	215	.065	.03	.4	.1	0.11984
43.	p13	11	2.8	2 0	20	5.0	11.04132
44.	p14	6.0	2.8	2 0	10.	5 0	6.00000
45.	p15	1.025	.14	.05	2.	2	1.09304
46.	p16	.3	.03	.1	.6	.15	0.36606
47.	p17	.0275	.014	.005	.05	.02	0.04983
48.	p18	4 5	1.8	1	8	3	2.32973
49.	p19	18.5	1 8	15.	22	20.	22.00000
50.	p20	40.	18.	20.	60.	40.	40.00000
51.	a111	3	18	.1	.5	.3	0.30000
52.	a112	09	.045	03	15	1	0.09000
53.	a113	3	18	.1	.5	..3	0.50000

Tabelle 1

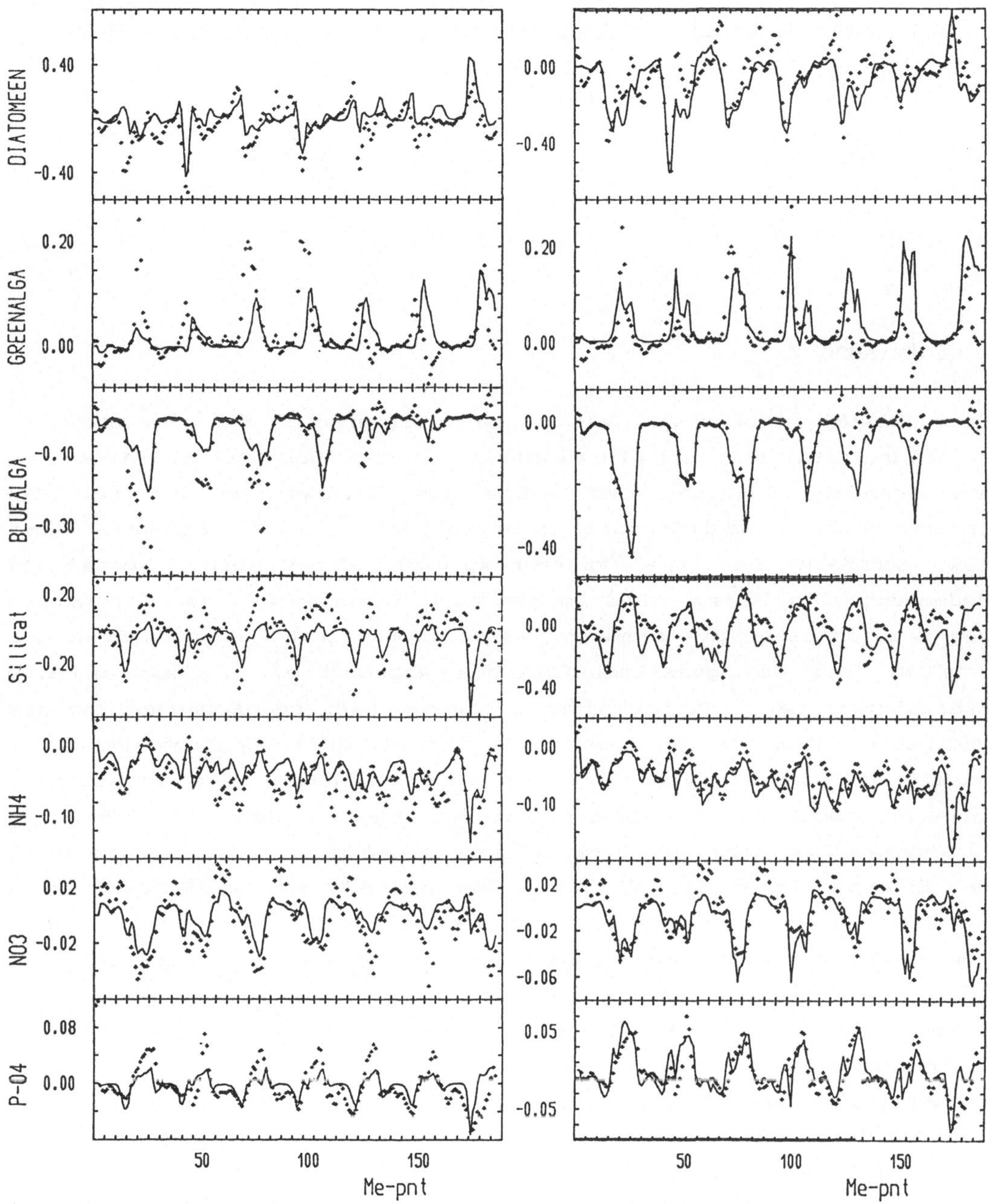

Abbildung 2 a
Das Modell ERNA mit den
empirisch gefundenen
Parameterkombinationen

Abbildung 2 b
Das Modell ERNA mit den
durch Optimierung ermit-
telten Parametern

ENTSCHEIDUNGSUNTERSTÜTZUNG FÜR DIE WASSERGÜTEBEWIRTSCHAFTUNG

Peter Rudolph
Zentralinstitut für Kybernetik und Informationsprozesse
Kurstraße 33
Belin
1086

1. Einleitung

Eine sowohl ökonomisch als auch ökologische Wassergütebewirt-
schaftung ist für die Bereitstellung des Rohstoffes Wasser in
ausreichender Qualität und Menge als Trinkwasser und für die
Industrie und Landwirtschaft unverzichtbar. Die Flußsysteme haben
als mehrfach genutzte Wasserresourcen dabei eine vorrangige
Bedeutung. Die Flüsse sind leider auch Sammelbecken der Abprodukte
der Gesellschaft. Sie sammeln, assimilieren und transportieren den
größten Teil der gesellschaftlichen Abprodukte, die als direkte
Einleitungen der Industrie, der Landwirtschaft und Kommunen und als
indirekte Kontamination über den Luftpfad, durch Erosion und durch
andere diffuse Einträge in die Gewässer gelangen. Der Zustand der
Gewässer macht den dringenden Handlungsbedarf deutlich. Bei dem
Flußsystem der Elbe ist dies besonders dringlich. Im Einzugsgebiet
der Elbe mit ihren Nebenflüssen leben ca. 82% der Gesamtbevölkerung
der DDR, der Wasserbedarf berträgt 8300 m^3/a, davon allein 1300 m^3/a
zur Bewässerung in der Langwirtschaft. Durch den gleichzeitigen
Anstieg der Abwasserlast (insgesamt 36 Mio Einwohnergleichwerte EGW)
führten durch neue Klärwerke erreichte Lastsenkung (ca 3,6 Mio EGW)
zu keiner wesentlichen Verbesserung ([1]). Die Konsequenzen von
notwendigen Bewirtschaftungsmaßnahmen, also Steuerungseingriffen,
für das Ökosystem müssen durch Computerstudien analysiert werden
([2],[3],[7]).
Die systemanalytischen Komponenten zum Studium der Wassergüte in
Flußsystemen sollen hier vorgestellt werden.

2. Das Beratungsystem

Das Ziel besteht in einem rechnergestützten Entwurf von
Sanierungsstrategien unter bestehenden ökonomischen Randbedingungen

und ökologischen Zielstellungen. Mit Computerhilfe werden derartige Entscheidungen entwickelt, deren Einfluß auf die Stabilität des Ökosystems und die Abhängigkeit von stochastischen Umwelteinflüssen untersucht. Das Anliegen des DSS ist es, moderne mathematische Methoden zur Analyse eines integrierten Modells bereitzustellen, die gleichzeitig unterschiedlichen Zielstellungen gerecht werden:

- eine detaillierte wissenschaftliche Analyse der ablaufenden dynamischen Prozesse, die entsprechend detaillierte Modelle voraussetzt,

–eine kurz-, mittel- und langfristige Vorhersage und

- eine langfristige Planung wofür statische Modelle ausreichend sind (Abb.1)

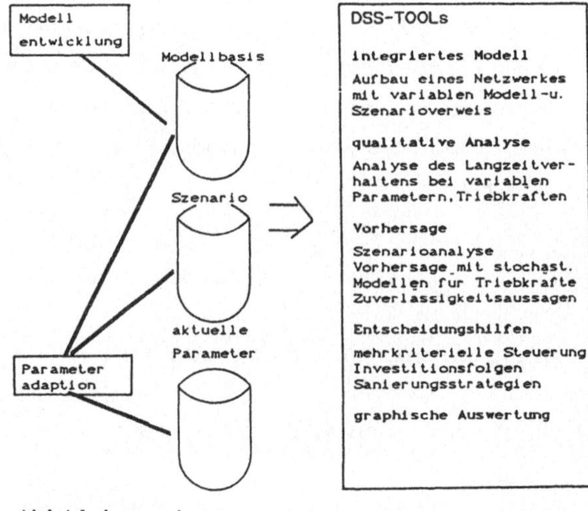

Abbildung 1

3. Das integrierte Modell

Basis der Untersuchungen ist ein integriertes Modell für das Gewässersystem. Die aquatischen Ökosysteme sind charakterisiert als nichtlineare dynamische Systeme, die energetisch und stofflich offen und stochastisch gestört sind. Eine Unterteilung des Gewässers in Teilbereiche mit als homogen vorausgesetzter Längsvermischung ist durch einen gerichteten Graphen repräsentiert, an dem unterschiedliche Zustandsmodelle angebunden werden können. Die Güteprozesse sind durch

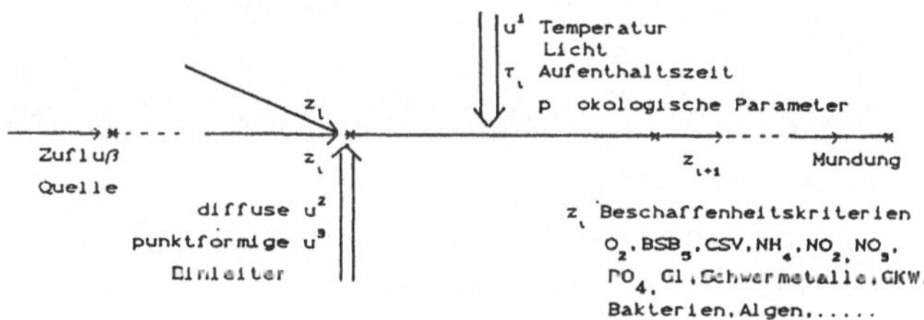

Abbildung 2

Systeme nichtlinearer Differentialgleichungen (DGS) beschrieben, die von ökologischen und geographischen Parametern p sowie treibenden Umwelteinflüssen u(t) abhängen. Unter den jeweiligen Annahmen an die Durchmischung in einem Flußsegment (eine Kante im Topologiegraph) genügen als äquivalente Beschreibung der Dynamik der ablaufenden Güteprozesse ein "Rührkessel"modell

$$\dot{z}_{i+1}(t) = f_i(t,u_i(t),p,z_{i+1}(t)) + c_i(z_{i+1}(t)-z_i(t))$$

bzw. ein "Pfropfen"modell

$$z_{i+1}(t) = z_i(t-\tau_i) + \int_{t-\tau_i}^{t} f_i(s,u_i(s),p,z_{i+1}(s))ds$$

f_i charakterisiert den autochthonen Stoffumsatz bzw. Wachstumsprozesse. Der 2. Term erfaßt die äußere Massenbilanz des betrachteten Flußsegments. Das dynamische Verhalten ist dabei beeinflußt durch die Zeitverläufe der Triebkräfte in typischen Situationen. Werden die Triebkräfte über einen Zeitraum als konstant bzw. periodisch angenommen, ergeben sich für diesen Zeitraum stationäre Modelle. Für die oben genannten Modelltypen ergeben sich bei zeitunabhängiger Dynamik

$$z_{i+1} = z_i + \int_{0}^{\tau_i} f_i(u,p,z_{i+1}(s))ds, \qquad (1)$$

die staionären Punkte des DGS

$$0 = f_i(u,p,z_{i+1}) + c_i(z_{i+1} - z_i), \qquad (2)$$

bzw. Grenzzyklen des DGS als die entsprechenden stationären Modelle. Das Softwaresystem CANDYS/CM unterstützt den interaktiven Aufbau von DGS im Vergleich mit Datenzeitreihen ([4]). Die rechnergestützt erarbeiteten Modelle als Systeme gewöhnlicher Differentialgleichungen, die von Parametern und Zeitverläufen der Triebkräfte abhängen, können z. B. in MODULA-2 in einer Modellbank abgelegt werden. Ein auf MINUIT ([5]) basierendes Softwaresystem unterstützt mit Methoden der Optimierung und des graphischen Vergleichs das Adaptieren der freien Parameter. Je nachdem, ob die Daten saisonale Mittel oder Zeitreihen sind, werden statische oder dynamische Modelle in das integrierte Modell einbezogen. In Abbildung 3 sind für den Elbestrom auf DDR-Gebiet die Ergebnisse mit einem Bilanzmodell (1) und in Abbildung 4 mit einem nichtlinearen Modell für den Stickstoffhaushalt (vgl. [6]) vom Typ (2) für einen Abschnitt der unteren Elbe dargestellt.

4. Vorhersage, Analyse, Entscheidungsvorbereitung

Die Modelle sind in Abhängigkeit von den treibenden Umwelteinflüssen $u(t)$ (z.B. Durchflußgeschwindigkeit, Lichtangebot, Temperatur) entsprechend der Aufenthaltszeit im Flußsegment für

Kurzzeitprognosen und Szenarienanalysen geeignet. Will man Prognosen über einen längeren Zeitraum mit Zuverlässigkeitsaussagen ableiten, müssen die durch Daten gegebenen Inputprozesse durch stochastische Modelle ersetzt werden. Dies geschieht dadurch, daß eine Komponente u(t) als stochastischer Prozeß (mit linearer Drift und nichtlinearer Diffusion) modelliert wird oder der Parameter p und u(t) durch Zufallsgrößen bzw. durch beschränkte stochastische Prozesse ersetzt werden. Für zufällige Parameter mit einer Dichte h_o berechnet sich die mehrdimensionale Zustandsdichte h aus der Jacobimatrix $J(t, x_o, p)$ des DGS

$$h(t_1, x_1, \ldots, t_n, x_n) = h_o(p) \left(\sum_{i=1}^{n} J(t_i, x_o, p)^2 \right)^{-\frac{1}{2}}$$

Daraus kann dann ein Vertrauensbereich zu verschiedenen Zeitpunkten berechnet werden. Sind die Triebkräfte u durch stochastische Ito-Differentialgleichungen ersetzt, ergibt sich ein Vertrauensbereich als Envelope aller Lösungen des DGS bei eingesetzten Realisierungen u aus einem Konfidenzschlauch. Diese Realisierungen können bei stochastischen Prozessen mit bekannten Verteilungen direkt bzw. werden mit stochastischen Integrationsverfahren iterativ berechnet.
Die in den Modellparametern abgebildeten Umwelteinflüsse, wie z. B. Wachstumsparameter von Algen, Bakterien, Pflanzen, Fischen usw. variieren in Grenzen bzw. sollen gezielt verändert werden, um das Ökosystem in eine gewünschte Entwicklung zu steuern. Die stationären und periodischen Lösungen, deren Anzahl und Stabilität charakterisieren das Langzeitverhalten der durch das Modell ausschnitthaft abgebildeten in Natur ablaufenden biologischen und chemischen Umwandlungsprozesse. Die Lösungen können sich bei wechselnden Umwelteinflüssen ändern. Die berechneten kritischen Parameterwerte (Bifurkationspunkte), die diese Veränderung möglicherweise mit Änderung der Stabilität verursachen, sind Ausdruck für kritische oder erstrebenswerte Änderung des Ökosystems. Die Analyse wird mit dem interaktiven Analysesystem CANDYS/QA ([6]) durchgeführt.
Einem weiteren besonders wichtigen Aspekt einer qualifizierten Entscheidungsvorbereitung zur Sicherung aller maßgeblichen Nutzungsansprüche an die Elbe wird mit der Entwicklung von optimalen Investitionsfolgen entsprochen, die Vorschläge zur Rang- und Reihenfolge beinhalten. Entscheidungen zur Steuerung von Prozessen in der Wasserwirtschaft werden in Abhängigkeit von subjektiven und

situationsbedingten Bewertungen durch einander widersprechender Ziele getroffen (z.B. beschränkte Investitionsmöglichkeiten, größtmögliche ökologische Stabilität , schnelle Verbesserung der Situation), zwischen denen ein Kompromiß gefunden werden muß. So erreichte Steuerungsvorschläge geben bei begrenzten Investitionsmitteln eine optimale Auswahl möglicher Reinigungstechnologien (verschiedene Ausbaustufen bei Klärwerken, Recycling , Schadstoffeliminierung in der Industrie selbst) für eine Reihe von Einleitern aus Industrie und Landwirtschaft an ([7]). Von den entlang des Flusses lokalisierten Elbverschmutzern werden diejenigen zur Rekonstruktion vorgeschlagen, die den größten Einfluß auf die Gesamtbelastung haben bzw. an einigen Meßpegeln (z.B. Boizenburg) dominierend sind. Abbildung 5 zeigt den Einfluß der Hauptverschmutzer entlang des Elbestromes bei 50%-iger bzw. bei 100%-iger Reduzierung der Einleitungen auf den Nitratgehalt und auf den chemischen Sauerstoffbedarf.

4. Literatur

[1]   Bericht über den ökologischen Zustand der Elbe Ministerium für Naturschutz,Umweltschutz und Wasserwirtschaft,1990
[2]   Rudolph,P.;Sydow,A.;Kaden,S.: Computer Aided Water Quality Managemenet,Kongreßband Der Hafen eine ökologische Herausforderung, Umweltbehorde Hamburg,September 1989,152-157
[3]   Braun,P.,Rudolph,P.: Ein Modell für die computergestützte Entscheidungshilfe bei der Wasserbewirtschaftung,msr, Verlag Technik 32,(7) 306-308
[4]   Funke,R.: Water Quality Forecast for a Lake Based on the Eutrophication Model ERNA and using the Ito Stochastic Calculus, Proceedings of the Int. Workshop Environmental Management in the Baltic Region,Leningrad, Nov. 1989
[5]   James,F.,Roos,M.: MINUIT,Computer Physics Comm.10 (1975)
[6]   Jansen,W.;Feudel,U.:The Interaction of Selfpurification and Nutrient Load in Water Quality Models Proceedings of the Int. Workshop Environmental Management in the Baltic Region,Leningrad, Nov. 1989
[7]   Model,N.;Rudolph,P.:River Pollution Control,ibid

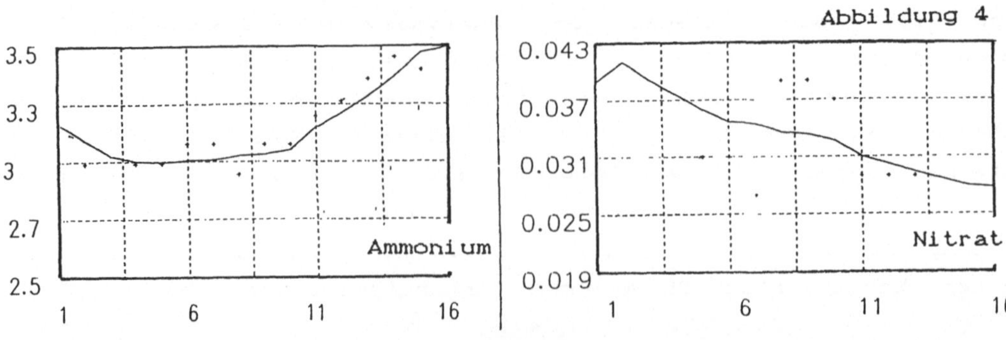

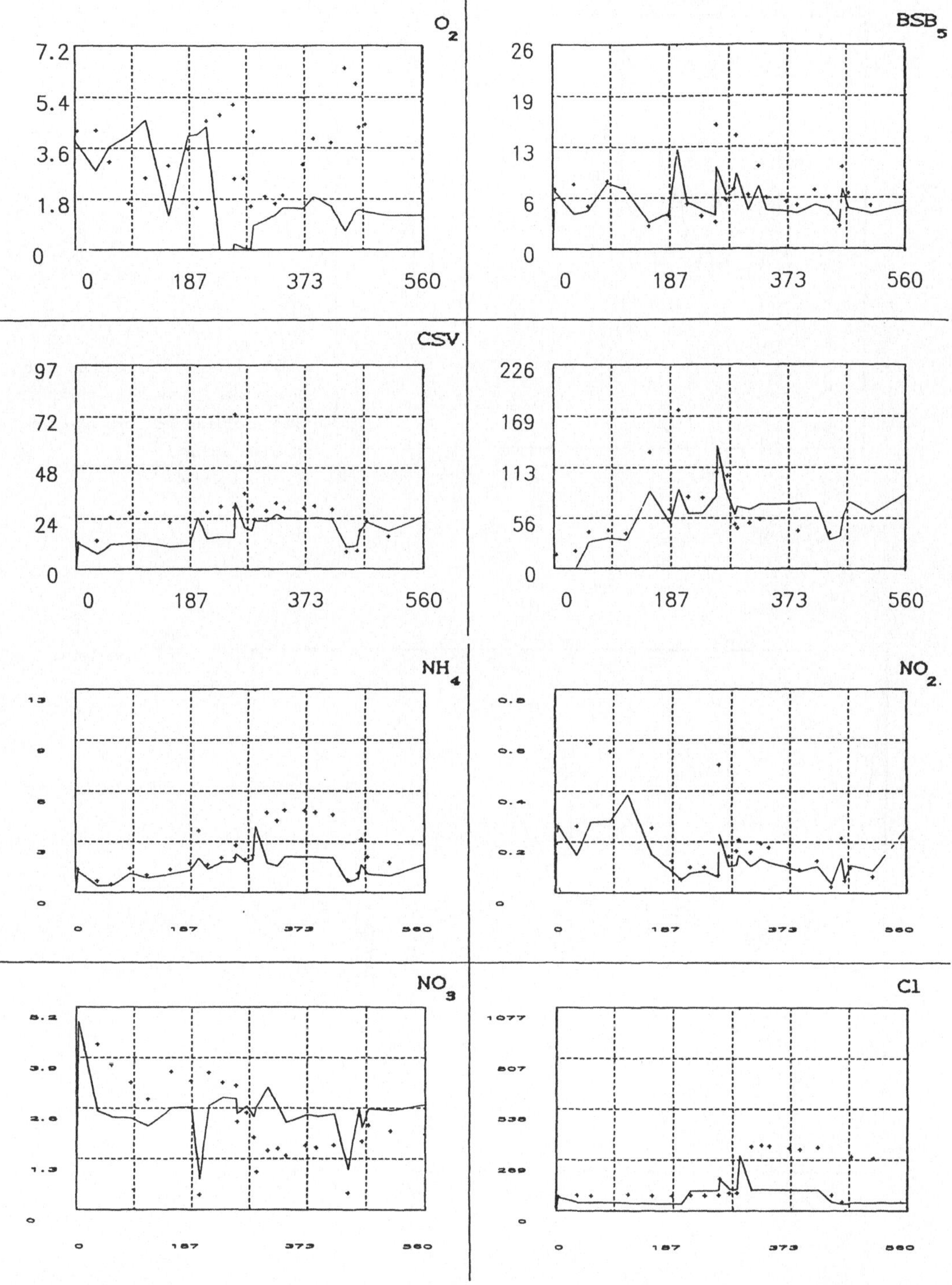

Abbildung 3

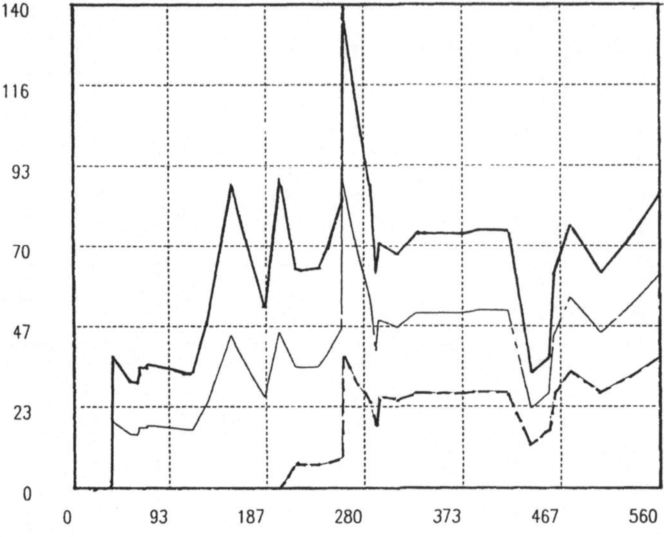

Abbildung 5
50%-ige bzw. 100%-ige Reduzierung der Hauptverschmutzer bei chemischen Sauer-stoffbedarf und Nitratgehalt

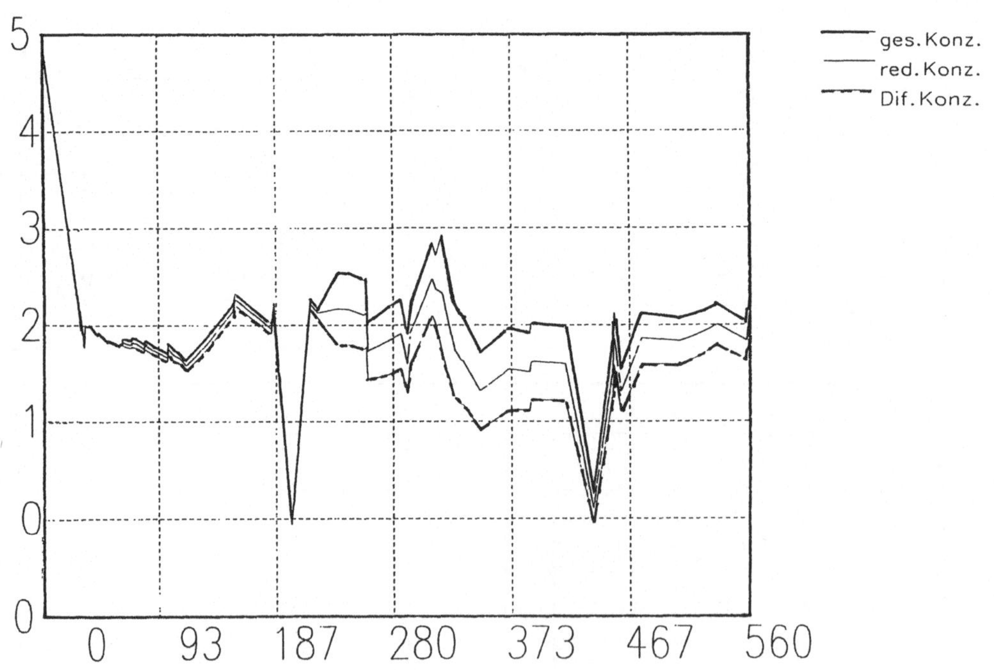

<u>Mehrkriterielle Sanierungsstrategien für Fluß- und</u>
<u>Luft- Schadstoffemittenten</u>

N. Model; P. Rudolph   *(Flussanierung)*
N. Model; P. Lasch; K. Bellmann   *(Luftsanierung)*

AdW der DDR, Zentralinstitut fuer Kybernetik und Informationsprozesse, 1086 Berlin, Kurstrasse 33

1. Einleitung

Vorgestellt werden Ergebnisse zur Ermittlung optimaler Rekonstruktionsfolgen für Fluß- und Luftverschmutzer.

Demonstriert werden die Berechnungsverfahren im Fall der Flußsanierung am Beispiel der Elbe (in Zusammmenarbeit mit dem Institut für Wasserwirtschaft) und im Fall der Luftsanierung am Beispiel der durch Luftemittenten (z.B. Kraftwerke) verursachten Schadstoffimmissionen (in Zusammenarbeit mit dem Meteorologischen Dienst der DDR und dem Zentrum für Umweltgestaltung).

Die erarbeiteten Computerprogramme sind sowohl für andere Flußsysteme als auch geographisch anders lokalisierte Luftsysteme einsetzbar.

2. Steueraufgabe

Im Steuervektor sind alle (steuerbaren) Emittenten zusammengefaßt. Jeder Komponente dieses Vektors wird dabei der Schadstoffausstoß eines Emittenten zugeordnet.

Der Steuereingriff erfolgt diskret. Die entsprechende Komponente des Steuervektors hat im vorliegenden Fall drei Varianten

   a) mittlerer Schadstoffausstoß ohne Rekonstruktion

   b) mittlerer Schadstoffausstoß nach Rekonstruktionsvariante 1

   c) mittlerer Schadstoffausstoß nach Rekonstruktionsvariante 2

Die bei den Rekonstruktionen (z.B. Bau von Klärwerken oder Rauchgasreinigungsanlagen) anfallenden Kosten sind in entsprechenden Kostenvektoren erfaßt.

Im Vergleich zur Variante 1 sind die in der Rekonstruktionsvariante 2 verwendeten Technologien wirkungsvoller, verursachen allerdings auch höhere Kosten.

Die für das Steuerproblem benötigten Zielfunktionale wurden entsprechend der zu realisierenden Aufgabenstellungen der beiden Sanierungsstrategien gewählt.

3. Flußsanierung

Ziel der Untersuchungen ist die Ausarbeitung von Sanierungsstrategien für Einleiter (Industrieeinleiter, kommunale Abwassereinleiter, Nebenflüsse u.a.) so, daß durch die Sanierungsmaßnahmen die Gewässergüte wesentlich erhöht wird. Dabei werden sechs Klassen der Wassergütebeschaffenheit unterschieden.

Die Grundlage der Klassifizierung sind regelmäßige Untersuchungen der in die Klassifizierungen an den Beschaffenheitsmeßstellen im Flußlängsschnitt einzubeziehende Kriterien. Dies sind u.a.

* gelöster Sauerstoff

* biochemischer Sauerstoffbedarf

* chemischer Sauerstoffverbrauch-Mn

* chemischer Sauerstoffverbrauch-Cr

* Ammonium

* Nitrit

* Nitrat

* Chlorid

Sanierungsstrategien sind dabei unterschiedlich aufwendige Schadstoffabbautechnologien (z.B. Klärwerke), die demzufolge auch unterschiedliche Investitionen erfordern.

Da die Investmittel beschränkt sind, müssen solche Rekonstruktionsfolgen ausgewählt werden, die der Forderung nach Verbesserung der Wassergüte am besten gerecht werden.

Entlang des Flußlaufes werden Bilanzprofile ausgewählt, an denen die Frachten bzw. Schadstoffkonzentratione auf vorgegebene Werte zu reduzieren sind. Diese Reduktionen sollten möglichst schnell erfolgen. Die Gesamtschadstoffeinleitung und die Kosten sollten möglichst gering sein.

Zur Realisierung dieser (sich durchaus widersprechenden Forderungen) wurde ein mehrkriterielles Optimierungsprogrammpaket CREH-rechnergestützte Entscheidungsfindung, ZKI) verwendet, mit dessen Hile pareto- optimale Lösungen berechnet werden.

Zur Simulation der Konzentrationen wird z.Zt. ein 22-komponentiges Bilanzmodell verwendet.

Der Flußlängsschnitt der Elbe wurde in 94 Segmente unterteilt, wobei im ersten Segment die Vorbelastung aus der ČSFR berücksichtigt wird.

c_{ij} sei die Konzentration des j-ten Kriteriums (z.B. Nitrat) am i-ten Segment.

Berechnet wurden die c_{ij} nach der Formel (lineare Abbaukinetik)

$$c_{ij} = (c_{i-1\,j} + u_{ij} + d_{ij}) \, e^{-\mu_j \tau}$$

c_{ij} - Konzentration des j-ten Kriteriums am Segment i

μ_j - Abbaubeiwert des j-ten Kriteriums

u_{ij} - punktförmiger Eintrag des j-ten Kriteriums am Segment i

d_{ij} - diffuser Eintrag des j-ten Kriteriums am Segment i

Der Einsatz eines gesteuerten 7-komponentigen nichtlinaeren Stickstoffhaushaltmodell ist vorgesehen.

Resultate:

- pareto-optimale Rekonstruktionsfolgen für Flußeinleiter
- Fracht- bzw. Konzentrationszeitverlauf (im gewählten Realisier-
  ungszeitraum) als Zeitreihe bzw. farbgraphisch an den
  Bilanzprofilen
- farbgraphische Darstellung der erreichten Fracht- bzw.
  Konzentrationsabsenkungen im Flußlängsschnitt bzw. Einzugsgebiet

4. Luftsanierung

Den Berechnungen liegt das 'Referenz-Emittentenkonzept' zugrunde. Das Schadstoffausbreitungsverhalten der vom Meterologischen Dienst der DDR (MD) ausgewählten Referenz-Emittenten ist typisch für das Schadstoffausbreitungsverhalten realer Emittenten der entsprechenden Bauhöhe und Geographie.

Vom MD wurden mit Hilfe eines dreidimensionalen Ausbreitungsmodells

$$\frac{\partial c}{\partial t} = \frac{\partial c\ v_x}{\partial x} - \frac{\partial c\ v_y}{\partial y} - \frac{\partial c\ v_x}{\partial z} + \frac{\partial K_x \frac{\partial c}{\partial x}}{\partial x} + \frac{\partial K_y \frac{\partial c}{\partial y}}{\partial y} + \frac{\partial K_z \frac{\partial c}{\partial z}}{\partial z} + Q - S$$

c   - Schadstoffkonzentration (z.B. SO_2)

v_x, v_y, v_z - Komponenten des Windvektors

K_x, K_y, K_z - Komponenten des Austauschvektors

Q, S   - Quelle bzw. Senke

normierte Schadstoffimmissionswerte in den Gitterpunkten eines flächendeckenden (25x25)-km Rasters berechnet.

Unter Verwendung der MD-Daten erfolgt die flächendeckende Immissionsberechnung im (10x10)-km Gitter bzw. an frei gewählten Rezeptor-Punkten durch zweidimensionale Spline-Interpolationen

$$f_{ij}(x,y) = \sum_{k,l=1}^{4\ 4} a_{ijkl}(x-x_i)^{k-1}(y-y_j)^{l-1}$$

Für den Fall, daß die Bauhöhen der realen Emittenten nicht mit den zugeordneten Referenz-Emittenten übereinstimmen, erfolgt eine kubische Spline-Interpolation

$$f_k(h) = \sum_{l=1}^{4} c_{kl}(h-h_k)^{l-1}$$

bezüglich der Bauhöhen.

Die Ermittlung des (10x10)-km Grundimmissionsrasters erfolgt mit der Gesamtheit der Schadstoffemittenten, an denen keine Rekonstruktion vorgesehen (bzw. durchführbar) sind.

Die flächendeckende Gesamtimmission ergibt sich mit Hilfe der für die Optimierung zugelassenen Emittenten zuzüglich der Grundimmission und eines Importanteils.

Die Ermittlung der geforderten Immissionsreduktion ergibt sich aus dem Produkt zwischen Gesamtimmisssion am gewählten Rezeptor-Punkt und Reduktionsfaktor.

Eine Absenkung der reduzierten Gesamtimmission unter die Grundimmission ist dabei nicht möglich.

Unter Emissionen betrachten wir Jahresmittel des entsprechenden Schadstoffes (z.B. SO_2) in 1000 Kt.

Das Ziel der Optimierungsrechnungen besteht in der Ermittlung optimaler Rekonstruktionsfolgen (und damit einer Verringerung des Schadstoffausstoßes) von Emittenten (z.B. Braunkohlenkraftwerke). Die geographische Lage dieser Emittenten im DDR-Gebiet kann dabei beliebig sein.

Zielstellungen:

- Maximierung der Anzahl von gewählten Rezeptor-Punkten, an denen eine geforderte Immisssionsreduktion erreicht wird (diese Punkte können beliebige Rasterelemente des benutzten DDR-flächendeckenden (10x10)-km Gitters sein

- Minimierung der zeitlich bewichteten Differenz des Immissionszeitverlaufes zwischen erreichter und geforderter Immissionreduktion an den gewählten Rezeptor-Punkten

- Minimierung des Gesamt-Schadstoffausstoßes der zur Optimierung zugelassenen Emittenten

- Minimierung der Gesamtkosten

- Minimierung des Realisierungszeitraumes

Resultate:

* pareto-optimale Rekonstruktionsfolgen für Emittenten

* Immissionszeitverlauf (im gewählten Realisierungszeitraum) als Zeitreihe bzw. farbgraphisch an den Rezeptorpunkten

* farbgraphische Darstellung der erreichten Immissionsreduktionen im gesamten DDR-Gebiet

Literatur:

(1) Model,N.; Rudolph,P.: River Pollution Control, Proceedings of the International Workshop 'Envirenmental Management in the Baltic Region ', Leningrad November 21-24, 1989

(2) Model, N.; Born, J.: Calculation of Air Pollution Immission Values by Using of Two-dimensional Spline-Approximations, Syst. Anal. Model. 6 (1989) 2, 103-108, Akademie-Verlag Berlin

(3) Lasch, P.; Model, N.; Bellmann, K.: The PEMU-Air Pollutant transport Model based on Emittent-Rezeptor-Point -Transmission Calculation, Proceedings of the International Symposium held in Berlin (GDR), September 12-16, 1988

(4) Discher, H.-J.; Damrath, M.: Das operative regionale Ausbreitungsmodel der DDR zur Bestimmung grenzüberschreitender Schadstoffströme. In: Abhandlungen des meterologischen Dienstes der DDR, 1986

(5) Späth, H.: Spline-Algorithmen zur Konstruktion glatter Kurven und Flächen, Oldenburg-Verlag, 1983

(6)   Rathke, P.: IGAP- A Graphics Systems for Interactive Visualization of Scientific Data. In: Proceedings of the 4-th Computer Graphics Conference held in Smolenice, CSFR 1989, pp. 106-109

(7)   Straubel, R.; Wittmüss, A.: Das Programmsystem REH zur rechnergestützten Entscheidungshilfe, ZKI-Informationen 2-86 Akademie der Wissenschaften der DDR, Berlin, 1986

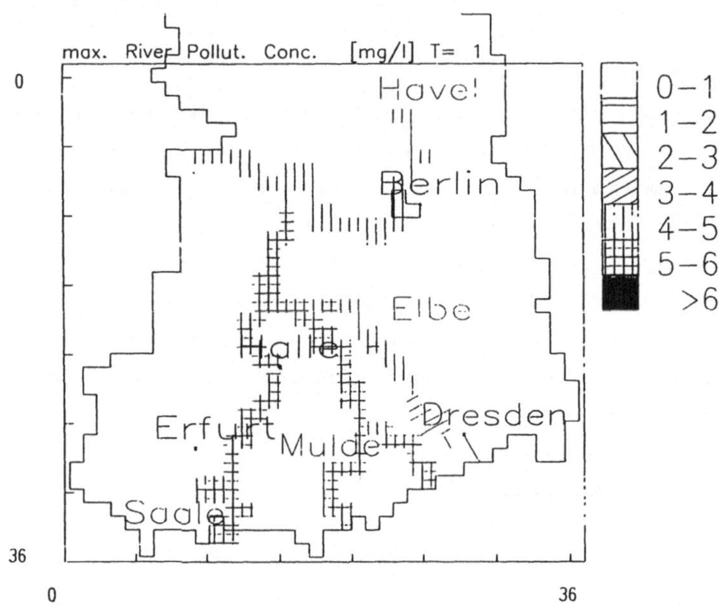

Konzentrationsverteilung im ELBE-Einzugsgebiet ohne Sanierung

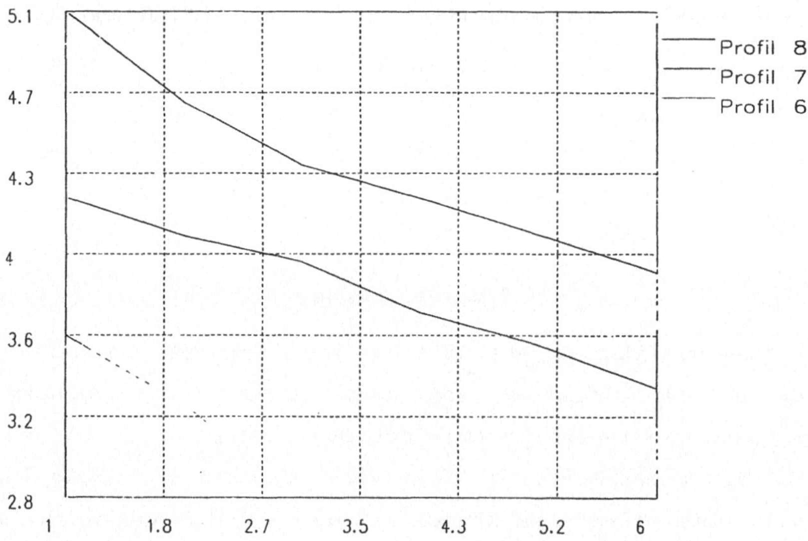

Zeitverhalten der Konzentration an gewählten Bilanzprofilen im Realisierungszeirraum

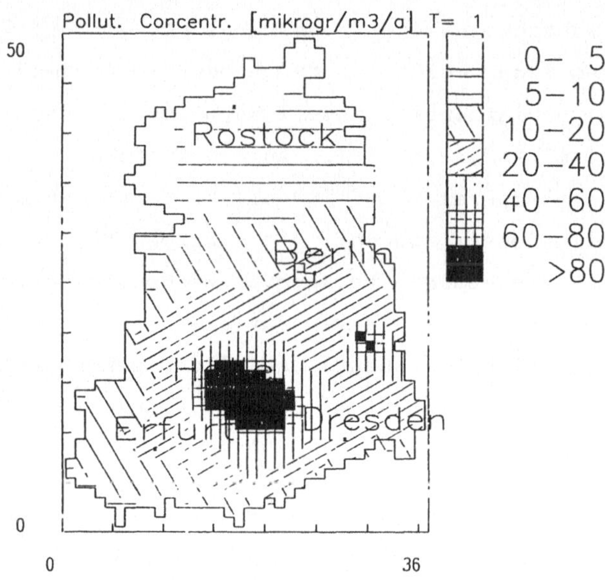

Immissionsverteilung von Luftschadstoffen im DDR-Gebiet ohne Sanierung

Vorhersage der Wasserqualität am Zu- und Abfluß von Seen

Reinhart Funke

Akademie der Wissenschaften der DDR

Zentralinstitut für Kybernetik und Informationsprozesse

Kurstraße 33, DDR-1086 Berlin,

Zusammenfassung

Die Wasserqualität am Abfluß eines Sees kann für den Durchfluß-
zeitraum auf der Grundlage von Meßdaten am Zufluß vorhergesagt
werden, wenn die innere Dynamik des Sees, d.h. die Prozesse der
Selbstreinigung des Sees, in Form von mathematischen Modellen bekannt sind. Diese Prozesse können mit dem von P.Braun entwickelten
Eutrophierungsmodell ERNA [1] beschrieben werden.

Um die Wasserqualität für Zeiträume vorhersagen zu können, die
die Durchflußzeit des Wassers durch den See übersteigen, muß eine
Vorhersage der wahrscheinlichen Wasserqualität am Zufluß getroffen
werden. Zu diesem Zweck wurde ein geschlossenes (gegenüber dem Modell ERNA etwas vereinfachtes) Modell entwickelt, das die Wasserqualität am Zufluß beschreibt [5]. Äußere Einflüsse, wie Globalstrahlung, Wassertemperatur und Durchflußrate werden unter Einbe-
ziehung von Diffusionsprozessen, die als Lösungen von stochastischen Differentialgleichungen betrachtet werden, vorhergesagt, auf
deren Grundlage dann Vorhersagen für die eigentlichen Wasser-
qualitätsmerkmale (Gehalt an Grünalgen, Blaualgen und Kieselalgen,
Detritusgehalt, Nährstoffgehalt an Silikat, Ammonium, Nitrat und
Phosphat) gemacht werden.

1.Die Modellierung der stochastischen Triebkräfte Globalstrahlung, Wassertemperatur und Durchfluß

Gegeben sei die stochastische (Ito-) Differentialgleichung (1)

$$(1) \quad \begin{aligned} d\,x(t) &= [p\,a_1(t) + a_2(t)\,x(t)\,]\,dt + \\ &+ (2\,a_1(t)\,x(t))^{1/2}\,d\,w(t), \quad x(0) = x_0 > 0, \end{aligned}$$

wobei $a_1(t) \geq 0$ und a_2 Lipschitz-stetig sind, $w(t)$ ein Standard-
Wienerprozeß und $p \geq 1$ ist. Falls eine für $t > 0$ positive Lösung
$f(t)$ der Gleichung (2)

(2) $\qquad f'(t) = a_1(t) + a_2(t)\, f(t),\ f(0) = 0$

und eine für $t \geq 0$ positive Lösung $g(t)$ der Gleichung (3)

(3) $\qquad g'(t) = a_2(t)\, g(t),\ g(0) = 1,$

existiert, dann ist die Übergangswahrscheinlichkeit des Prozesses $x(t)$ explizit durch die folgende Gleichung (4) gegeben:

$$(4) \qquad p(t,x) = \sum_{k=0}^{\infty} \left[e^{-h(t)}\, h^k(t)\, \frac{1}{k!} \right] q_{p+k,\,1/f(t)}(x),$$

$h(t) = x_0\, g(t)/f(t)$, wobei $q_{\varkappa,\lambda}(x)$ die Dichte der Gamma-Verteilung (5) ist (siehe [4]):

$$(5) \qquad q_{\varkappa,\lambda}(x) = \begin{cases} 0, & x \leq 0 \\[2mm] \dfrac{\lambda^{\varkappa}}{\Gamma(\varkappa)}\, x^{\varkappa-1}\, e^{-\lambda x}, & x > 0. \end{cases}$$

Der Prozeß $x(t)$ besitzt den folgenden Erwartungswert (6)

(6) $\qquad m(t) = \mathbf{E}\, x(t) = p\, f(t) + x_0\, g(t)$

und die folgende Streuung (7)

(7) $\qquad s^2(t) = \mathbf{D}\, x(t) = p\, f^2(t) + 2\, x_0\, f(t)\, g(t).$

Damit können für $x(t)$ zeitabhängige Vertrauensgrenzen berechnet werden. Im Falle $p \geq 15$ kann die Hyper-Gamma-Verteilung (4) durch eine Gauss-Verteilung mit den Parametern $m(t)$ und $s^2(t)$ approximiert werden. Dann können die 99%-igen Vertrauensgrenzen nach den folgenden Gleichungen berechnet werden:

$\qquad c_o(t) = m(t) + 2.57583\, s(t), \qquad c_u(t) = m(t) - 2.57583\, s(t).$

Für $1 \leq p < 15$ ist die Hyper-Gamma-Verteilung (4) unsymmetrisch. Für $p = 7.5$, z.B., sind die 99%-igen Vertrauensgrenzen durch die folgenden Gleichungen gegeben:

$\qquad c_o(t) = m(t) + 3.24982\, s(t), \qquad c_u(t) = m(t) - 1.89832\, s(t).$

Es sei jetzt angenommen, daß die Globalstrahlung $x_1(t)$ und der Durchfluß $x_3(t)$ der stochastischen Differentialgleichung

$\qquad dx_i(t) = [p_i a_{i1}(t) + a_{i2}(t) x_i(t)]dt + (2\, a_{i1}(t)\, x_i(t))^{1/2}\, dw_i(t),$

$x_i(0) = x_{i0} \geq 0$, $i = 1;3$, mit $a_{i2}(t) \equiv a_{i2} = \text{const} < 0$, den positiven periodischen Funktionen $a_{i1}(t)$, $i = 1;3$ mit der Periode 364 (= ein Jahr) und den beiden unabhängigen Wienerprozessen $w_1(t)$ und $w_3(t)$ genügt. Dann gilt

$\qquad f_i(t) = A_i(1 - e^{a_{i2}t}) + B_i(\cos \omega t - e^{a_{i2}t}) + C_i \sin \omega t$

und $g_i(t) = e^{a_{i2}t}$ mit $\omega = 2\pi/364$. Für $t \to \infty$ verschwindet $g_i(t)$ und $f_i(t)$ nähert sich der periodischen Funktion

$$A_1 + B_1 \cos \omega t + C_1 \sin \omega t.$$

Die Wassertemperatur $x_2(t)$ genüge der Gleichung

$$dx_2(t) = [a_{21} x_1(t) + a_{22} x_2(t) + a_{23}]\, dt, \quad x_2(0) = x_{20},$$

wobei $a_{21} > 0$ und a_{22}, $a_{23} < 0$. Dann liefert die Gleichung (8)

$$(8) \qquad \dot{m}_2(t) = a_{21}\, m_1(t) + a_{22}\, m_2(t) + a_{23}$$

für $m_i(t) = E\, x_i(t), \quad i = 1,2$

$$m_2(t) = A_2 + B_2 \cos \omega t + C_2 \sin \omega t + D_2\, e^{a_{12}t} + E_2\, e^{a_{22}t},$$

wobei A_2, B_2, C_2, D_2 und E_2 von a_{11}, a_{21}, x_{10}, x_{20}, p_1 und ω abhängen.

Die entsprechenden Vertrauensgrenzen können aus Gleichung (8) durch Ersetzen von $m_1(t)$ durch die obere bzw. untere Vertrauensgrenze $c_o(t)$ bzw. $c_u(t)$ berechnet werden.

2. Die Modellierung der Algen- und Schadstoffdynamik am Zufluß

Es seien y_1 die Diatomeen, y_2 Chlorophyceen, y_3 Cyanophyceen und y_4 Detritus. Weiter seien y_5 Silikat, y_6 Ammonium, y_7 Nitrat und y_8 Phosphat – die entsprechenden Nährstoffkomponenten, und u_1, u_2, u_3 (konstanter) Ammonium-, Nitrat- und Phosphateintrag.

Wir definieren $q_i(x_2) = a_i x_2^2 + b_i x_2 + c_i$ als quadratische Funktionen der Wassertemperatur und $d_i(y_j) = y_j/(k_i + y_j)$ Michaelis-Menten-Kinetiken der Nährstoffkomponenten.

Dann wird die Algen- und Nährstoffdynamik beschrieben durch die folgenden Differentialgleichungen:

$$\dot{y}_1 = y_1[x_1 q_1(x_2)d_1(y_5)d_2(y_8) - q_2(x_2) + \gamma_1 \exp(-l_1 x_3)]$$

$$\dot{y}_2 = y_2[\alpha_1 x_1 q_3(x_2)d_3(y_6)d_4(y_8) + (1-\alpha_1)x_1 q_4(x_2)d_5(y_7)d_6(y_8)$$
$$- q_5(x_2) + \gamma_2 \exp(-l_2 x_3)]$$

$$\dot{y}_3 = y_3[\alpha_2 x_1 q_6(x_2)d_7(y_6)d_8(y_8) + (1-\alpha_2)x_1 q_7(x_2)d_9(y_7)d_{10}(y_8)$$
$$- q_8(x_2) + \gamma_3 \exp(-l_3 x_3)]$$

$$\dot{y}_4 = \beta_1 q_2(x_2)y_1 + \beta_2 q_5(x_2)y_2 + \beta_3 q_8(x_2)y_3 - q_9(x_2)y_4$$

$$\dot{y}_5 = q_{10}(x_2) - \beta_4 x_1 q_1(x_2)y_1 d_1(y_5)d_2(y_8) + \beta_5 q_2(x_2)y_1$$

$$\dot{y}_6 = u_1 - \alpha_1 \beta_6 x_1 q_3(x_2)y_2 d_3(y_6)d_4(y_8)$$
$$- \alpha_2 \beta_7 x_1 q_6(x_2)y_3 d_7(y_6)d_8(y_8)$$
$$+ \beta_8 q_9(x_2)y_4 - q_{11}(x_2)y_6 + q_{12}(x_2)y_7$$

$$\dot{y}_7 = u_2 - \beta_9(1-\alpha_1)x_1 q_4(x_2)y_2 d_5(y_7)d_6(y_8)$$
$$- \beta_{10}(1-\alpha_2)x_1 q_7(x_2)y_3 d_9(y_7)d_{10}(y_8)$$
$$+ \beta_{11}q_{11}(x_2)y_6 - \beta_{12}q_{12}(x_2)y_7$$

$$\dot{y}_8 = u_3 - \beta_{13}x_1 q_1(x_2)y_1 d_1(y_5)d_2(y_8)$$

Konfidenzschlauch für die Globalstrahlung
386
307
227
148
68
840
863
887
910
Konfidenzschlauch für den Durchfluß
109
88
67
46
25
840
863
887
910
Konfidenzschlauch für die Wassertemperatur
14
11
9
6
3
840
863
887
910
Zeitachse in Tagen

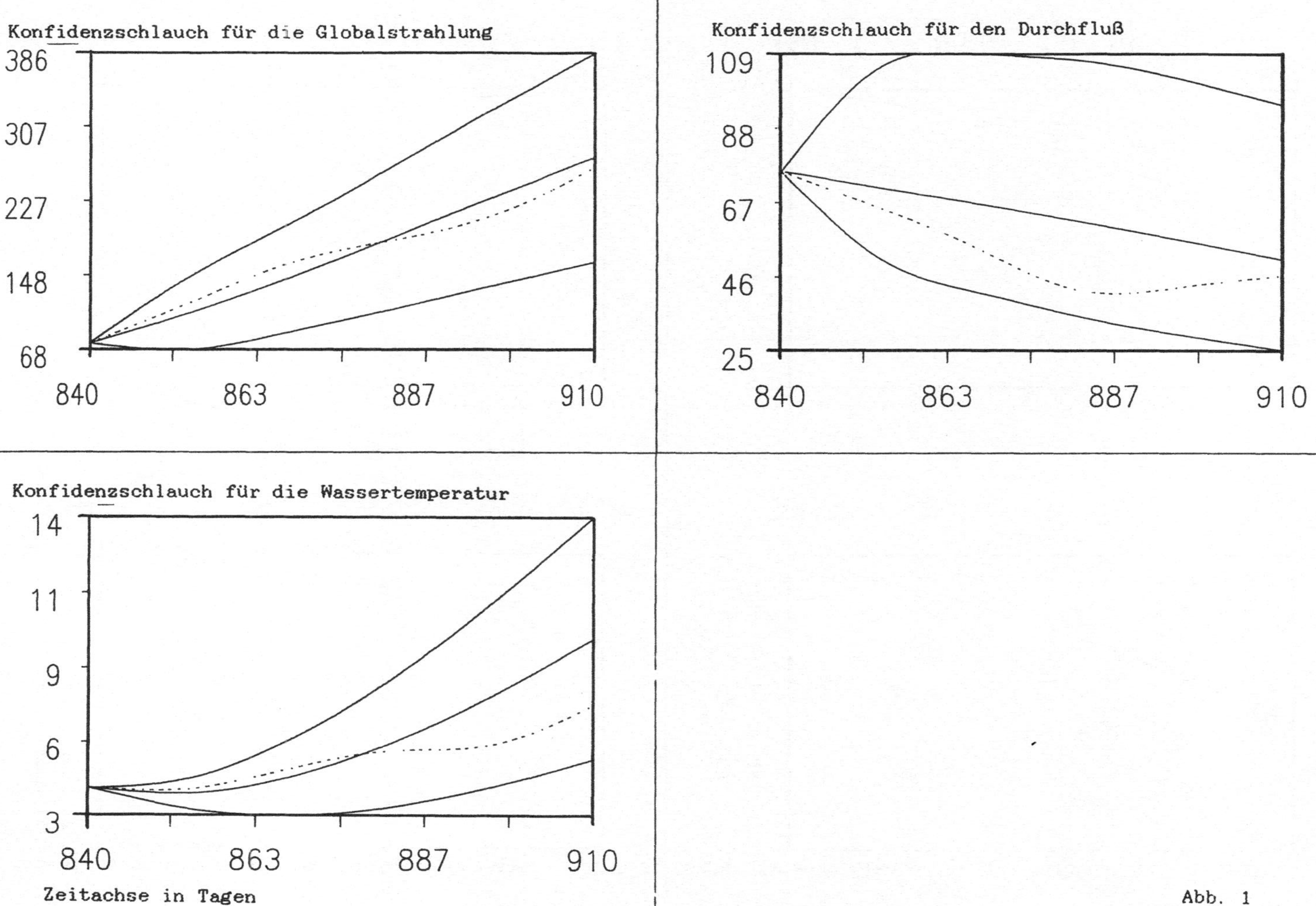

Abb. 1

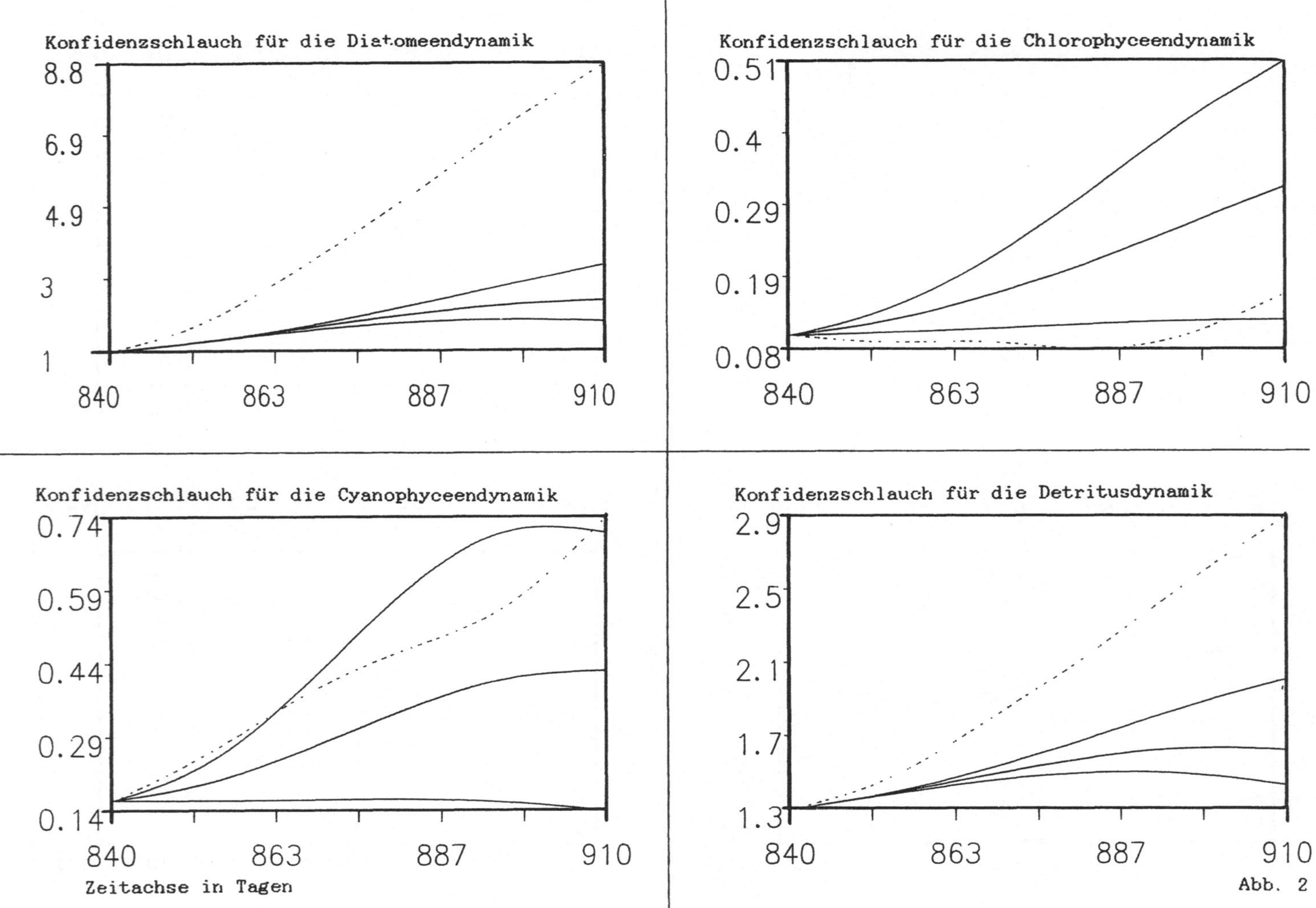

Konfidenzschlauch für die Diatomeendynamik
8.8
6.9
4.9
3
1
840
863
887
910
Konfidenzschlauch für die Chlorophyceendynamik
0.51
0.4
0.29
0.19
0.08
840
863
887
910
Konfidenzschlauch für die Cyanophyceendynamik
0.74
0.59
0.44
0.29
0.14
840
863
887
910
Zeitachse in Tagen
Konfidenzschlauch für die Detritusdynamik
2.9
2.5
2.1
1.7
1.3
840
863
887
910
Abb. 2

Konfidenzschlauch für die Silikatdynamik
13.6
11.5
9.5
7.5
5.4
840 863 887 910
Konfidenzschlauch für die Ammoniumdynamik
2.06
1.86
1.65
1.45
1.25
840 863 887 910
Konfidenzschlauch für die Nitratdynamik
3
2.6
2.2
1.8
1.4
840 863 887 910
Zeitachse in Tagen
Konfidenzschlauch für die Phosphatdynamik
1.55
1.39
1.24
1.08
0.92
840 863 887 910

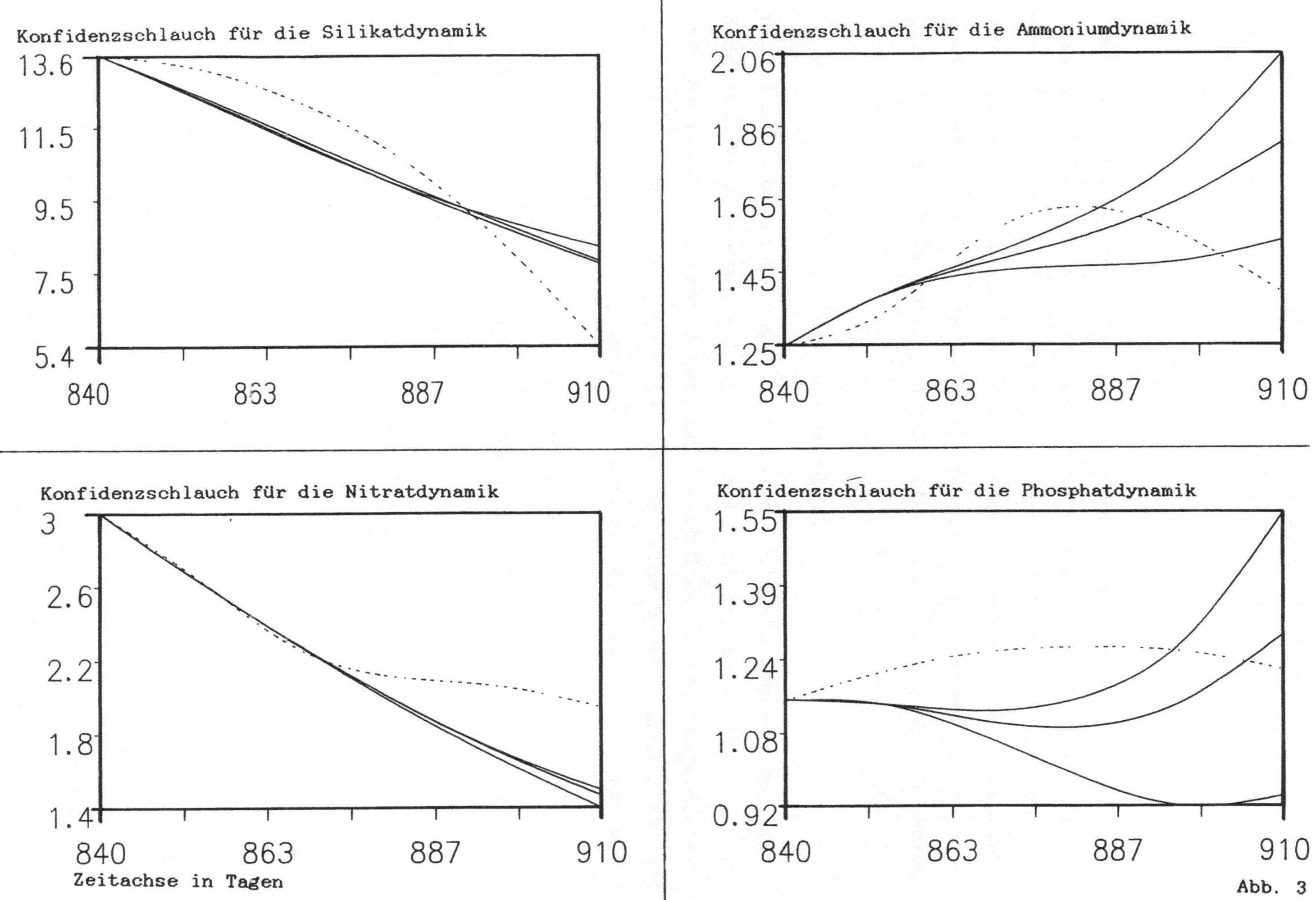

Abb. 3

$$- \beta_{14} y_2 \alpha_1 x_1 q_3(x_2) d_3(y_6) d_4(y_8)$$

$$- \beta_{15} y_2 (1-\alpha_1) x_1 q_4(x_2) d_5(y_7) d_6(y_8)$$

$$- \beta_{16} y_3 \alpha_2 x_1 q_6(x_2) d_7(y_6) d_8(y_8)$$

$$- \beta_{17} y_3 (1-\alpha_2) x_1 q_7(x_2) d_9(y_7) d_{10}(y_8)$$

$$+ \beta_{18} q_9(x_2) y_4 .$$

Im Modell wurden solche Prozesse wie Assimilation, Respiration und Algenablagerung, Nitrifikation und Denitrifikation berücksichtigt. Die Assimilation der Chlorophyceen und Cyanophyceen erfolgt bevorzugt unter Verbrauch von Ammonium (solange vorhanden), da Ammonium energetisch besser verwertbar ist als Nitrat. Bei der Modellierung wurden das Eutrophierungsmodell ERNA [1] und das Stickstoffmodell [6] berücksichtigt.

3. Resultate

Vertrauensgrenzen für Globalstrahlung, Wassertemperatur und Durchfluß sind in Abb. 1 gezeigt. Abb. 2 zeigt die Algendynamik und Abb. 3 die Nährstoffdynamik am Zufluß. Bei der Modellierung, der Parameterbestimmung, der Simulation und der Prognose am Zufluß ist das Programmsystem CANDYS/CM verwendet worden.

4. Literatur

[1] Braun, P.: Das Eutrophierungsmodell ERNA, Dissertation A, TU Dresden, 1984.

[2] Braun, P.; Rudolph, P.; Albrecht, K.-F.: Computer Aided Decision for Water Quality Management, In: Sydow, A.; Tzafestas, S.G.; Vichnevetsky, R.: Proceedings of the 3. Symposium on Systems Analysis and Simulation 1988, Mathematical Research, Vol. 47, Akademie-Verlag Berlin, 1988, 75-78.

[3] Funke, R.: CANDYS/CM - A Dialogue System for Modelling Continuous Dynamical Systems with Chain Structure by Differential Equations, Mathematical Research, Vol. 46, 169-171.

[4] Funke, R.: A Nonlinear Diffusion with Hyper-Gamma Distribution, Stochastic Analysis and Applications, 7(1989)1, 19-33.

[5] Funke, R.: Water Quality Forecast for a Lake Based on the Eutrophication Model ERNA and Using the Ito Stochastic Calculus, Proceedings of the International Workshop Environmental Management in the Baltic Region, Leningrad, 21.-24. Nov. 1989.

[6] Hajek, P.-M.: Stickstoffoxidation in Fließgewässern, Berichte aus Wassergütewirtschaft und Gesundheitsingenieurwesen, TU München, Nr. 52.

ARASIM

Eine Modellbank zur Simulation von Kläranlagen mit Simplex II

verfaßt von Hubert Hoffmann
Universität Passau

Inhaltsübersicht

1. Simplex II allgemein

2. Simplex II speziell
   - Modellbanken
   - hierarchische Modellbildung
   - Klassenkonzept
   - Versionenkonzept

3. Kläranlagen allgemein

4. Modellbank ARASIM

1. Simplex II allgemein

Das Simulationssystem Simplex II besteht in seiner Grundstruktur aus drei Teilen:

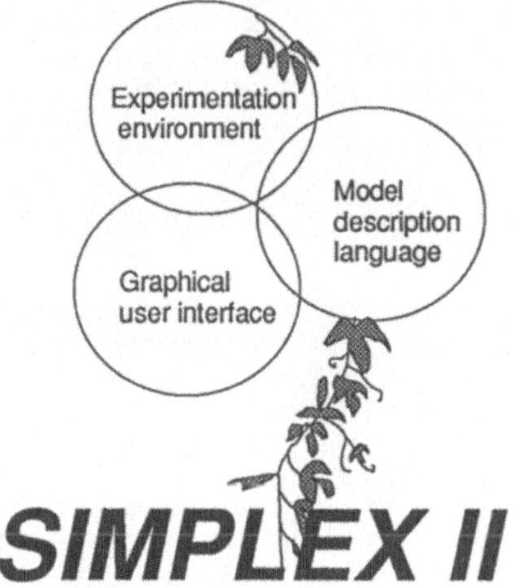

1. einer graphischen Benutzerschnittstelle, die der graphischen Modellerstellung und der Darstellung von Simulationsergebnissen durch Präsentationsgraphiken und Animation dient,
2. einer Modellbeschreibungssprache (Simplex-MDL) als höherer Programmiersprache, die durch spezielle, an die Problemstellungen der Simulation angelehnte, programmiersprachliche Konstrukte zur Beschreibung einer Modellkomponente dient sowie
3. einer Experimentierumgebung, die zur Verwaltung der Simulationsmodelle und deren Experiment- und Modelldaten dient.

2. Simplex II speziell

Ausgehend von der Überlegung, daß jedes anspruchsvollere (Simulations-)Modell zwar eine logische Einheit bildet, dennoch aber in verschiedene (Teil-)Komponenten zerfällt, gelangt man zum Begriff der Modellbank.
Die Modellbank beinhaltet folglich sämtliche Komponenten eines (Simulations-)Modelles. Durch logische Verbindungen, sogenannte Connections, der einzelnen Komponenten untereinander können diese zu einer höheren Komponente zusammengefaßt werden. Derartig verbundene Komponenten können wiederum mit anderen einzelnen, sog. Basiskomponenten, oder ebenfalls zusammengefaßten Komponenten zu einer neuen höheren Komponente vereinigt werden, so daß eine hierarchische Modellierung möglich ist.
Soll jedoch eine (Sub-)Komponente mehrmals in einer (hierarchisch) höheren Komponente vorkommen, so ist dies durch das sogenannte Klassenkonzept möglich. Dabei bildet jede Komponente eine eigene Klasse. Bei der Modellierung einer höheren Komponente ist es daher möglich mehrere Ausprägungen einer Komponentenklasse mit Ausprägungen anderer Komponentenklassen zu vereinigen. Eine Mehrfachhaltung gleicher Komponenten entfällt somit.
Ebenfalls ist es möglich in einer Modellbank mehrere Versionen einer Modellkomponente zu halten. Dabei wird dem Simulationssystem beim Installieren einer Komponente lediglich mitgeteilt, welche der möglichen Versionen die momentan aktuelle ist. Diese Version wird nun in allen höheren Komponenten, die diese Komponente als Subkomponente enthalten, als die z. Zt. aktuelle betrachtet. Durch dieses Versionenkonzept lassen sich, ohne die Schnittstellen einer Komponente nach oben hin zu ändern, beliebig komplizierte Versionen ein und derselben Komponente im Modellentwurfszyklus austesten.

3. Kläranlagen allgemein

Die Reinigungsleistung einer Kläranlage erfolgt, grob eingeteilt, in zwei Stufen:

- In der mechanischen Reinigungsstufe transportiert das Hebewerk das im Sammler zusammengelaufene Abwasser in die Anlage. Hier durchläuft es zunächst eine Rechenanlage, in der der Grobschmutz durch ein automatisch gesteuertes Rechen nach oben geräumt und in Behältern gesammelt wird. Dieses Rechengut wird zur Müllverbrennung gebracht und dort weiterbehandelt. Das Abwasser strömt nun in ein belüftetes Sandfangbecken, in dem sich im Abwasser enthaltene Feinsande absetzen, die mit Hilfe einer mechanischen Räumeinrichtung entfernt und zur Ablagerung zu einer Deponie gebracht werden. Die letzte Station dieser Stufe ist die <u>Vorklärung</u>, ein Längsbecken mit einem mechanischen Räumer, in der sich relativ große, partikuläre Stoffe in verhältnismäßig kurzer Zeit (2 bis 3 Stunden) absetzen. Damit sind bereits ca. 25 % der ursprünglichen Verschmutzung entfernt.
- Die biologische Stufe umfaßt mehrere sog. <u>Belebungsbecken</u>, deren Reinigungsverfahren auf der Freßtätigkeit von frei schwimmenden Organismen beruht. Die Nahrung für diese Lebewesen wird vom einfließenden, hier schon stark vorgereinigten Abwasser mitgeführt. Je nach Belebungsbeckentyp unterscheidet man Nitrifikations- und Denitrifikationsbecken.

Bei der Nitrifikation soll im Abwasser enthaltenes Ammonium (NH_4) - ein Fischgift - zu Nitrat (NO_3) oxidiert werden. Deshalb wird Sauerstoff benötigt, der dem Belebungsbecken mittels Belüftungsanlagen in Form von Luft oder reinem Sauerstoff zuzuführen ist.
Die Denitrifikation ist die Reduktion von Nitrat zu Stickstoff (N), das bei zu starker Abgabe die Gewässer überdüngt. Dieser Prozeß wird nur unter Mangel an freiem Sauerstoff (O_2) von bestimmten Bakterien durchgeführt. Daher darf hier nicht belüftet werden. Die Denitrifikation wird jedoch von den meisten Klärwerken noch nicht angewendet.
In den <u>Nachklärbecken</u> werden vorhandene partikuläre Stoffe im jetzt gut gereinigten Abwasser so weit wie möglich gesammelt, teilweise in das Belebungsbecken zurückgepumpt bzw. als Überschußschlamm der Schlammbehandlung übergeben. Die Reinigungsleistung dieser Station, gemessen an der ursprünglichen Verschmutzung, beträgt 13,3 %.

4. Die Modellbank Arasim

Gemäß dem Komponentenprinzip aus Kapitel 2 ist die Modellbank ARASIM in Simplex II aufgebaut. Dabei sollen die Komponenten den realen Teilkomponenten einer Kläranlage weitgehend entsprechen.

Zufuhr: Diese Komponente stellt die Quelle des Systems dar. Sie bestimmt den täglichen Zufluß des Abwassers in die Kläranlage

Vorklärung: Sie dient der Modellierung der mechanischen Reinigung einer Kläranlage. Auf die Modellierung der Reinigung des Abwassers von groben Bestandteilen in der Rechenanlage und im Sandbecken wurde verzichtet.

Belebungsbecken: Komponente, die das Herzstück der biologischen Reinigung einer Kläranlage modelliert.

Nachklärung: In dieser Komponente wird die Trennung des Belebungsschlamms vom gereinigten Abwasser durch Sedimentation modelliert. Gleichzeitig wird hier die Schlammenge berechnet, die als Rückflußschlamm ins Belebungsbecken fließt und jene, die als Überflußschlamm das System verläßt.

Zur Bestimmung der Verschmutzung des Abwassers werden folgende 13 Inhaltsstoffe des Abwassers und des Schlamms berücksichtigt:

- zur CSB - Bilanz:
  X_A : Autotrophe Biomasse
  X_H : heterotrophe Biomasse
  X_P : biologisch inerte, partikuläre organische Produkte
  X_I : biologisch inerte, partikuläre organische Stoffe
  X_S : biologisch langsam abbaubare organische Stoffe
  S_I : biologisch inerte, gelöste organische Stoffe
  S_S : biologisch rasch abbaubare, gelöste organische Stoffe
        (Substrat)
  S_O : Sauerstoff

- zur Stickstoffbilanz:
  S_NH : Ammonium
  S_NO : oxidierter Stickstoff
  S_ND : gelöster biologisch rasch abbaubarer organischer
         Stickstoff
  X_ND : biologisch langsam abbaubarer Stickstoff

  S_A : Alkalinität

Wobei partikuläre Substanzen durch ein vorangestelltes X und
gelöste Substanzen durch ein vorangestelltes S verdeutlicht
werden.

Die Komponenten der Modellbank ARASIM im einzelnen:

<u>Die Komponente Zufluß</u>

- Blockbild

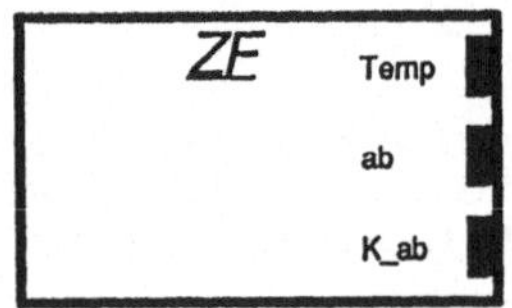

- Eingangsgrößen: %
- Ausgangsgrößen
  Temp   : Temperatur des Abwassers
  ab   : Zuflußvolumenstrom ins System
  K_ab   : Stoffkonzentration der 13 im Modell enthalte-
                nen Inhaltsstoffe des Abwassers
- Versionen
  zf-konst   : konstanter Wert der Ausgabegrößen
  zf-varzu   : variabler Zuflußvolumenstrom
  zf-varkonz : dynamischer Zuflußstrom mit variabler Konzen-
                tration der Inhaltsstoffe

<u>Die Komponente Vorklärung</u>

- Blockbild

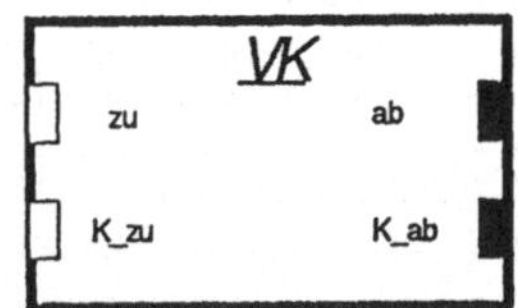

- Eingangsgrößen
  zu: : Zuflußvolumenstrom
  k_zu   : Zuflußkonzentration
- Ausgangsgrößen
  ab   : Abflußvolumenstrom
  K_ab   : Abflußkonzentration
- Versionen
  vk-mix : einfließende Substanzen verteilen sich ohne zeit-
           liche Verzögerung auf das gesamte Becken
  vk-rohr: zwischen zeitlich hintereinander einfließendem
           Abwasser findet keine Durchmischung statt.

Die Komponente Belebungsbecken

Blockbild:

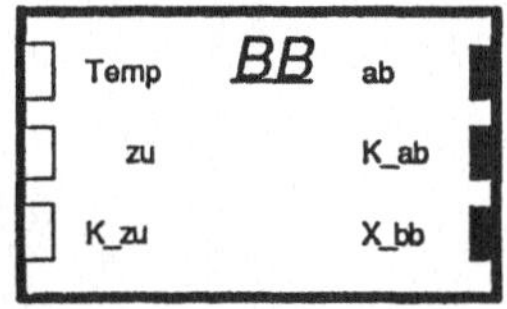

- Eingangsgrößen
  zu   : Zuflußvolumenstrom
  k_zu : Zuflußkonzentration
  Temp : Abwassertemperatur im Becken
- Ausgangsgrößen
  ab   : Abflußvolumenstrom
  K_ab : Abflußkonzentration
  X_bb : Gesamtschlammasse im Becken

Die Komponente Nachklärung

Blockbild:

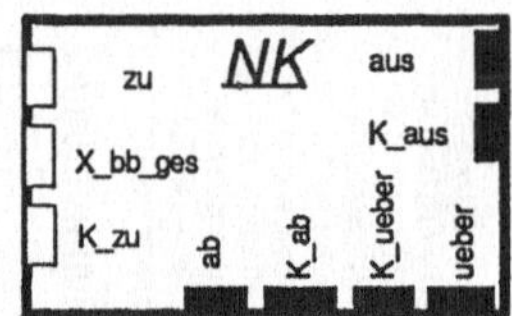

- Eingangsgrößen
  zu   : Zuflußvolumenstrom
  k_zu   : Zuflußkonzentration
  X_bb_ges : Abwassertemperatur im Becken
- Ausgangsgrößen
  aus   : Ausflußvolumenstrom
  K_aus   : Abflußkonzentration
  ab   : Rückflußschlammvolumen
  K_ab   : Rückflußschlammkonzentration
  über   : Überschußschlammvolumen
  K_über   : Überschußschlammkonzentration
- Versionen
  nk-mix   : volldurchmischtes Nachklärbecken
  nk-flux   : Nachklärung als Schichtenmodell

Um dem Leser die Beschreibung der Dynamik einer Modellkomponente in SIMPLEX - Mdl zu demonstrieren, möchte ich auf
die nachfolgende Seite verweisen. Aufgrund des eingeschränkten
Platzangebotes für diesen Beitrag kann dies aber nur an einem
einfachen Beispiel vorgeführt werden. Bei der Simulation des
Absetzvorganges in einer Schicht eines Nachklärbeckens kann
der Anwender unter den folgenden Integrationsverfahren wählen:
    Euler-Verfahren,   Runge-Kutta-Verfahren   4.Ordnung,   Runge-
    Kutta- Fehlberg-Verfahren 5./6. Ordnung,   Implizites Runge-
    Kutta-Verfahren   6.Ordnung,   Extrapolationsverfahren   nach
    Bulisch-Stör 6.   Ordnung und für steife Differential-
    gleichungen das Runge-Kutta-Verfahren mit varianten Koeffi-
    zienten nach Rosenbrock-Wanner.
Zum Abschluß sei noch auf eine Graphik verwiesen, die als
Beispielimplementierung einer Kläranlage als Simulationsmodell
in SIMPLEX II dient.

```
#############################################
#                                           #
#          D O K U M E N T A T I O N        #
#                                           #
#############################################
#                                           #
#  Projekt:      Klaeranlagensimulation     #
#                                           #
#  Komponente:   sch2                       #
#                                           #
#  Version:      V0                         #
#                                           #
#  Autor:        Klaus Helderer             #
#                                           #
#  Erst.-datum:  Apr. 1989                  #
#                                           #
#  Letzte Aend.: 20. Juni 1989              #
#                                           #
#  Funktion:   * Nachklaerbeckenschicht fuer 2 Stoffe
#              * Entscheidung, ob Zuflussschicht, obere oder untere
#                Schicht, in der Komponente selbst auf Grund der
#                angeschlossenen Sensorgroessen
#              * Berechnung der aktuellen Schichtkonzentration fuer
#                2 Stoffgruppen, partikulaere und geloeste
#              * Berechnung der aktuellen Sinkgeschwindigkeit
#                                           #
#############################################

BASIC COMPONENT sch2

#############################################
#                                           #
#   M O D E L L E L E M E N T E   D E K L A R I E R E N   #
#                                           #
#############################################

DECLARATION OF ELEMENTS

  CONSTANTS

    oberfl(REAL) := 320,      #**2    Beschickungsflaeche der Schicht
                              #       des Nachklaerbeckens
    hoehe (REAL) := 0.25,     #       Hoehe der Nachklaerb.schicht
    v_max (REAL) := 7.80,     #  /h   maximale Sinkgeschwindigkeit
    SVI   (REAL) := 100       #=1 /g  Schlamm-Volumen-Index

  STATE VARIABLES
  CONTINUOUS
    partikulaere Stoffe gesamt
    X_tot (REAL) := 1860,     #g CSB/m**3  Gesamtschlammkonzentration
    geloeste Stoffe gesamt
    S_tot (REAL) := 40        #g CSB/m**3


DEPENDENT VARIABLES
CONTINUOUS
  Volumenstroeme innerhalb des Schichtenmodells
    durch (REAL) := 2000,   #m**3 /d   Durchflussmenge dieser Schicht
    auf   (REAL) := 2000,   #m**3 /d   Ausflussmenge in die darueber-
                                       liegende Schicht
    ab    (REAL) := 2000,   #m**3 /d   Ausflussmenge in die darunter-
                                       liegende Schicht
  Volumen der Schicht des Nachklaerbeckens
    vol   (REAL) := 800,    #m**3

  Konzentration in dieser Schicht
    ARRAY [2] konz (REAL):=700,  #g CSB/m**3  Stoffkonzentrationen der
                                 #g N /m**3   einzelnen Fraktionen

  Sinkgeschwindigkeit der partikulaeren Substanzen in dieser Schicht
    v   (REAL) := 100      #m /d   (0.09 m/s)

SENSOR VARIABLES
CONTINUOUS
  darueberrliegende Schicht
    v_o (REAL) := 0,             #m /d        Sinkgeschwindigkeit
    ARRAY [2] k_o (REAL) := 0,   #g CSB/m**3  Stoffkonzentrationen der
                                             einzelnen Stoffe
  darunterliegende Schicht
    v_u (REAL) := 0,             #m /d        Sinkgeschwindigkeit
    ARRAY [2] k_u (REAL):= 0,    #g CSB/m**3  Stoffkonzentrationen der
                                             einzelnen Stoffe
  Zufluss aus darueber- oder darunterliegender Nachklaerbeckenschicht
    zu_u  (REAL) := -1,     #m**3 /d   Zuflussmenge pro Tag von unten
    zu_o  (REAL) := -1,     #m**3 /d   Zuflussmenge pro Tag von oben
  Zufluss aus Belebungsbecken (nur bei einer Schicht aktiv)
    ARRAY [2] k_zu (REAL):= C,  #g CSB/m**3  Stoffkonzentrationen der
                                            einzelnen Stoffe
    zu_bb (REAL) := -1      #m**3 /d   Zuflussmenge pro Tag


#############################################
#                                           #
#   D Y N A M I K   D E R   K O M P O N E N T E   #
#                                           #
#############################################

DYNAMIC BEHAVIOUR

# ALGEBRAIC EQUATIONS

# Volumen der Schicht bestimmen
  vol := oberfl * hoehe;

# Sinkgeschwindigkeit in dieser Schicht bestimmen
# aus G.T. Daigger u. R.E. Roper Jr.: "The Relationship between SVI ..."
#   in der Originalgleichung: X_tot in g/l, v in m/h, SVI in ml/g
#   daher Umwandlungen erforderlich: X_tot / 1000, v * 24
  v := 24 * v_max * EXP ( (-0.148 - 0.0021 * SVI) * X_tot / 1000);

# aktuelle Konzentrationen der Fraktionen in ARRAY konz uebertragen
  konz [1] := X_tot;
  konz [2] := S_tot;


# Falls Zufluss vom Nachklaerbecken kommt, Durchfluss in die oberen und
# unteren Schichten aufteilen
  IF zu_bb >= 0       #Zufluss von BB: wird aufgeteilt
    DO
      durch := zu_bb;
      ab    := zu_o;
      auf   := zu_u;

  DIFFERENTIAL EQUATIONS

#   partikulaere Stoffe: Wasserbewegung
#                            + herabsinkende Feststofffracht
#                            - absinkende Feststofffracht
      X_tot' := durch * (k_zu[1] - X_tot) / vol
                     + ( MIN (v_o * k_o[1],v * X_tot)
                       - MIN (v * X_tot,v_u * k_u[1]) ) / hoehe;
#   geloeste Stoffe: Wasserbewegung
      S_tot' := durch * (k_zu[2] - S_tot) / vol;

    END
  END
  ELSIF zu_u >= 0    #Zufluss von unterer Schicht: hydraulischer Auftrieb
    DO
      durch := zu_u;
      ab    := 0;
      auf   := zu_u;

  DIFFERENTIAL EQUATIONS

#   partikulaere Stoffe: Wasserbewegung in obere Schicht
#                            + herabsinkende Feststofffracht
#                            - absinkende Feststofffracht
      X_tot' := durch * (k_u[1] - X_tot) / vol
                     + ( MIN (v_o * k_o[1],v * X_tot)
                       - MIN (v * X_tot,v_u * k_u[1]) ) / hoehe;
#   geloeste Stoffe: Wasserbewegung
      S_tot' := durch * (k_u[2] - S_tot) / vol;

    END
  END
  ELSE            #Zufluss von oberer Schicht: hydraulischer Antrieb
    DO
      durch := zu_o;
      ab    := zu_o;
      auf   := 0;

  DIFFERENTIAL EQUATIONS

#   partikulaere Stoffe: Wasserbewegung
#                            + herabsinkende Feststofffracht
#                            - absinkende Feststofffracht
      X_tot' := durch * (k_o[1] - X_tot) / vol
                     + ( MIN (v_o * k_o[1],v * X_tot)
                       - MIN (v * X_tot,v_u * k_u[1]) ) / hoehe;
#   geloeste Stoffe: Wasserbewegung
      S_tot' := durch * (k_o[2] - S_tot) / vol;

    END
  END
END OF sch2
```

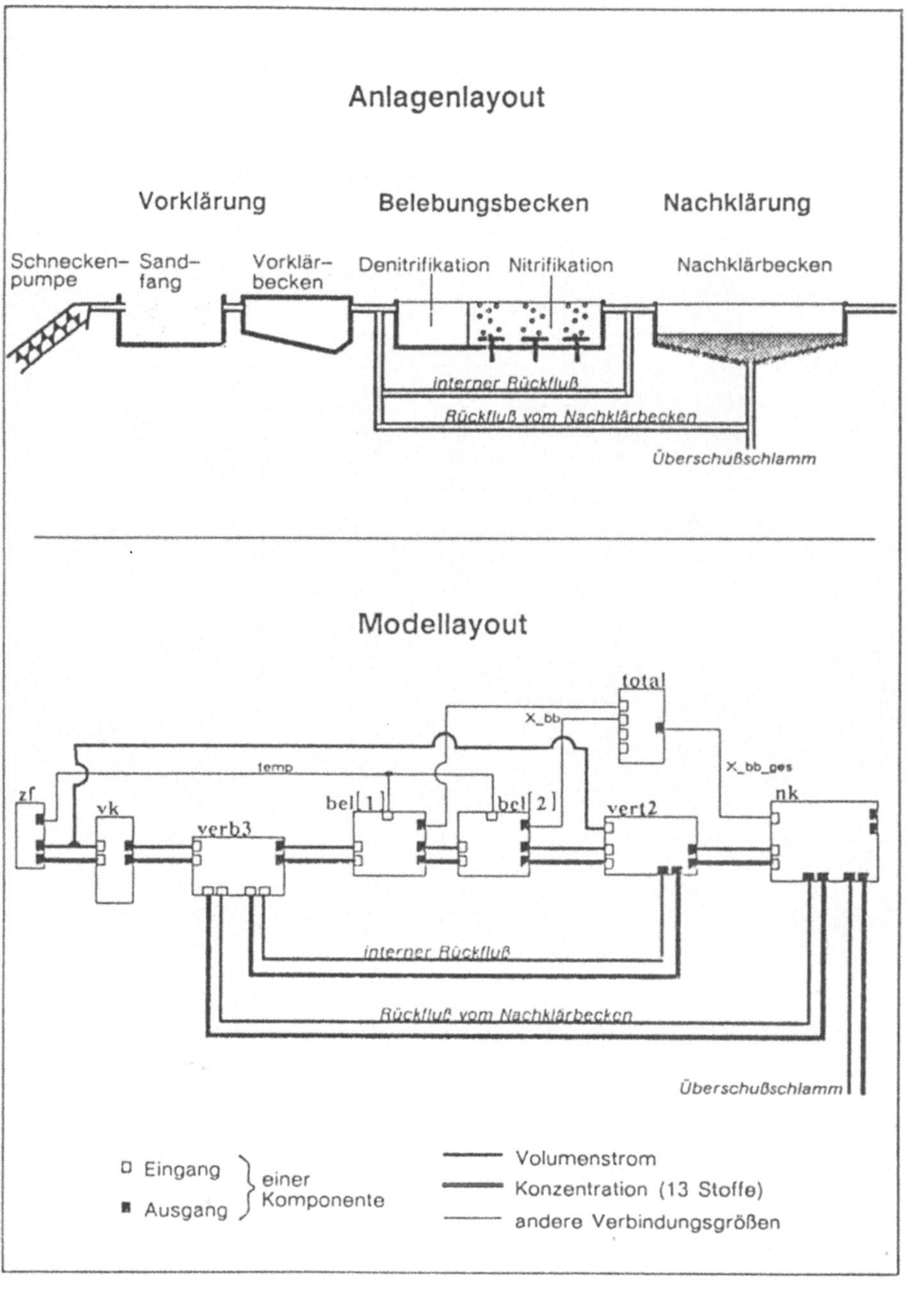
Anlagenlayout
Vorklärung
Belebungsbecken
Nachklärung
Schnecken-
pumpe
Sand-
fang
Vorklär-
becken
Denitrifikation
Nitrifikation
Nachklärbecken
interner Rückfluß
Rückfluß vom Nachklärbecken
Überschußschlamm
Modellayout
total
X_bb
X_bb_ges
temp
zf
vk
verb3
bel[1]
bel[2]
vert2
nk
interner Rückfluß
Rückfluß vom Nachklärbecken
Überschußschlamm
Eingang
Ausgang
einer
Komponente
Volumenstrom
Konzentration (13 Stoffe)
andere Verbindungsgrößen

MATHEMATIK

STOCHASTIC MODELING OF REACTION-MIGRATION SYSTEMS

Gottfried Jetschke

Sektion Mathematik
Friedrich-Schiller-Universität Jena
DDR-6900 Jena

In modern chemistry, biotechnology or ecology often the following
situation has to be modeled: There is a spatially extended system consisting of a lot of individuals (or particles). These individuals may
reproduce themselves or die (= "reaction") and they are able to carry
out spatial movement (= "migration"). Typical examples are chemical or
microbial processes in large vessels (f.e. bioreactors or sewage
treatment plants) or biological populations living in large areas.

Since reproduction and movement of the individuals appear to be of
stochastic nature two main questions have to be aswered: (1) How can
the macroscopic behaviour of the system can be derived from the microscopic rules? (2) Which stochastic effects occur due to the very
large but finite number of individuals?

For simplicity we will only model diffusion-like movement showing the
tendency of spreading out the population and to smooth steep density
profiles.

1. Types of models

Two main classes of models can be distinguished:

(1) Spatially discretized models

To apply the methods describing homogeneous systems the area is arti-
ficially divided into many small cells or boxes of equal size. The
reactions are modeled by birth-and-death processes within each cell as
known for uniform systems. Diffusion-like migration can be described
in this setting by stochastic jumps between neighbouring cells. As a
result we get a high-dimensional Markovian birth-and-death process
$(X_1(t),...,X_N(t))$, $t \geq 0$, where $X_k(t)$ denotes the number of particles
(of one or of several species) in the k-th cell.
To be more explicit let us consider the very simple case of one
species living in a one-dimensional habitat of length L which is
divided into N cells of length $1 = L/N$. The change of particle
numbers due to reproduction (where at least one particle is created or

removed at the same time) and diffusion is modeled by the phenomeno-
logical ansatz for the transition probabilities:

(i) For all $k = 1,\ldots,N$ we put

$$\text{Prob}\{X_k(t+\tau) = n_k+1 \mid X_i(t) = n_i,\ i=1,\ldots,N\} = 1 \cdot \lambda(n_k/1) \cdot \tau + o(\tau) \ ,$$

$$\text{Prob}\{X_k(t+\tau) = n_k-1 \mid X_i(t) = n_i,\ i=1,\ldots,N\} = 1 \cdot \mu(n_k/1) \cdot \tau + o(\tau) \ ,$$

$$\text{Prob}\{X_k(t+\tau) = n_k \mid X_i(t) = n_i,\ i=1,\ldots,N\} = 1 - 1 \cdot [\lambda(n_k/1)+\mu(n_k/1)] + o(\tau) \ .$$

(ii) For all $k,k' = 1,\ldots,N$ we put

$$\text{Prob}\{X_{k'}(t+\tau) = n_{k'}+1 ,\ X_k(t) = n_k-1 \mid X_i(t) = n_i ,\ i=1,\ldots,N\} =$$

$$= \begin{cases} d \cdot n_k \cdot \tau + o(\tau) & k' = k\pm1 \ , \\ 0 & \text{otherwise} \ , \end{cases}$$

where for X_0 and X_{N+1} some boundary conditions have to be introduced
(for a closed system for example zero-flux conditions $X_0(t) \equiv X_1(t)$,
$X_{N+1}(t) \equiv X_N(t)$ for all t).

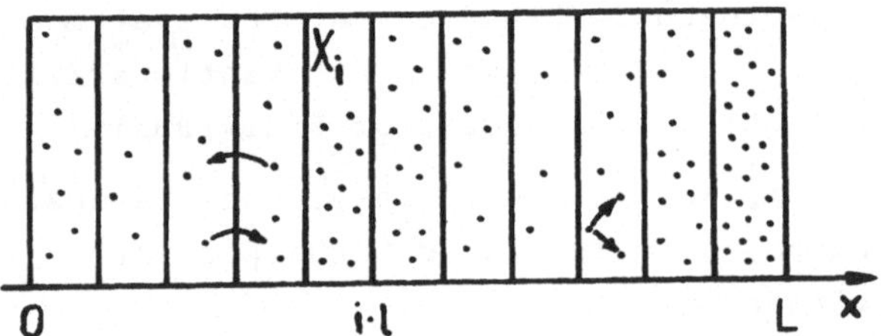

Fig. 1: Cell model of reaction and migration

The probability distribution and other characteristic properties of
this Markovian birth-and-death process (see Fig. 1) can be derived
from the corresponding high-dimensional Master equation (for details
see [12], [10], [5], [7], [1]).

(2) Spatially continuous models

To avoid discrete cells diffusional transport can be modeled by independent Brownian motions for each individual (assuming some boundary
conditions, for example in a closed system reflection at the boundaries). Birth and death of individuls are then stochastic events depending on the local configuration of the system. For example, if prey
and predator come close enough together then the predator will eat the
prey, hence that individual has to be removed. If two individuals of
the same species meet (at a small distance) then the probability to
produce offsprings will be lowered due to intraspecific competition.
In this way we obtain some kind of spatio-temporal branching process

$(x_1(t),\ldots,x_{n(t)}(t))$, $t \geq 0$, where $x_k(t)$ denotes the position of the k-th individual, taking values in the set of all possible particle configurations with a random total number n(t) of individuals (Fig. 2, for details see [6], [2], [3]).

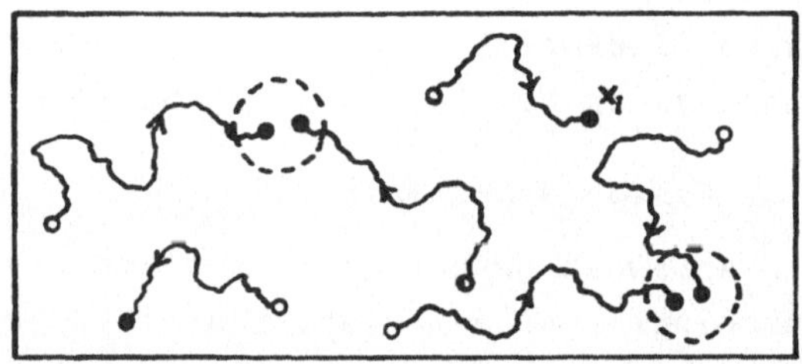

Fig. 2: "Encounter" model of reaction and migration

(3) Approximations

Since real systems are large and often contain a very high number of individuals several approximations, deterministic and stochastic ones, can be made (see [1]).

a) Deterministic limit

If the (relative) system size and the (local) particle number tend to infinity such that the mean particle density remains finite we obtain in the deterministic limit a macroscopic reaction-diffusion equation.

To be more specific consider the cell model with the transition rates given above, which are (up to an extra factor 1) functions of the particle density alone. Since the cells have to grow (1→∞) but the limit shall be spatially continuous the number of cells also has to grow (N→∞) but even faster. This forces us to rescale the coordinate such that on the new x-axis the system is defined on the interval $[0,\tilde{L}]$ with new cell size $\tilde{l} = \tilde{L} \cdot 1/L = \tilde{L}/N$ where necessarily $\tilde{l} \to 0$. The N-dimensional process $(X_k(t))_{k=1}^{N}$ corresponds to a step function $U(t,x) = X_k(t)/1$ on $[(k-1)/N, kN)$, $k=1,\ldots,N$.

If we assume that for N→∞, L→∞ we have $U(0,\cdot) \to u_0$ in $\mathcal{L}^2(0,\tilde{L})$, $N^2/1 \to \infty$ and $d=DN^2$ (entailing 1→∞ and d→∞) then it can be proved that on $\mathcal{L}^2(0,\tilde{L})$ the process $U(t,\cdot)$ for every t converges in probability as N→∞, L→∞ to the solution $u(t,\cdot)$ of the deterministic equation

$$\frac{\partial u}{\partial t}(t,x) = D \cdot \frac{\partial^2 u}{\partial x^2}(t,x) + f[u(t,x)] \ , \quad u(t,0)=u_0 \ , \quad t \geq 0 \ , \quad x \in [0,\tilde{L}] \qquad (1)$$

with $f(u) := \lambda(u) + \mu(u)$ together with appropriate boundary conditions (here $(\partial u/\partial x)(t,x) = 0$ for x=0 and $x=\tilde{L}$ and all t).

b) Gaussian approximation

A first stochastic approximation is given by

$$U(t,x) = u(t,x) + V(t,x)/\sqrt{L} \;,$$

where the random Gaussian field $V(t,x)$ solves a linear stochastic partial differential equation. That means that the evolution is approximately given by Gaussian fluctuations around the deterministic path.

c) Diffusion approximation

In a somewhat better approximation the evolution can be described by the solution of a nonlinear stochastic partial differential equation driven by a space-time Gaussian white noise with a state-dependent intensity. The mathematics of these equations is still under research.

For the branching-type encounter models similar approximations can be done.

2. A simple ecological model: Spread of a species

A population usually is found only within a finite geographic region. Whereas the living conditions in the middle of the area are rather good they gradually become worser in the outer parts of the habitat. Therefore, survival of the species at the boundaries is only maintained by immigration from the central part. Quite naturally questions concerning the size of the habitat and its fluctuations have to be answered.

Again we will consider a rather simple situation, namely one species in a one-dimensional space (f.e. at a coast line), which we will describe by the cell model of section 1. In one half-space ($x<0$) we assume constant favourable conditions, hence the established population density will fluctuate around its natural capacity. At the other side ($x>0$) the death rate always shall exceed the birth rate due to constant unfavourable conditions. As a consequence of migration (with rate d) into the positive half-space and negative mean reproduction rate ($\lambda<\mu$) there will be a largest cell number $N(t)$ containing at least one individual. This is the actual length of the habitat.

Analytic investigations are under performance, but computer simulations show interesting effects. Near the right end of the habitat the mean population density is decreasing exponentially, but the realized length $L(t) = N(t)\cdot 1$ is a random variable with high dispersion. Moreover, local extinction can create isolated regions of occurrence. If migration is high and mortality only slightly above natality the popu-

lation can spread out and survive in an area being large compared with the "core" region where positive net reproduction occurs.
If the unfavourable domain has a finite size its length can be estimated such that the population with a sufficiently high probability can (or can not!) cross this strip and reach the other "good" region.

3. Population genetics: Speciation due to spatial genetic drift

Random genetic drift is a stochastic effect due to the finite size of a population causing deviations from the mean genetic dynamics. If we consider a population living in a spatially extended region and assume that selection may have several optima of mean fitness then the genetic drift can produce interesting global effects.

Let us consider a very simple situation, namely a continuous character under disruptive selection with several stable genetic equilibria. If migration is modeled by diffusion then under some further assumptions (f.e. fast recombination and a large number of involved loci) we arrive at the partial differential equation (1), where u now denotes the mean of the character (being normally distributed),

$$f(u) = \frac{d}{du} \log \overline{W}_0(u) \qquad ,$$

with $\overline{W}_0(u)$ as the (local) mean fitness of the population and D as its dispersal rate (see [11], [9], [4]).

Equation (1) can be written as a gradient system in a function space,

$$\frac{du}{dt} = - \frac{\delta S(u)}{\delta u} \quad , \qquad S(u) = \int_0^L \{\frac{D}{2}|u'(x)|^2 - F[u(x)]\} \, dx \quad ,$$

where $\delta/\delta u$ denotes the functional derivative and $F'(u) \equiv f(u)$. For all continuous initial functions $u_0(\cdot)$ a unique solution $u(t,\cdot)$ exists for all $t>0$, and as $t\to\infty$ any trajectory reaches a fixed point being a stationary point of the potential S (i.e. $\delta S(u)/\delta u=0$). Hence almost all trajectories are attracted by the stable equilibria being local minima of S. All other fixed points are of saddle type.

Due to the finite size of the population inevitably stochastic deviations of this deterministic behaviour occur. This sampling drift is often modeled by adding in (1) a space-time Gaussian white noise ξ:

$$\frac{\partial U}{\partial t} = D \cdot \frac{\partial^2 U}{\partial x^2} + f(U) + \sigma \cdot \xi(t,x) \qquad , \qquad (2)$$

$$E \, \xi(t,x) = 0 \, , \qquad E \, \xi(t,x)\xi(s,y) = \delta(t-s) \cdot \delta(x-y).$$

The parameter σ measures the "noise" intensity and is proportional to the inverse square root of the population size (see [4], [9], [11]).

A precise meaning to this formal stochastic PDE has been given ([7], [8]), and the solution is a Markovian process taking values in the space of continuous functions (of the x variable). Hence the population can be thought of as moving across an adaptive landscape in an infinite-dimensional function space. Since the deterministic part of (2) is of gradient type disruptive selection together with diffusion pushes U towards one of the minima of S (being the states with maximal total mean fitness) whereas random drift enables the system to climb uphill and to cross crest lines in a mountaineous landscape.

As $t \to \infty$ the system reaches a stationary probability distribution which is formally given by the density

$$p^{st}[q(\cdot)] = N \cdot \exp\,(2S(q)/\sigma^2) \tag{3}$$

implying that the minima of S are the most probable states ([7], [8]).

If the effective population size is large the sampling drift is weak and the dynamics can be described more detailed. According to (3) most of its time the system spends near one of the stable equilibria. But in the long run sometimes especially favourable fluctuations occur which allow the system to leave one basin of attraction and to enter a neighbourhood of another stable equilibrium. (In physics such a transition between alternative ground states is called "tunneling".)

Let us assume that we have only two stable equilibria u_1 and u_2 and that their basins of attraction are separated by the set Σ. The minimum of $S(q)$ on Σ is reached in the lowest saddle point v. The following assertions can be proved:

(i) For sufficiently small σ the mean transition time from u_1 to u_2 is roughly given by $\exp(-2 \cdot \Delta S/\sigma^2)$ with $\Delta S := S(v) - S(u_1)$.

(ii) For small noise tunneling most probably occurs via the lowest saddle point v on a path consisting of an uphill movement opposite to the gradient of S (i.e. nearly the reverse of the deterministic motion) until it reaches the neighbourhood of the saddle and a subsequent rapid movement to the other equilibrium.

A straightforward interpretation of (ii) states that the weakly noisy system most likely chooses the way of least resistance - a common behaviour in social life.

Let us discuss the cubic nonlinearity

$$f(u) = -c(u - u_1)(u - u_3)(u - u_2) \quad , \quad c > 0 \quad , \quad u_1 < u_3 < u_2 \quad ,$$

which is the simplest case with two alternative stable equilibria. Suppose that $u_3 - u_1 < u_2 - u_3$, hence the situation is slightly asymmetric. For no-flux boundary conditions the states $u(x) \equiv u_1$ and $u(x) \equiv u_2$ are the

253

only stable fixed points. If L exceeds a critical length the lowest
saddle point v is a density profile with a peak near L/2 (Fig. 3). We
see that a stochastic shift from u_1 to u_2 is possible and most likely
occurs via v. Therefore, the transition between these two stable equilibria exhibits a kind of nucleation mechanism: Most probably in the
interiour of the system a region with high values of U is formed spon-
taneously. If this core is large enough both fronts spread throughout
the whole system (Fig. 4). Of course a subsequent transition from u_2
to u_1 is possible but less probable.

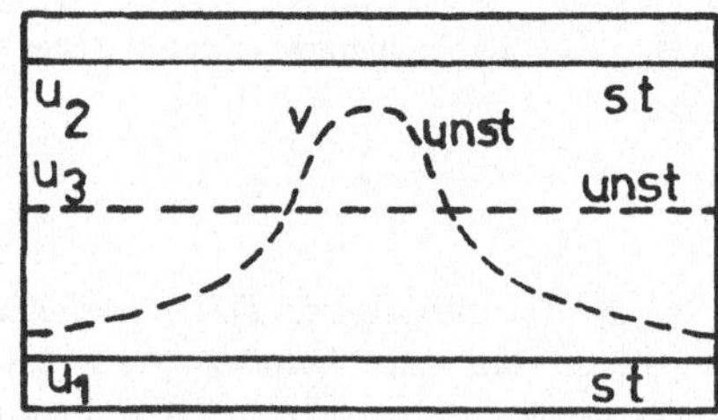

Fig. 1: Stable and unstable
fixed points for (1)

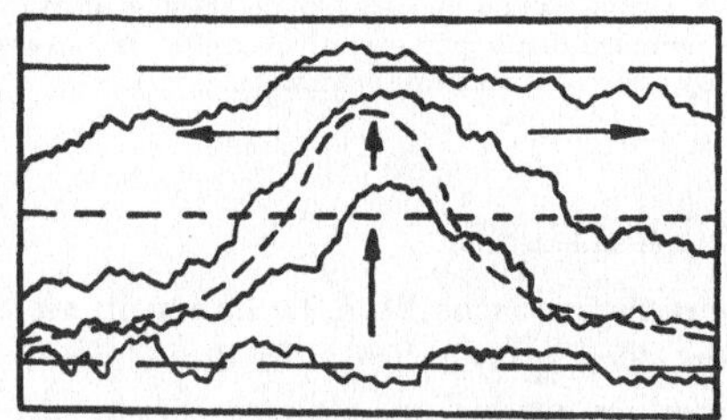

Fig. 2: Most probable transition
path from u_1 to u_2

These considerations show that a stochastic process of speciation is
quite feasible within a continuous population. It is not restricted to
strongly isolated demes or to a founder effect.

References

[1] ARNOLD, L.: On the consistency of mathematical models of chemical
    reactions. In: Dynamics of Synergetic Systems, Springer Series in
    Synergetics, Vol. 6, Berlin/Heidelberg/ New York 1980.
[2] DITTRICH, P.: A stochastic model of a chemical reaction with
    diffusion. Prob. Theor. Rel. Fields 79 (1988), 115-128.
[3] FEISTEL,R., EBELING,W.: Evolution of Complex Systems, Berlin 1989.
[4] FELSENSTEIN,J.: Excursions along the interface between disruptive
    selection and stabilizing selection. Genetics 93 (1979), 773-795.
[5] HAKEN, H.: Synergetics. An Introduction. 3rd Ed., Springer,
    Berlin/Heidelberg/New York 1983.
[6] ITO, K.: Distribution-valued processes arising from independent
    Brownian motions. Math. Zeitschr. 182 (1983), 17-33.
[7] JETSCHKE, G.: Mathematik der Selbstorganisation. DVW Berlin und
    Vieweg Braunschweig/Wiesbaden 1989.
[8] JETSCHKE, G.: Unendlichdimensionale Diffusionsprozesse als Lösun-
    gen stochastischer partieller Differentialgleichungen, Disser-
    tation B, Jena 1989.
[9] LANDE, R.: Genetic variation and phenotypic evolution during
    allopatric speciation. Amer. Natur. 116 (1980), 463-479.
[10] MALEK-MANSOUR, M., VAN den BROECK, C., NICOLIS, G., TURNER, J.W.:
    Asymptotic properties of Markovian master equations, Ann. Phys.
    131 (1981), 283-313.
[11] ROUHANI,S., BARTON,R.: Speciation and the "shifting balance" in a
    continuous population. Theor. Popul. Biol. 31 (1987), 465-492.
[12] VAN KAMPEN, N.G.: Stochastic processes in physics and chemistry.
    North-Holland, Amsterdam/New York/Oxford 1981.

Zufallsgraphen in der Soziometrie

Erhard Godehardt, Düsseldorf

Zusammenfassung. In der Soziometrie, wo Gruppenstrukturen wie zum Beispiel Cliquenbildung in Schulklassen analysiert werden, kann man mit kombinatorischen Strukturmodellen arbeiten. Da hier im allgemeinen keine Relationen vorliegen, sondern unsymmetrische Eigenschaften wie die Wahlen der einzelnen Personen Gegenstand der Untersuchung sind, werden sogenannte gerichtete Graphen (oder: Digraphen) zur Beschreibung und Modellbildung benutzt. Hierfür werden Testverfahren vorgestellt, mit deren Hilfe man die Relevanz einer Gruppenstruktur ähnlich einfach statistisch absichern kann wie in der Clusteranalyse.

Summary. In sociometric research, group structures like cliques in school classes have to be analyzed. Here, we can use combinatorial models like directed graphs (or digraphs) to describe, e.g., the result of a questionnaire where persons choose other persons in a group. We present test procedures to test the randomness of group structures found by such a procedure of choices.

1. Einleitung

Graphentheoretische Modelle haben in vielen Bereichen, wo die Strukturen von Daten exploriert werden, Einzug gehalten. In der Clusteranalyse können sie sowohl dazu benutzt werden, um Gruppen zu entdecken, als auch dazu, um deren Relevanz statistisch abzusichern. Hier benutzt man sogenannte "ungerichtete" Graphen, bei denen die Meßwerte als Knoten interpretiert werden und die Kanten durch eine Ähnlichkeitsrelation definiert werden ([3], [4]).

Auch in der Soziometrie, wo Gruppenstrukturen wie zum Beispiel Cliquenbildung in Schulklassen analysiert werden, kann man mit kombinatorischen Strukturmodellen arbeiten. Da hier im allgemeinen keine Relationen vorliegen, sondern unsymmetrische Eigenschaften wie die Wahlen der einzelnen Personen Gegenstand der Untersuchung sind, werden sogenannte gerichtete Graphen (oder: Digraphen) zur Beschreibung und Modellbildung benutzt. Hierfür werden Testverfahren vorgestellt, mit deren Hilfe man die Relevanz einer Gruppenstruktur ähnlich einfach statistisch absichern kann wie in der Clusteranalyse ([1], [5]).

2. Ein Beispiel

In einem Ferienlager mit $n = 40$ Teilnehmern (Jungen und Mädchen zwischen zehn und 14 Jahren) wurde nach drei Wochen eine Umfrage abgehalten: Jeder Teilnehmer sollte $m = 5$ andere (nicht sich selbst) nennen, mit denen er oder sie am liebsten zusammen waren. Stimmenkumulierung war nicht erlaubt ([2]). Das Ergebnis kann als Digraph dargestellt werden. Dabei sind die Personen als indizierte Punkte (oder Knoten, englisch: vertices) repräsentiert und die Wahlen als Pfeile (oder gerichtete Kanten, englisch: arcs); ein Pfeil von Person A zu Person B wird gezeichnet, wenn B von A gewählt wurde). In Abbildung 1 ist das Ergebnis der Wahlen graphisch dargestellt.

In unserem Beispiel mit $N = m\,n = 200$ Pfeilen fanden wir $k = 2$ isolierte Personen (ein Junge und ein Mädchen), d.h. Personen, die von keinem Teilnehmer gewählt wurden (diese werden von Soziologen auch Igel genannt), und drei populäre Personen im Sinne von [1] (je eine mit 14, 15 und 18 Stimmen, vgl. Abbildung 2). Aus Abbildung 1 entnehmen wir, daß 35 Paarbildungen (gegenseitige Wahlen) vorkommen. Die Frage ist, ob solche Konstellationen, also zwei oder mehr isolierte Personen, 35 oder mehr Paare bzw. drei oder mehr Personen, die mindestens vierzehnmal gewählt werden, unter der Annahme zufälliger Wahlen erwartet werden dürfen.

Die Annahme zufälliger Wahlen als Nullhypothese bedeutet hier, daß wir eine homogene Gruppe von Personen vor uns haben, die weder in einzelne Teilgruppen aufgrund von Präferenzen oder Antipatien bei den zerfällt noch auffällige Personen — also solche, die entweder zu oft oder

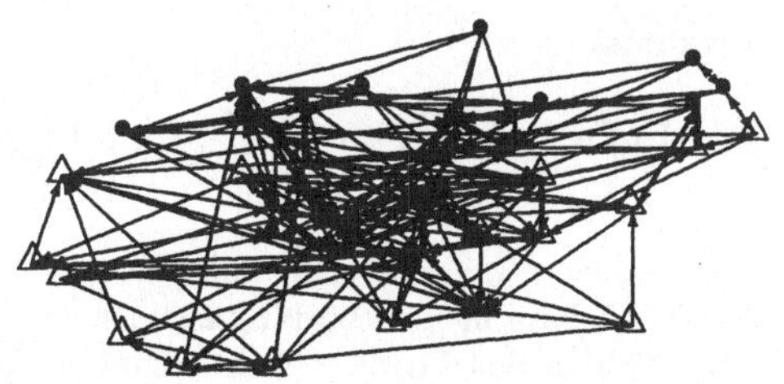

Abbildung 1: *Digraph der Wahlen des Ferienlager-Beispiels;*
$\triangle$*: Jungen,* $\bullet$*: Mädchen; Namensinitialen oder Nummern sind weggelassen.*

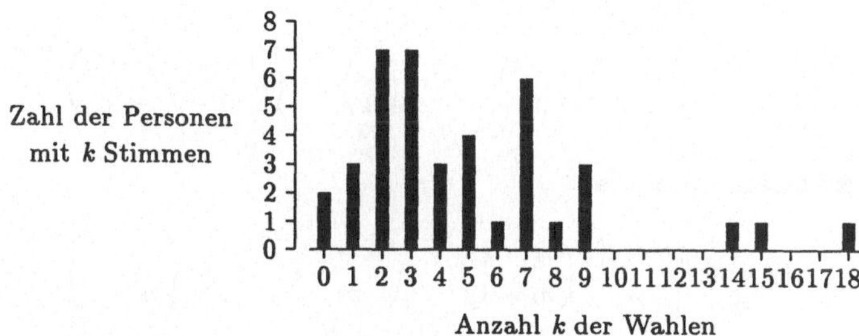

Abbildung 2: *Wahlergebnis des Ferienlager-Beispiels*

gar nicht gewählt werden — enthält. Die Zufallsvariablen "Zahl der nichtgewählten Personen", "Zahl der Paare" (gegenseitigen Wahlen, englisch: matchings), "Zahl der Personen, die am meisten Stimmen auf sich vereinen" oder "Zahl der Personen, die von allen gewählt wurden" können als Teststatistik dienen, wenn ihre Verteilung unter der Nullhypothese bekannt ist.

3. WAHRSCHEINLICHKEITSMODELL FÜR DIGRAPHEN

Neben den Begriffen "Knoten" und "gerichtete Kante" sind zwei weitere Termini aus der Graphentheorie im folgenden wichtig: Unter dem Ausgrad (oder Außengrad, englisch: outdegree) verstehen wir die Zahl der von einem Knoten ausgehenden gerichteten Kanten; der Eingrad (oder Innengrad, englisch: indegree) eines Knotens ist die Zahl der auf ihn gerichteten Kanten. Bei unserem Wahlbeispiel hat der Ausgrad jedes Knotens den Wert 5. Wir haben zwei Knoten mit Eingrad 0 (zwei Kinder, die keinmal gewählt worden sind), drei Knoten mit Eingrad 1 (drei Kinder, die je einmal gewählt worden sind) usw., wie Abbildung 2 zeigt.

Wenn wir annehmen, daß keine besonderen Sympathien oder Antipatien die n Personen in Untergruppen aufspaltet, so können wir annehmen, daß jede Person ihre m Stimmen "zufällig" aufgeteilt hat, also die Wahrscheinlichkeit dafür, eine bestimmte andere Person zu wählen, für jede dieser Personen $1/(n-1)$ ist (in unserem Beispiel ist $n = 40$ und $m = 5$). Der Digraph, der das Wahlverhalten bildlich darstellt, kann somit als Realisation eines Zufallsdigraphen mit n Knoten angesehen werden, bei dem mn Pfeile zufällig so zwischen den Knoten verteilt werden, daß von jedem Knoten m Pfeile ausgehen.

Unter dieser Annahme lassen sich die Verteilungen der Zufallsvariablen X, der Anzahl der Personen, die nicht gewählt wurden, sowie W, der Anzahl der Personen, die von allen anderen gewählt wurden, elementar herleiten. Mit Hilfe der Sieb-Formel (formula of inclusion and exclusion) erhalten wir für die Verteilung der Anzahl der Personen, die nicht gewählt wurden (also den Eingrad 0 besitzen),

$$P(X = k) = \binom{n}{k} \sum_{\mu=k}^{n-1-m} (-1)^{\mu-k} \binom{n-k}{\mu-k} \binom{n-\mu}{m}^{\mu} \binom{n-1-\mu}{m}^{n-\mu} \binom{n-1}{m}^{-n}.$$

Der Erwartungswert dieser Anzahl ist

$$\mathrm{E}_{t,n,N}X = n\binom{n-1}{m}\binom{n-2}{m}^{n-1}\binom{n-1}{m}^{-n} = n\left\{\frac{n-1-m}{n-1}\right\}^{n-1}.$$

Die Verteilung der Anzahl der Personen, die von allen anderen gewählt wurden (also den Eingrad $n-1$ besitzen), ergibt sich durch Komplementbildung aus der vorigen Verteilung. Wir zeichnen statt des Digraphen für die Wahlen den für die "Nichtwahlen". Jede Person hat $n-1-m$ andere nicht gewählt, somit gehen von jedem Knoten $n-1-m$ Pfeile für die nicht gewählten aus, und die Personen, von allen gewählt wurden, haben in diesem komplementären Digraphen den Eingrad 0. Somit ergibt sich für die Verteilung der Anzahl der Personen, die von allen anderen gewählt wurden (also den Eingrad 0 besitzen),

$$P(W = k) = \binom{n}{k}\sum_{\mu=0}^{n-k}(-1)^{\mu}\binom{n-k}{\mu}\binom{n-k-\mu}{m-k-\mu+1}^{k+\mu}\binom{n-k-\mu-1}{m-k-\mu}^{n-k-\mu}\binom{n-1}{m}^{-n}.$$

Der Erwartungswert dieser Anzahl ist

$$\mathrm{E}_{t,n,N}W = n\binom{n-1}{m}\binom{n-2}{m-1}^{n-1}\binom{n-1}{m}^{-n} = n\left\{\frac{m}{n-1}\right\}^{n-1}.$$

In unserem Beispiel aus dem letzten Abschnitt haben wir $X = 2$ Personen, die nicht gewählt wurden, und erhalten aus obiger Formel

$$P(X = 0) = 0.82473\,, \qquad P(X \le 1) = 0.98614\,,$$
$$P(X \le 2) = 0.99931\,, \qquad P(X \le 3) = 0.99998\,.$$

Das bedeutet, daß wir zu einer Irrtumswahrscheinlichkeit von $\alpha = 0.05$ die Nullhypothese zufälliger Wahlen innerhalb dieser 40 Personen ablehnen dürfen. Wir können also annehmen, daß sich innerhalb des Verlaufs des Ferienlagers Untergruppen herausgebildet haben.

Die exakte, finite Verteilung der Zufallsvariablen V, der Zahl der gegenseitigen Wahlen oder Paare, ist bekannt, jedoch recht kompliziert und für numerische Berechnungen bei größerem Stichprobenumfang nicht geeignet. Wir geben hier nur den Erwartungswert und die Standardabweichung an:

$$\mathrm{E}_{t,n,N}V = \frac{n}{n-1}\frac{m^2}{2} = \frac{m^2}{2}\left(1 + O\left(1/n\right)\right),$$
$$\mathrm{Var}_{t,n,N}V \sim \mathrm{E}_{t,n,N}V.$$

Es ist ferner bekannt, daß die Verteilung von V gegen eine Poisson-Verteilung mit Parameter $m^2/2$ strebt.

In unserem Beispiel mit $m = 5$ würden wir $25/2 = 12.5$ Paare erwarten. Wir erhalten aber tatsächlich 35 Paare. Die Wahrscheinlichkeit, soviele oder noch mehr Paare unter der Nullhypothese zufälliger Wahlen zu erhalten, ist praktisch Null, so daß wir auch bei dieser Teststatistik von signifikanten Untergruppen ausgehen können.

4. VERGLEICH VON WAHRSCHEINLICHKEITSMODELLEN

Exakte, finite Verteilungen sind in der Kombinatorik bzw. der Theorie der Zufallsgraphen ziemlich kompliziert und meist durch alternierende Summen oder speicherintensive Rekursionsformeln gegeben. Hier ist es von Nutzen, die asymptotischen Verteilungen zu kennen und zu wissen, ab welchem Stichprobenumfang n sie genau genug sind (vgl. [3], [4]). Weiter ist es von Nutzen zu wissen, ob die beiden folgenden Wahrscheinlichkeitsmodelle — zumindest asymptotisch — dieselben Ergebnisse für die oben genannten Zufallsvariablen liefern.

(1) Jede Person hat eine festgelegte Anzahl m von Wahlen.

(2) Jede Person wählt eine zufällige Anzahl anderer Personen (von 0 bis n, d.h., jede Person macht zuerst eine Zufallsentscheidung für die Anzahl der von ihr gewählten Personen und wählt dann diese Zahl einzelner Personen; Stimmenkumulierung, also mehr als eine Stimme je gewählter Person, ist allerdings verboten).

Das zweite Modell ist ersichtlich unstrukturierter als das erste, wo die Zahl der abgegebenen Stimmen je Person konstant ist. Wenn n Personen insgesamt N Stimmen abgegeben haben, so gilt jetzt nur noch, daß der Erwartungswert der abgegebenen Stimmen je Person N/n ist (also $E_{t,n,N} M = N/n$ mit einer Zufallsvariablen M anstelle einer Konstanten m). Dieses zweite Modell ist äquivalent zum Spezialfall $t = 2$ des z.B. in [3] und [4] diskutierten Wahrscheinlichkeitsmodells für ungerichtete, vollständig indizierte Zufallsmultigraphen $\Gamma_{t,n,N}$. Ein (ungerichteter) Multigraph $\Gamma_{2,n,N}$ besteht aus zwei ungerichteten Graphen Γ_{n,N_1} und Γ_{n,N_2}, die dieselben n Knoten haben und aufeinandergelegt worden sind (die Kanten müssen nicht identisch sein, auch die Kantenzahlen N_1 und N_2 nicht; wir setzen $N = N_1 + N_2$ als Gesamtzahl der Kanten des Multigraphen, siehe [3], [4]). Die Kanten in einem ungerichteten Graphen oder Multigraphen werden als einfache verbindende Linien zwischen zwei Knoten gezeichnet; eine Orientierung wie bei den Wahlen ist nicht definiert.

Einen Digraphen $\Delta_{n,N}$ mit n Knoten und N Pfeilen, etwa den Digraphen für die Wahlen mit $N = m\,n$, kann man auch als Multigraphen $\Gamma_{2,n,N}$ mit denselben n Knoten, zwei Schichten und N Kanten wie folgt darstellen. Für die Personen Nummer i und Nummer j ($i < j$) wird eine Kante $\kappa_{i,j,1}$ in der unteren (ersten) Schicht des Multigraphen gezeichnet, wenn der Pfeil $\delta_{i,j}$ von Person j nach Person i im Digraphen vorhanden ist; eine Kante $\kappa_{i,j,2}$ ist in der zweiten (oberen) Schicht von $\Gamma_{2,n,N}$, wenn der Pfeil $\delta_{j,i}$ im Digraphen vorhanden ist, d.h., wenn die Person j von der Person i gewählt wurde. Wenn zwei Personen i und j sich gegenseitig gewählt haben, so zeigt im Digraphen $\Delta_{n,N}$ je ein Pfeil von i nach j und von j nach i. In der Darstellung als Multigraph $\Gamma_{2,n,N}$ liegt entsprechend je eine Kante in der unteren Schicht und eine Kante in der oberen Schicht; wir sprechen von einer 2-gesättigten Belegung zwischen den Knoten i und j (englisch: 2-saturated connection; für die entsprechenden exakten Definitionen im Multigraphen-Modell siehe [3], [4]). Eine 2-gesättigten Belegung im Multigraphen entspricht also einem sich gegenseitig wählenden Paar im Digraphen.

Unter der Hypothese, daß die N Kanten "zufällig" zwischen die Knoten von $\Gamma_{2,n,N}$ verteilt werden (in beiden Schichten) haben wir

$$P(\Gamma_{2,n,N}) = \binom{2\binom{n}{2}}{N}^{-1}.$$

als Wahrscheinlichkeit, einen bestimmten Multigraphen $\Gamma_{2,n,N}$ nach der Verteilung der Kanten zu erhalten. Jeder nach diesem Modell gewonnene Multigraph heißt Zufallsmultigraph. Dieses Wahrscheinlichkeitsmodell entspricht dem, welches wir aus (2) unter der Annahme erhalten, daß die Wahlen zufällig erfolgen (sowohl bezüglich der Anzahl der von jeder Person abgegebenen Stimmen, als auch bezüglich der gewählten Personen); die einzige Nebenbedingung ist, daß insgesamt genau N Stimmen abgegeben werden.

Die asymptotische Verteilung der Anzahl der Paare unter der Annahme (2) brauchen wir somit nicht mehr neu herleiten: Wir können sie wegen der Analogie zwischen Digraphen $\Delta_{n,N}$ und Multigraphen $\Gamma_{2,n,N}$ z.B. als Spezialfall aus [3] direkt übernehmen (Satz 5-16 mit $t = s = 2$ und $c = m^2$). Es gilt somit der folgende Satz.

Satz: *In Folgen $(\Gamma_{t,n,N})_{n\to\infty}$ von Zufallsmultigraphen mit 2 Schichten, n Knoten und $N(n) = m\,n$ Kanten strebt der Erwartungswert der Anzahl der 2-gesättigten Belegungen gegen einen positiven Grenzwert für $n \to \infty$: $E_{t,n,N} V \to m^2/2$. Die Anzahl der 2-gesättigten Belegungen ist in der Grenze poissonverteilt mit dem Parameter $\lambda = m^2/2$.*

Die erwartete Zahl der Stimmen je Person ist nach diesem Satz $E_{t,n,N} M = N/n = m$, und es ergibt sich dieselbe Grenzverteilung wie unter der Annahme (1), daß jede Person gleichviele andere

Personen wählt. Der obige Satz für die Verteilung der Zahl der Paare erlaubt es uns, recht einfach mit Hilfe dieser Zufallsvariablen die Nullhypothese der Zufälligkeit beim gegenseitigen Wählen in einer Gruppe unter den Annahmen (1) und (2) zu testen. (Dabei ist es sinnvoller, in einem realen Test nicht wie in unserem Fereinlager-Beispiel vorzugehen sondern zuzulassen, daß die einzelnen Personen nicht gleichviele andere wählen, also mit Modell (2) zu arbeiten).

Die Gleichheit der Grenzverteilungen unter den beiden Annahmen (1) bzw. (2) ist nicht selbstverständlich. Bei den Zufallsvariablen X und W erhalten wir z.B. verschiedene Grenzverteilungen. Wir wollen das hier nur kurz an den Erwartungswerten demonstrieren. Damit die erwartete Zahl der Personen mit Eingrad 0 endlich bleibt, muß unter (1) gelten: $m = \log n + c$. In diesem Fall gilt $E_{t,n,N} X \to \lambda = e^{-c}$. Für Modell (2) ist wieder die Analogie zum Multigraphenmodell ausnutzbar und Satz 5-9 aus [3] mit $t = s = 2$ anwendbar. Hier ergibt sich $N \sim n \left(n \left(\log n + c \right) \right)^{1/2}$ als notwendige Kantenzahl, damit wir asymptotisch denselben Erwartungswert für die erwartete Zahl der isolierten Personen erhalten. Für die erwartete Anzahl der abgegebenen Stimmen muß also jetzt $E_{t,n,N} M = \left(n \left(\log n + c \right) \right)^{1/2}$ gelten. Diese Anzahl ist wesentlich stärker wachsend als die für m unter (1).

Damit der Erwartungswert der Personen mit Eingrad $n - 1$ (also der Personen, die von allen anderen gewählt wurden) positiv und endlich bleibt für $n \to \infty$, muß unter Modell (1) $m \sim n$ gelten; d.h., jeder muß unter der Nullhypothese zufälliger Wahlen praktisch alle $n - 1$ anderen wählen, ehe wir damit rechnen müssen, daß jemand von allen gewählt wird. Unter Voraussetzung (2) können wir Satz 5-15 aus [3] anwenden und erhalten hieraus $N \sim e^{-1/2} n^2$ als notwendige Kantenzahl dafür, daß $E_{t,n,N} W$ gegen denselben Grenzwert strebt. Damit ergibt sich $E_{t,n,N} M = e^{-1/2} n$ als notwendige Bedingung an den Erwartungswert der abgegebenen Stimmen je Person. Hier ist also unter dem weniger strukturierten Modell (2) eine proportional geringere Stimmenzahl zur Erreichung des gleichen Ziels notwendig unter Modell (1).

Literatur

[1] Capobianco, M., Palka, Z.: The Distribution of Popular Persons in a Group. *Social Networks* 5 (1983), 383–393

[2] Godehardt, E.: *Graphen in der Soziometrie*. Seminarvortrag 1974 (Vortrag und Daten bisher nicht veröffentlicht)

[3] Godehardt, E.: *Graphs as Structural Models: The Application of Graphs and Multigraphs in Cluster Analysis* (Advances in Systems Analysis, Vol. 4). Friedr. Vieweg & Sohn, Braunschweig – Wiesbaden 1988

[4] Godehardt, E.: Multigraphs as a Tool for Numerical Classification. In: Möller, D.P.F. (Ed.): *Erwin-Riesch Arbeitstagung: Systemanalyse biologischer Prozesse: 3. Ebernburger Gespräch (Advances in Systems Analysis, Vol. 5)*. Friedr. Vieweg & Sohn, Braunschweig – New York 1988, pp. 59–68

[5] Godehardt, E.: *Random Multigraphs and Digraphs in Sociometry*. 4th International Seminar on Random Graphs, Poznań, August 7–12, 1989